AUTOMOTIVE REFINISHING

THIRD EDITION

AUTOMOTIVE REFINISHING

Harry T. Chudy

Upper Saddle River, New Jersey
Columbus, Ohio

Library of Congress Cataloging-in-Publication Data

Chudy, Harry T.
    Automotive refinishing / Harry T. Chudy.—3rd ed.
      p. cm.
    Rev. ed. of: The complete guide to automotive refinishing / Harry T. Chudy. 1987, c1988.
    Includes index.
    ISBN 0-13-010073-0
    1. Automobiles—Painting. I. Chudy, Harry T. Complete guide to automotive refinishing. II. Title.

TL255.2 .C49 2001
629.2'6—dc21   00-027424

Vice President and Publisher: Dave Garza
Editor in Chief: Stephen Helba
Executive Editor: Ed Francis
Production Editor: Christine M. Buckendahl
Design Coordinator: Robin Chukes
Cover Designer: Tanya Burgess
Cover art: Neal Moss
Production Manager: Pat Tonneman
Marketing Manager: Jamie Van Voorhis

This book was set in ITC Century Book by Carlisle Communications, Ltd., and was printed and bound by The Banta Company. The cover was printed by Phoenix Color Corp.

Copyright © 2001, 1988, 1982 by Prentice-Hall, Inc., Upper Saddle River, New Jersey 07458. All rights reserved. Printed in the United States of America. This publication is protected by Copyright and permission should be obtained from the publisher prior to any prohibited reproduction, storage in a retrieval system, or transmission in any form or by any means, electronic, mechanical, photocopying, recording, or likewise. For information regarding permission(s), write to: Rights and Permissions Department.

10 9 8 7 6 5 4 3 2 1
ISBN: 0-13-010073-0

*This book is dedicated to my wife, Connie;
to my children, Rick and Sharon; daughter-in-law, Cindy;
to Jason and Benjamin Chudy, grandsons;
and to all young persons who aspire to become
automotive refinishing technicians.*

PREFACE

This book may be used by various groups for several purposes:

1. **By all students and apprentices** entering the automotive refinishing trade who want to learn how to apply automotive finishes.
2. **By paint instructors** in high schools, vocational schools, and colleges who need updated information about refinishing.
3. **By experienced painters** who want updated paint information and who want to increase their skills in refinishing.
4. **By all personnel connected with automotive refinishing** who want to become familiar with the latest developments in refinish products and their application.
5. **By anyone who wants to gain a working understanding** of the latest government clean air regulations and a thorough coverage of HVLP spray equipment.

For easy reference, the book is divided into self-descriptive chapters covering applicable portions of the refinish operation. **Chapter 1, "Description of the Automotive Refinishing Trade,"** describes those businesses and organizations that are members of the trade. Also included is a description of how paint shops are regulated.

Chapter 2, "How Vehicles Are Painted at the Factory and in the Refinish Trade," gives the painter insight into repair operations. Featured topics are paint shop safety rules, paint identification, and ordering paint from a paint jobber.

Chapter 3, "Paint Spray Guns and Paint Cups," explains how all spray guns are divided into two categories:

1. Conventional high-pressure spray guns
2. High-volume low-pressure (HVLP) spray guns

The chapter covers two high-pressure spray guns and several HVLP spray guns and their features. Also, VOC and transfer efficiency are described. Operating air pressures of HVLP and conventional guns are compared.

Chapter 4, "Spraying Techniques," explains how a novice, by learning and practicing the fundamentals in this chapter, can adjust a spray gun and apply an excellent spray fan. The chapter explains how to adjust and use each of the spray guns described in Chapter 3. It also includes a series of practice exercises.

Chapter 5, "Spray-Painting Equipment and Facilities," explains the role of OSHA. Paint shop safety and the use of respirators are emphasized. Air-supplied respirators feature a safe, oil-free pump that is NIOSH- and OSHA-approved and 100 percent carbon monoxide–free. Air requirements for HVLP and shop equipment are explained. The following topics are also described here:

1. Types of spray booths, with an emphasis on down-draft types
2. Respirator fit and maintenance
3. Vacuum sandblaster
4. Surface prep stations
5. Automatic gun cleaners and color-matching tools
6. Drying equipment

Chapter 6, "Surface Preparation," covers all operations included in preparing surfaces for application of topcoats. All paint repair systems require comparable preliminary workmanship. The following topics are featured in this chapter:

1. Use of grade 2500 Ultra Fine sandpaper for repair of basecoat and clearcoat finishes.
2. The latest in abrasives for power and hand sanding.

3. The latest in masking materials, including door aperture refinish tape (DART) and pre-folded drape masking sheets.
4. Tips on the use of masking tape.
5. A thorough coverage of featheredging.
6. Efficient use of paint removers.
7. Safety tips on the use of power tools.

Chapter 7, "Automotive Refinishing Solvents," describes the importance and purpose of refinishing solvents in the trade and how solvents have been affected by the Clean Air Act. The chapter describes VOC content and explains how to determine VOC content. Also included are procedures for checking paint viscosity, reducing a multicomponent paint material, and paint shop recordkeeping in a compliant area.

Chapter 8, "Undercoat Materials and Application," describes the purpose and application of undercoat materials. Procedures are given for applying high- and low-VOC undercoats, including waterbase types. Coverage of two-component undercoat systems is included. Also included is a coverage of polyester glazing putties, which are more efficient in repair systems. Care must be taken to apply special primers on bare metal before applying waterbase finishes.

Chapter 9, "Automotive Topcoats," familiarizes the student or painter with the description and behavior of automotive finishes. The chapter emphasizes the following topics:

1. The components of a paint finish.
2. Types of automotive topcoats.
3. Car factory paint use.
4. G.M. paint evolution.
5. Description of metallic colors.
6. Description of multicoat paint finishes.
7. Description of field paint repair materials.
8. Availability of waterbase field repair colorcoats.
9. Drying of enamel finishes.
10. Paint compatibility.
11. Test to determine the type of paint.
12. Paint thickness limit.

Chapter 10, "Complete Vehicle Refinishing," includes all the necessary procedures for preparing a car and a review of the topcoats for application. The tools and equipment needed in a paint mix room are listed. A procedure for using a graduated paint mix paddle is included. Special techniques are outlined for the application of color to the roof, hood, and the rear compartment for best results. The chapter has a checklist of rule reminders that a shop needs to follow when performing a complete refinish. Two sequential methods of a complete refinish job are also included.

Chapter 11, "Panel and Sectional Panel Refinishing," is completely revised. Sectional panel repairing, with great emphasis on blending, has become a popular repair method. The key to successful color matching is proper blending with the proper blending agents. The chapter includes special procedures for the repair of pearl or tricoat finishes, which also includes blending. A procedure for applying fine line striping tape is included.

Chapter 12, "Spot and Spot/Partial Panel Repair," has been completely revised. Basecoat/clearcoat and single-stage repairs are explained, with great emphasis on blending. The latest techniques in handling the blending of the sail panel to the roof panel are included. Also featured in the blending of a door with the side of the car on a tricoat finish.

Chapter 13, "Compounding and Polishing," has also been completely revised. New technologies in multicoat finishes have brought about new technologies in compounding and polishing. The chapter introduces a new approach to making repairs on any type of finish without the need for repainting. The procedure is simple and revolutionary. The 3M company has available a paint polishing system called PERFECT-IT II, which stresses simplicity in going from the finest abrasive to a more coarse one to make a repair. Risks are eliminated because repairs progress from finer abrasives to coarser ones as the situation demands. Maguiar's and other polish companies have equivalent products available.

Chapter 14, "Color-Matching Fundamentals and Techniques," covers topics carefully assembled by experts from car companies and paint manufacturers. This chapter lays a solid groundwork for the person who wants to learn the art of tinting and color matching. To do acceptable paint repairs on a competitive basis, each paint shop must be able to tint a color at the time a car is being painted. A shop should be equipped with a complete set of tinting colors, tinting guides, and color-mixing equipment from leading paint suppliers.

Chapter 15, "Painting Interior Plastic Parts," covers two systems of painting interior plastic parts that are in use:

1. A low-VOC system (with HVLP equipment) is required in compliant areas.

PREFACE

2. A conventional refinish system continues to be used in noncompliant areas.

Check with your local paint jobber or with your local air pollution control (APC) district to determine which products and application procedures are permitted in your area.

Chapter 16, "Painting Flexible Plastic Parts," covers two systems of painting exterior plastic parts that are in use:

1. A low-VOC system (with HVLP equipment) is required in compliant areas.
2. A conventional refinish system continues to be used in noncompliant areas.

Check with your local paint jobber to determine which products and application procedures are required.

Chapter 17, "Paint Conditions and Remedies," summarizes the most common paint problems, their causes, and how they are repaired. This chapter is a compilation of information from leading car companies and paint manufacturers.

Chapter 18, "Rust Repairs and Prevention," explains how to treat rust. The chapter includes a procedure for making a patch repair for a perforation condition and procedures for applying chip resistant coatings.

NOTE: *Equipment or product descriptions and recommendations found in this textbook are for general reference purposes only as teaching aids. They do not necessarily constitute an endorsement of any specific type or brand of equipment or products by the author or publisher.*

ACKNOWLEDGMENTS

The author wishes to express his sincere thanks to a number of people without whose special help this book could never be assembled. Special thanks go to the following people:

Louis McClain, paint instructor, General Motors Training Centers
Ken Davis, PPG Refinishing
Robert Yearick, Dupont Refinishing
Andy Ladak, BASF Refinishing
John Zoia, 3M Automotive Aftermarket Division
Michele Vale, California Air Resources Board
John Fitzgerald, Safety Kleen Corporation
John Meyers, S & H Industries
Paul Lenzen, Mattson Spray Equipment
Ken Marg, AccuSpray Incorporated
Tony Larimer, Sata Spray Equipment
Linda Samons, ITW-Binks-DeVilbiss
Dan Dryke, General Motors Corporation
Tom Lamberg, ATCO Precision Tool, Inc.

BRIEF CONTENTS

1 Description of the Automotive Refinishing Trade   1

2 How Vehicles Are Painted at the Factory and in the Refinish Trade   9

3 Paint Spray Guns and Paint Cups   21

4 Spraying Techniques   37

5 Spray-Painting Equipment and Facilities   67

6 Surface Preparation   89

7 Automotive Refinishing Solvents   121

8 Undercoat Materials and Application   131

9 Automotive Topcoats   141

10 Complete Vehicle Refinishing   153

11 Panel and Sectional Panel Refinishing   167

12 Spot and Spot/Partial Panel Repair   185

13 Compounding and Polishing   195

14 Color Matching Fundamentals and Techniques   207

15 Painting Interior Plastic Parts   223

16 Painting Flexible Plastic Parts   233

17 Paint Conditions and Remedies   245

18 Rust Repairs and Prevention   261

Glossary   271

Index   277

CONTENTS

CHAPTER 1
Description of the Automotive Refinishing Trade ... 1

Need for the Trade 1
General Description of the Trade 1
Description of a Qualified Painter 1
How a Typical Paint Shop Operates 2
Refinish Paint Companies 3
Federal, State, and Local Government Regulation 5
Review Questions 8

CHAPTER 2
How Vehicles Are Painted at the Factory and in the Refinish Trade ... 9

Description of the Production Line 9
Factory Painting 9
Refinish Trade Painting 11
Paint Shop Safety Rules 12
Paint Identification 14
Color Charts 17
Ordering Colors from a Paint Jobber 19
Review Questions 20

CHAPTER 3
Paint Spray Guns and Paint Cups ... 21

Introduction 21
Conventional High Pressure Spray Guns 21
Fluid Containers: General Description 28
High-Volume, Low-Pressure (HVLP) Spray Guns 32
Conventional High-Pressure Spray Painting Versus HVLP Spray Painting 33
Review Questions 36

CHAPTER 4
Spraying Techniques ... 37

Introduction 37
Spray Gun Adjustments 37
Spray Painting Stroke 46
Types of Surfaces to Be Sprayed 50
Spray Gun Maintenance 50
Spray Gun Cleaners 52
Cleaning a Mattson Spray Gun 53
Spray Gun Exercises 58
Review Questions 65

CHAPTER 5
Spray-Painting Equipment and Facilities ... 67

Introduction 67
Role of OSHA and NIOSH 67
Environmental Protection Agency 68
Respirators 68
Dusting Gun 72
Air Compressors 73
Compressed Air Filter and Regulator 75
Refinish Hoses 76
Spray Booths 78
Drying Equipment 82
Paint Mixing Room and Equipment 83
Surface Preparation and Work Station 85
Spray Gun Washer and Solvent Recycler 86
Review Questions 87

CHAPTER 6
Surface Preparation — 89

Introduction 89
Car Washing 89
Materials Needed for Surface Preparation 90
The Art of Sanding 96
Hand Sanding 96
Power Sanders and Power Files 100
Paint Removers 105
Sandblasting and Mediablasting 108
Masking 110
Determine the Surface Condition 119
Review Questions 120

CHAPTER 7
Automotive Refinishing Solvents — 121

Purpose of Solvents 121
Types of Solvents 121
Automotive Refinishing Rules Limit VOC Concentration 122
Types of Nonexempt Solvents 122
Paint Viscosity Cup 124
Percentage of Reduction and Mixing Ratio 126
Recordkeeping 128
Review Questions 129

CHAPTER 8
Undercoat Materials and Application — 131

General Types of Undercoats 131
History 131
Purpose of Undercoats 131
Undercoat Categories 132
Undercoat Application 135
Review Questions 138

CHAPTER 9
Automotive Topcoats — 141

Purpose of Color 141
What Is Color? 141
Types of Automotive Topcoats 143
Topcoat Classifications 144
American Car Factory Paint Use 146
Low VOC Paint Systems 147
Description of Single-Stage Metallic Colors 147
Paint Compatibility 148
Test to Determine the Type of Finish 148
Paint Thickness Limits 149
Paint Thickness Gauges 149
Review Questions 152

CHAPTER 10
Complete Vehicle Refinishing — 153

Introduction 153
Tools and Equipment Needed in the Paint Mixing Room 153
Checklist for Preparing a Car Before Color Coating 154
Preparation of Color 154
Color Reduction (Conventional Method) 155
Use of Hardener (Catalyst) in Acrylic and Urethane Enamel 155
Check the Spray Outfit Before Painting 156
Hose Control When Spraying Extra-Wide Surfaces 158
Paint Application Systems 158
Paint Application Procedure 158
Differences Between Acrylic Lacquer and Enamel Paint Systems 161
Applying Single-Stage Acrylic and Urethane Enamel Paint Systems 161
Applying the Single-Stage Acrylic Lacquer Paint System 162
Repairing Mottling During Single-Stage Lacquer and Enamel Paint Application 163
Usage of Silicone Additive 163
Review Questions 164

CHAPTER 11
Panel and Sectional Panel Refinishing — 167

Introduction 167
Description of Complete Panel Repair 167
Need for Color Accuracy 167
Tricoat Paint Repairs 174
Sectional Panel Repair 178
Paint Striping Replacement 181
Review Questions 183

CHAPTER 12
Spot and Spot/Partial Panel Repair 185

Introduction 185
Successful Spot and/or Partial Panel Repair 186
Surface Preparation for Spot Repairs 186
Single-Stage Finish Spot Repair Technique 188
Single-Stage Acrylic Lacquer Spot Repairs 188
Single-Stage Acrylic Enamel Spot Repairs 189
Spot/Partial Panel Repairs on Multicoat Finishes 190
Review Questions 194

CHAPTER 13
Compounding and Polishing 195

Introduction 195
Compounding in Refinishing 195
Polishing 196
3M PERFECT-IT™ II Paint Finishing System 198
Runs and Sags 201
Dirt 201
Orange Peel 202
Fisheyes 202
Damage from Acid Rain and Industrial Fallout 203
Review Questions 205

CHAPTER 14
Color Matching Fundamentals and Techniques 207

Introduction 207
Variables Affecting Color Match 207
Other Causes of Color Mismatches 209
Lighting Conditions and Color Inspection 209
Concept of Black and White Spray-Out Panels 211
Theory of Color 211
How We See Color and Why 212
The BRYG Color Wheel 212
Munsell Color Tree Concept 213
How to Solve Color-Matching Problems 214
Understanding Paint Formula Ingredients 217
How to Use a Color Tinting Guide 217
Hints for Color Matching 218
Color Plotting 218
Review Questions 220

CHAPTER 15
Painting Interior Plastic Parts 223

Introduction 223
How Parts Are Serviced 223
How Trim Parts Are Color Keyed 223
Federal and Factory Paint Standards 224
Testing Quality and Durability of Paint Products 225
Painting Interior Plastic Parts 226
Products for Painting Body Interior Vinyl Trim Parts 228
Luggage Compartment Finishing 231
Review Questions 231

CHAPTER 16
Painting Flexible Plastic Parts 233

Painting Flexible Plastic Parts 233
Repair of Bumper Cover Before Painting 237
Painting Vinyl Tops 240
Review Questions 242

CHAPTER 17
Paint Conditions and Remedies 245

Introduction 245
Acid and Alkali Spotting (Lacquer and Enamel) 245
Blistering (Lacquer and Enamel) 246
Blushing (Lacquer) 246
Bull's Eye (Lacquer and Enamel) 246
Chalking (Lacquer) 247
Checking and Cracking (Lacquer and Enamel) 247
Micro-Checking (Lacquer) 247
Crazing (Lacquer) 248
Dirt in the Finish (Lacquer and Enamel) 248
Dry Spray (Lacquer and Enamel) 249
Etching (Lacquer) 249
Fisheyes (Lacquer and Enamel) 249
Lifting of Enamels (Lacquer and Enamel) 250
Mottling (Lacquer and Enamel) 250
Off-Color (Lacquer and Enamel) 251
Excessive Orange Peel (Lacquer and Enamel) 251
Overspray (Lacquer and Enamel) 252
Peeling (Lacquer and Enamel) 252

Pinholing (Lacquer and Enamel) 252
Rub-Through (Lacquer) 253
Runs or Sags (Lacquer and Enamel) 253
Rust Spots and Rusting (Lacquer and Enamel) 254
Sand or File Marks (Lacquer and Enamel) 254
Sandscratch Swelling (Lacquer) 255
Shrinking and Splitting of Putty (Lacquer and Enamel) 255
Streaks in the Finish (Lacquer and Enamel) 256
Sweat-Out or Bloom (Lacquer) 256
Water Spotting (Lacquer) 256
Wet Spots (Lacquer and Enamel) 257
Wheel Burn (Lacquer) 257
Wrinkling of Enamels 257
Binks Summary on Dirt in the Finish 258
Review Questions 259

CHAPTER 18
Rust Repairs and Prevention 261

Introduction 261
"Stop Rust" Products 261
Causes of Rust 262
Types of Rust 263
Determining Rust Repair Costs 264
Rust Removal 264
Chip-Resistant Coatings 266
Review Questions 269

Glossary 271

Index 277

CHAPTER 1

Description of the Automotive Refinishing Trade

NEED FOR THE TRADE

America is a nation on wheels. Every part of our society and all our industries depend on wheels of one form or another. There are hundreds of millions of cars, vans, and trucks on American roads.

The appearance of a car is important to every car owner. The single most important thing that makes a car appealing to the owner is the beauty and gloss of the paint finish. People know that a good paint finish is essential to the lasting beauty and durability of a car.

Between 30 and 40 million cars pass through collision and paint repair shops each year for major repairs. Additional millions of vehicles pass through bump and paint shops for minor collision damage and general paint repairs. According to insurance companies, the total cost for all this repair work runs into billions of dollars. The need for automotive painters and metal repair persons never was so great, and the need grows each year.

GENERAL DESCRIPTION OF THE TRADE

The purpose of the automotive refinishing trade (see Figure 1–1) is to keep the paint finish of all cars on the road in good repair. To accomplish this purpose requires many thousands of painters and automotive-related businesses and organizations such as:

1. Federal, state and local governments
2. Paint shops
3. Refinish paint companies
4. Refinish paint spray gun companies
5. Refinish tool and equipment companies
6. Miscellaneous refinish material companies.
7. Insurance companies.

While all automotive-related paint businesses and the people working in them play a vital role in the refinish trade, the painter is the single most important person.

By design, the automotive refinishing trade is so widespread all over the country that a person traveling in any state, in any type of production car, can have paint repairs done on any color by a qualified painter and the color match will be commercially acceptable. To make such a condition possible requires a great amount of teamwork and cooperation among many thousands of people in the refinishing trade. Millions of cars are being painted every day all over the country as a result of collision damage repair or for other reasons.

The center of the refinishing trade is the paint shop. Three types of paint shops are most prominent in the trade:

1. **Conventional paint shops** do most of the paint work in the refinish trade because they greatly outnumber all other paint shops.
2. **High-volume, low-cost paint shops** do a good share of the refinish business nationally.
3. **Custom paint shops** have the ability to do the best quality work. They are also the most expensive.

DESCRIPTION OF A QUALIFIED PAINTER

To a bystander watching a painter at work, painting may appear to be an easy, routine operation. What the bystander is really seeing is a qualified painter

FIGURE 1-1. Auto refinish trade businesses and organizations.

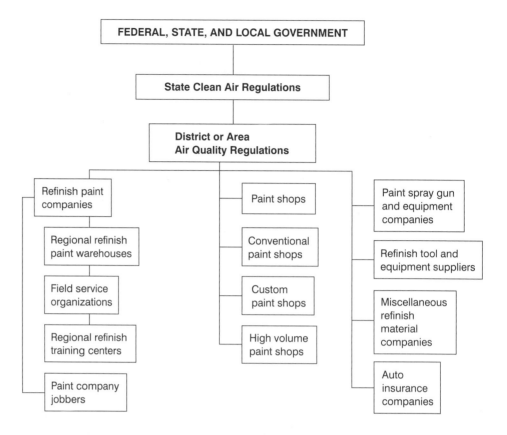

with artistic skills developed after countless hours of patient work and practice at his or her trade. To be qualified, a painter must have the following abilities:

1. Be expert at handling a paint spray gun.
2. Be expert at refinish fundamentals, including paint problems and repairs.
3. Be thoroughly familiar with production paint systems.
4. Be thoroughly familiar with refinish paint systems.
5. Be expert at color matching with refinish paint systems.
6. Be thoroughly familiar with all applicable federal, state, and local safety regulations.

It takes a certain amount of teamwork among all the people in the refinishing trade to achieve quality results. If the paint color is not right, if the paint equipment is faulty, or if the painter is not qualified, paint repairs will not be commercially acceptable.

HOW A TYPICAL PAINT SHOP OPERATES

When a person drives into a paint shop for repairs, a qualified person, usually a painter or estimator, writes up the **repair order.** The estimator inspects the car thoroughly for work to be done and writes up each paint operation. In one column on the repair order, the amount of time (in tenths of an hour) is noted opposite each operation. This is known as **labor.** All labor time is totaled at the bottom of the column. The total labor time is then multiplied by the flat-rate hourly scale for the area. This is the total labor charge. In another column, the amount of cost for each item of **material** used on the car is listed. This includes all paint, sandpaper, masking tape, and other items used in painting the car. All material costs are totaled at the bottom of the column. The total of the labor charges and material costs represents the total cost of a paint job. When extra work must be done on a car before painting, such as rust repairs and dent repairs, the charge is determined on a straight-time hourly basis and the extra charge is added to the cost of the paint job.

A common rule to determine the cost of a paint job in the trade is the one-third plus two-thirds formula. The first third represents the total cost of all materials for the paint job. The two-thirds represents the total cost of labor. Thus, if the total cost for materials is $50, the total cost for labor would be $100, and the total cost of the paint job would be $150. Rust repairs and dent repairs are determined on a straight-time basis, and the costs are added to determine the final cost of the paint job.

DESCRIPTION OF THE AUTOMOTIVE REFINISHING TRADE

Generally, **the responsibilities of a painter in a paint shop are to:**

1. Write repair orders.
2. Order all necessary paint materials.
3. Supervise painter's helpers and apprentices to prepare each car for paint.
4. Supervise and help complete all masking.
5. Prepare paint materials as required.
6. Apply paint materials on cars as required.
7. Supervise all masking removal and final car cleanup.
8. Supervise car delivery to owner.
9. File repair orders.
10. Keep required records of all paint work done.

Generally, **the responsibilities of a painter's helper or apprentice are to:**

1. Move cars into and out of the shop.
2. Wash and clean up cars before and after paint operations.
3. Help with car parts removal and installation.
4. Clean and sand paint surfaces as required.
5. Apply plastic filler and metal finish as required and when qualified to do so.
6. Clean and mask cars before the painting operation.
7. Apply undercoats when qualified to do so.
8. Help sand undercoats as required.
9. Have the painter check all completed surface preparation operations.
10. Help to clean up the shop at end of each day or as often as required.

REFINISH PAINT COMPANIES

All major paint companies who supply paint to car factories also supply paint to the refinish trade. Additional paint companies who do not supply paint to car factories also supply refinish paint to the trade. All refinish paint suppliers are obligated to match and to repair all colors produced by the car factories. To achieve this objective, paint companies:

1. **Supply color and undercoat materials** to the refinish trade through paint jobbers.
2. **Make available color charts, tinting guides, color programs for computers, color measuring tools (spectrophotometers), special formulas to match various shades of the same color, and special body shop support programs** that cover all phases of modern refinishing.
3. **Provide training to all who need it** for proper use of refinish materials.
4. **Provide sales literature, brochures,** and material safety data sheets (MSDSs) that cover all safety hazards faced by the painter.
5. **Hire service representative assistance** to handle any and all refinish problems related to company products.
6. **Provide safety equipment needed by the painter.**
7. **Develop, test, and improve new lines of refinish products** continually to keep up with field requirements.
8. **Guarantee their products in writing** when properly applied by certified painters and paint shops.

Paint Company Regional Refinish Training Centers

Most paint companies operate numerous special **paint training centers** located in strategic, high-population areas all over the country (see Figure 1–2). **Special paint training courses are conducted continuously for painters and paint jobbers** who handle their refinishing products. These professionally prepared programs result in more highly skilled painters and more efficient paint jobbers.

The training centers have been organized at a most opportune time. The Clean Air Act passed by Congress has prompted the refinish industry to re-invent and re-establish completely new lines of paint technologies that do not contaminate the environment and reduce air pollution. **The more sophisticated the automotive refinishing industry gets, the greater is the need for specialized training.**

The objective of schools is to teach painters and apprentices the know-how of the new generation of refinish products, HVLP spray painting equipment, and application techniques. Painters learn firsthand why the most durable and best-looking finishes are applied in three steps: first, the primer coats; then the midcoats; and finally the topcoats. The best paint systems are designed with components that are interdependent. (For information about enrollment into these Training Centers, check with your local paint jobber or with your local paint company sales representative.)

FIGURE 1–2. Typical paint class in session at BASF Refinish Training Center. (Courtesy of BASF Corporation.)

Paint Jobbers

Paint jobbers perform several tasks. They:

1. **Make all color** and **undercoat systems available** to paint shops.
2. **Make** recent model **(factory package), ready for use colors** available to paint shops.
3. **Formulate past model colors** on request.
4. **Provide** all necessary **spray painting equipment.**
5. **Provide** all necessary **surface preparation materials.**
6. **Conduct new product training programs** among customer paint shops as required.
7. **Distribute material safety data sheets** and refinish product brochures to customers who request them.
8. **Provide a delivery service** for paint shops who need it.
9. **Provide** to painters a complete line of **safety equipment** such as **respirators, goggles,** etc.
10. **Provide** apprentices and do-it-yourselfers with **refinish information and special guidance** on request.
11. **Provide all customers with the latest information regarding government regulations.**

Paint Spray Gun and Equipment Companies

Certain spray gun and equipment companies are mentioned in this book. There are many additional companies in the trade with products similar to or exactly like them. Space does not allow coverage of all spray guns. Spray gun companies:

1. **Supply and service** complete lines of spray guns and equipment to all paint jobbers and paint shops.
2. **Conduct special paint training programs** for painters and paint shops.
3. **Provide complete service information and technical assistance** to prospective paint shop owners who plan to organize, equip, and maintain up-to-date paint shops.
4. **Conduct research and development programs for** the continued **improvement of all spray guns and equipment** in cooperation with the refinish industry.

To contact a spray gun company, consult your local paint jobber.

Miscellaneous Refinish Material Companies

The following is a partial list of items available through your local paint jobber (additional items are also available):

1. Masking tape and masking paper
2. Striping tape
3. Sheet and disc sandpaper
4. Ultra fine and micro fine sandpaper
5. Cleaning solvents for paint, metal, glass, and plastic
6. Rubbing compounds
7. Car polishes

DESCRIPTION OF THE AUTOMOTIVE REFINISHING TRADE

8. Cements, compounds, and sealers
9. Plastic filler for metal repair
10. Plastic filler for flexible plastic repair

Paint Shops

Three popular types of paint shops are described in the following.

CONVENTIONAL PAINT SHOPS

Conventional paint shops:

1. Make up the greatest number of paint shops in the country.
2. Use refinish materials recommended by the factory.
3. Apply refinish material according to label directions as recommended by paint suppliers.
4. Work closely with insurance companies and charge for repair work on a mutually agreed on, flat-rate basis.

Almost all metal repair shops are connected with a paint shop that does repair work in accordance with factory recommendations. Some paint shops specialize in both **custom** and **conventional** paint work and keep busy without the aid of a collision repair shop.

CUSTOM PAINT SHOPS

Custom paint shops:

1. Hire painters who combine **artistic and refinish skills** that are the most advanced in the trade.
2. Spend more time on paint jobs because custom painting requires more time than conventional forms of painting.
3. Use exotic color materials and systems not used on factory cars.
4. Charge for custom paint work on an individual job basis. Rates are usually the highest in the refinish trade.

Some paint shops specialize in custom paint work only. Other custom paint shops are tied in with a conventional paint shop and with a collision repair shop.

HIGH-VOLUME, LOW-COST PAINT SHOPS

High-volume, low-cost paint shops are set up in a special manner, like a small assembly line, to do many paint jobs per day at a very low cost:

1. They have special paint made for them at very low cost and in only a limited set of colors.
2. They have production-line-type specialists to do surface preparation and masking.
3. Normally, they color-coat at a very fast production line speed.

These shops have the capability, on special order, to do conventional-type paint work. They can also use the conventional method of pricing. Also, these shops specialize in paint work and generally are not tied in with a collision repair shop.

INSURANCE COMPANIES

Insurance companies tie into the refinish trade by:

1. Making all driving on the roads possible. Without insurance, most drivers could not afford to be on the road, which is why driving without insurance is unlawful.
2. Using predetermined flat-rate and repair schedules established by vehicle manufacturers and automotive flat-rate study organizations to establish the cost of repairs. These schedules are updated periodically to reflect changes in the industry.
3. Competing with one another for insurance business. However, they work with one another in establishing groundwork for all metal and paint repairs.

Insurance companies pay for most of the collision repair, which runs in the billions of dollars each year.

FEDERAL, STATE, AND LOCAL GOVERNMENT REGULATION

The refinish trade has been targeted as an important contributor to air pollution because of the great amount of volatile organic compounds (VOCs) it produces each year. **VOCs are essentially the solvents used in the manufacture and application of refinish products.** The government's plan to reduce VOCs and to implement clean air regulations is carried out by the Environmental Protection Agency (EPA) and may be divided into two parts:

- Part One of the plan features a national rule on VOCs, which establishes minimum VOC limits for the entire country.
- Part Two of the plan is made up of clean air rules and regulations that must be enacted by each state based on local pollution conditions.

Table 1–1 shows how VOCs are limited (in pounds per gallon) in all refinish products for the entire country.

TABLE 1-1 National Rule VOC Limits

Product Category	VOC Limit
Pretreatment wash primer	6.5
Primer and primer-surfacer	4.8
Primer sealer	4.6
Single stage and basecoat/clearcoat topcoats[a]	5.0
Topcoats of three or more stages	5.2
Specialty coatings	7.0

[a]VOC average for the system

How the National Rule Affects Paint Shops

1. **Responsibility for compliance rests with the manufacturer.** All paint companies are prohibited from making noncompliant products since adoption of the rule in 1998.
2. **All products made before 1998 may be sold and used.** Paint companies can sell these products until inventories are exhausted. This provision allows the industry to use up the inventory of noncompliant material in the supply chain and to prevent large amounts of paint waste.
3. **Most lacquer-based products will not be compliant for general refinishing.** However, some lacquer products may be used as **specialty coatings only** in the repair of antique cars. This usage is limited to 10 percent of total production per month.
4. **The use of HVLP equipment is not required.**
5. **The rule does not regulate surface or plastic cleaners.**
6. **The rule does not require enclosed spray gun cleaners.**
7. **The rule does not require record keeping.**

See Table 1-2 for a comparison between the national rule and regional regulations.

Part two of the clean air plan includes rules and regulations that must be enacted by each state based on local pollution problems. The first state with a clean air program was California. Since California has the most severe pollution problems, it has enacted the strictest regulations since the early 1990s. Other states with an early clean air program were New York, New Jersey, Texas, and Washington.

All states are required to enact a clean air program based on local pollution problems. States can make their regulations more strict than the national rule, but they cannot make them more lenient. Because pollution problems are so different from one area to another, all districts or areas will be affected by different sets of clean air laws. No two clean air programs will necessarily be alike.

State regulations will take precedence if they are more strict than the national rule. The national rule will take effect in nonregulated states and in those states with less restrictive regulations.

In California (see Figure 1-3), groups of counties with common problems have formed their own clean air districts. (APCD stands for air pollution control district.) Each district shows its own area code and telephone number. Most states will probably organize their pollution fighting strategies similar to California's.

Compliant Areas

About 31 areas in the United States (made up of one or more counties) representing about 190 million people have been designated as having the most serious air pollution by the EPA. These areas are mostly metropolitan and the fast-growing urban areas in the West. (See Table 1-3.) These areas are known as **nonattainment areas** because they do not meet minimal air pollution standards. Because these areas need regulation, they are also known as **compliant areas.**

Least Regulated Areas

Parts of the country with no or minimal pollution problems are known as **least regulated areas.** Since pollution is not a problem, paint shops face less clean air regulations. When a paint shop owner applies for a license with the proper government authorities, the owner promises to follow all applicable rules and regulations governing the shop. To be sure that clean air laws in effect for the area are being followed, all paint

TABLE 1-2 National Rule Versus Regional Regulations

	National	Regional
VOC limits on products?	Yes	Yes
HVLP spray equipment required?	No	Yes (required in some areas)
Enclosed spray gun washers required?	No	Yes (required in some areas)
VOC recordkeeping at shop level required?	No	Yes (required in some areas)

DESCRIPTION OF THE AUTOMOTIVE REFINISHING TRADE

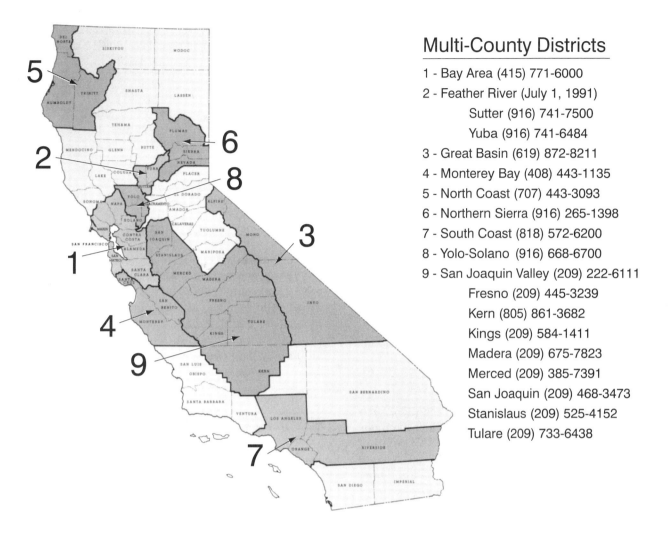

Multi-County Districts

1 - Bay Area (415) 771-6000
2 - Feather River (July 1, 1991)
  Sutter (916) 741-7500
  Yuba (916) 741-6484
3 - Great Basin (619) 872-8211
4 - Monterey Bay (408) 443-1135
5 - North Coast (707) 443-3093
6 - Northern Sierra (916) 265-1398
7 - South Coast (818) 572-6200
8 - Yolo-Solano (916) 668-6700
9 - San Joaquin Valley (209) 222-6111
  Fresno (209) 445-3239
  Kern (805) 861-3682
  Kings (209) 584-1411
  Madera (209) 675-7823
  Merced (209) 385-7391
  San Joaquin (209) 468-3473
  Stanislaus (209) 525-4152
  Tulare (209) 733-6438

County APC Districts

Amador (209) 223-6406	Lassen (916) 257-8311 x110	San Diego (619) 694-3307
Butte (916) 891-2882	Mariposa (209) 966-3689	San Luis Obispo (805) 549-5912
Calaveras (209) 754-6400	Mendocino (707) 463-4354	Santa Barbara (805) 961-8800
Colusa (916) 458-5891	Modoc (916) 233-3939 x401	Shasta (916) 225-5674
El Dorado (916) 621-5897	No.Sonoma (707) 433-5911	Siskiyou (916) 842-8029
Glenn (916) 934-6500	Placer (916) 889-7130	Tehama (916) 527-4504
Imperial (619) 339-4606	Sacramento (916) 386-6650	Tuolumne (209) 533-5693
Lake (707) 263-7000	San Bernardino (619) 243-8920	Ventura (805) 654-2806

FIGURE 1–3. California air pollution control districts. (Courtesy of California Air Resources Board and California EPA Compliance Division.)

shops in **most regulated areas** must be ready for an inspection by state, county, or district officers at all times. Paint shops that violate clean air laws are subject to severe monetary penalties. To determine in which category your area or city is classified, check with your local paint jobber or local clean air district.

Who Regulates Paint Shops?

All automotive refinish paint shops are regulated in four major areas:

1. **Fire prevention**
2. **Health and emergency services**

TABLE 1–3 Number of Unhealthful Air Days per Year for Selected Cities

The average number of days per year with unhealthful air from 1991 through 1995, according to monitoring by the Environmental Protection Agency.

City	Days
Seattle	0.6
Portland	2.6
Denver	4.2
Salt Lake City	8.8
New York	9.6
Phoenix	9.8
Las Vegas	10.2
Houston	116.0
Los Angeles	150.8

  3. OSHA
  4. **Air pollution control districts** (also known as **air quality management districts**)

More than a business permit is required to operate a paint shop. Other government agencies that protect paint shops, painters, and the environment enforce laws on specific health, safety, and environmental concerns. All paint shops should contact all agencies that oversee paint shop operations. **Learn and know all the requirements for your paint shop. Not knowing the law is a poor excuse. Keep informed.**

  1. Fire departments focus on fire safety and require safe storage and use of flammable materials. They require you to:
     a. Use spray booths.
     b. Use safety cans.
     c. Store flammable solvents in fire-safe cabinets. Contact your local fire department for their requirements.
  2. Health services and Office of Emergency Services (OES) ensure the safe disposal of hazardous wastes. They look for:
     a. Disposal of hazardous waste in approved sealed containers.
     b. Use of certified waste disposal services. Contact your local health services or OES office for their requirements.
  3. OSHA works to reduce workers' exposure to harmful chemicals. To protect the worker, OSHA may require:
     a. Use of respirators.
     b. Good ventilation in spray booths and mixing rooms. Contact your local OSHA agency for your industry's requirements.
  4. Air pollution control districts (APCDs) and air quality management districts (AQMDs) attempt to reduce evaporating VOCs and respond to complaints about paint odors and overspray. Their regulations require:
     a. Closed containers.
     b. Reduced solvent use.
     c. Maintenance of booth exhaust system to prevent overspray. Contact your local APCD or AQMD for a permit and local requirements.

The best advice for all paint shops is follow the law. By doing so, the paint shop will save money and reduce pollution. To determine how you are affected by clean air or safety regulations in any area of the country, check with your paint jobber. All paint companies and paint jobbers are advised continually by government agencies regarding clean air regulations.

REVIEW QUESTIONS

1. List several businesses and organizations that make up the refinishing trade.
2. Describe how a paint shop operates.
3. List four major skills a painter must master to become a qualified painter.
4. Name six responsibilities of a painter.
5. Name six responsibilities of a painter's helper.
6. Name five tasks the paint jobber does for the refinish trade.
7. Describe the difference between a conventional paint shop and a custom paint shop.
8. Explain how paint companies familiarize paint shop owners and painters with new paint systems when they are developed.
9. In addition to supplying paint materials, list four tasks a major paint company does for the trade.
10. How can a painter or apprentice become enrolled in a paint training program at a paint company training center?
11. List four or more items provided for the refinish trade by companies like 3M.
12. Which regulation of the federal government most directly affects the refinish trade?
13. List two major tasks performed by insurance companies for the refinish trade.
14. What is the quickest way to learn which refinish regulations are in effect in any given area at any specific time?

CHAPTER 2

How Vehicles Are Painted at the Factory and in the Refinish Trade

The purpose of reviewing how vehicles are painted at the factory is to familiarize apprentices entering the automotive refinishing trade with factory paint systems and their characteristics. **Painters should know how factory paint behaves when field repairs are made.**

DESCRIPTION OF THE PRODUCTION LINE

A production line is a specially planned and engineered assembly line, moving continually at a particular speed, on which many assembly operations are performed for the purpose of building vehicles. Production-line methods of transporting vehicle bodies are adapted to the work operations to be done. Sometimes vehicle bodies go through production on body assembly trucks, and sometimes vehicle bodies are switched to an overhead conveyor line.

FACTORY PAINTING

Factory automotive paint systems and products are the best available for painting vehicles at factories. As better systems are developed, they are added immediately. Factory paint systems are geared to the speed of the production line. Because of line speed, all materials are designed to be **high-baked** so that they can dry in time for vehicle assembly. Production paint systems are not interchangeable with field repair systems. A production paint material requiring a high bake temperature would not work satisfactorily in the refinish trade because it is not designed to **air-dry** satisfactorily.

Paint Mixing Room

All paint materials to be sprayed on a production line are received and kept in large drums in a special room called the **paint mixing room.** Paint materials are mixed and prepared in drums at a specified **viscosity.** Viscosity is defined as the resistance of a paint material to flow. The heavier bodied a material, the slower it flows. The viscosity of a paint material is decreased as the amount of solvent in the material is increased. The viscosity of a paint material is measured by a **viscosity cup** and by how long it takes **(in seconds)** for a filled cup to empty through a hole at the bottom of the cup. How to check the viscosity of a paint material is discussed in Chapter 7. All paint materials are then pumped through paint system lines to assembly-line locations, where they are applied. Once a drum of paint material is connected to factory paint lines and is pumped through the lines, paint materials are recirculated continually even when not in use at the end of the shift. This prevents **settling** of essential paint ingredients.

Next, we describe the stages through which a production line passes when going through a typical car factory paint department.

Metal Conditioning

As vehicle bodies arrive for metal conditioning, two important operations are done initially:

1. Bodies are washed with a solvent to remove chalk marks, oil, and grease.
2. Body floors and all corners are then vacuum cleaned thoroughly to remove all dirt and metal-filing particles originating in the metal finishing department.

A metal conditioning booth is like a very long car wash booth. The booth is filled with many special spray nozzles located about every 3 feet in all directions on the floor, walls, and ceiling to surround each body. Nozzles are designed to produce a **full-jet** spray pattern, in which every square inch of space is filled with droplets of liquid chemicals or water. When all nozzles are turned on, they literally soak everything in the booth. Chemically treated and heated water is pumped through the nozzles under high pressure to saturate every car body. The stages of metal conditioning are as follows:

Stage 1: An acid cleaning of the body, followed by a clear water rinse.

Stage 2: The first **chemical treatment** (usually a **phosphate coating**) is sprayed on the body. This is followed by a clear water rinse. When body metal is contacted by the chemical, a chemical coating builds instantaneously.

Stage 3: A second chemical treatment, also a phosphate coating, is sprayed on the body. This is followed by a clear water rinse.

NOTE

The number and type of chemical treatments depend on the factory and the trade name of the metal conditioning system used.

At the conclusion of metal conditioning, the chemically treated bodies proceed through a heated oven to be dried.

Primer Application

After metal conditioning, the production line passes through the primer application department. Primer is applied to car bodies in two ways:

1. Some car factories are equipped with a **dip priming system** called "ELPO."
2. Other factories are equipped with **conventional spray priming systems.**

The **dip priming system** is made possible by the use of a special water-base primer. The dip system is made up of a long, large tank filled with water-base primer. The primer is electrically charged like a negative. A moving conveyor line runs overhead. Car bodies, hanging from the conveyor line, pass through the primer and are immersed completely or as required. The car bodies are electrically charged like a positive. As car bodies pass through the dip tank, they are primed automatically to the correct film thickness. The car bodies then proceed through a high-bake oven to dry.

The **conventional spray priming system** involves the use of pressure-feed spray equipment and lines. Several types of primer, and sometimes sealer, are generally used on passenger cars. The choice depends on the car factory.

Factory primer–surfacer application is done in two ways:

1. **Manual spraying is done by hand.** This system is used to paint all hard-to-get-at areas, which includes:
   a. Body interiors
   b. Door openings
   c. Hinge pillars
   d. Rear compartment interiors
   e. Rocker panel areas
   f. Front end panels
   g. Rear end panels

2. **Robotic spraying is done by automatic equipment.** As an electric eye is actuated by the vehicle body, the robotic spray equipment begins to operate. When the robotic priming operations are completed, the vehicle is passed through a high-bake oven to dry.

Wet Sand Operations

Bodies pass through the **wet sand department** for sanding of exterior body surfaces. All bodies are sanded to specifications. When sanding is completed, all of the residue from the sanding operation is washed off and the bodies are passed through a high-bake oven to dry.

Color Application

The color coats on all production vehicles are applied with a robotic spray painting method, which involves pressure equipment and paint lines. Color coats are applied in two or more applications depending on the paint system. There are two different types of high-solid acrylic enamel paint materials used on production vehicles: **single-stage enamel colors and the base color for the two-stage basecoat/clearcoat paint.**

Spraying color coats is done in two ways:

1. **Some manual spraying is done by hand.** This method is used to paint hard-to-get-at areas, which includes:
   a. Body interiors

b. Body door openings and door facings
   c. Rear compartment interiors
2. **Robotic spraying is done with special equipment.** This method is used to paint all other areas of the vehicle that are not manually sprayed.

Most vehicle assembly plants use a **wet-on-wet spray paint process** so that there is no **overspray** when all areas of the vehicle are painted. At assembly plants where **single-stage high-solids acrylic enamel colors** are applied, the vehicles proceed through a bake-oven system for drying. At vehicle assembly plants where the **two-stage high-solids basecoat/clearcoat acrylic enamels** are sprayed, the high-solids acrylic enamel base color application is applied first. Then the vehicles proceed directly to the **high-solids acrylic enamel clear line** where two or more coats of a **high-solids acrylic enamel or polyurethane clearcoat** are applied. From there vehicles proceed through the bake-oven for drying.

Color Systems Used by Vehicle Manufacturers

1. Before 1984, Ford, Chrysler, and American Motors vehicles, including trucks, were painted with a **single-stage acrylic enamel system.** Since then, these companies have used the **high-solids basecoat/clearcoat acrylic enamel system** on most vehicles. Some trucks are still painted with **single-stage acrylic enamels.**
2. Before 1984, all GM vehicles were painted with **single-stage acrylic lacquers or enamels,** about half and half. Since then, **high-solids basecoat/clearcoat acrylic enamel** finishes have been used increasingly. GM has now achieved its goal of 100 percent **basecoat/clearcoat high-solids acrylic enamel** usage at all of its car assembly paint finish lines. At some of its truck assembly plant paint finish lines, **single-stage high-solids acrylic enamel** paints are still being applied. At all of its other truck assembly plants, **high-solids basecoat/clearcoat acrylic enamel** paint finishes are applied on the paint line.
3. Before mid-1984, most imported vehicles, including Canadian vehicles, were painted with **single-stage acrylic enamels. High-solids basecoat/clearcoat acrylic enamel** finishes have been used on most imported vehicles since 1984. **Single-stage acrylic enamel** finishes are still being used on some imported truck production.

REFINISH TRADE PAINTING

The purpose of a general review about how vehicles are painted in the refinish trade is to familiarize the apprentice with:

1. What is expected of the apprentice in the way of quality.
2. How to make field repairs.

An apprentice should know:

1. How the clean air rules affect refinishing in his or her area.
2. What types of refinish materials are permitted for use in his or her area: low VOC or conventional?
3. What spray painting equipment is required.
4. General surface preparation and general spray painting procedures.

Surface Preparation

Surface preparation is an area where painters need the most help. These jobs are normally assigned to an apprentice or a painter's helper. Performing surface preparation is the most difficult but the best way for apprentices to learn the trade. Surface preparation operations range in **degree of difficulty** from surfaces that are **very easy to prepare** to surfaces that are **difficult to prepare.** Surfaces that are **easy to prepare** are new cars that simply require color coating or compounding. Paint problems that require the most labor in surface preparation involve **rust repairs,** which is what an apprentice or painter's helper can expect a great deal of as he or she learns the trade. Rust repairs are among the most difficult to repair and to **guarantee.** However, by following proper repair fundamentals, all paint repairs, including rust repairs, can be guaranteed by a good painter for a reasonable period of time: anywhere from one to three years or more. These repairs can be done if an owner is advised of the extra parts and labor that are required.

In general, refinish systems do a very good job of repairing cars, but they are not as durable as factory finishes. This is particularly true of the factory high-bake primer and/or primer–surfacer. **Factory primer should never be removed** unless necessary.

Color Application

The painter inspects the surface preparation and masking done by the apprentice. The painter prepares, applies, and blends the color to the car. Most color application operations are done with a **suction-feed**

production-type spray gun (described in Chapter 3) in a special paint area with a good air exhaust system or in a paint spray booth, depending on the size of the repair and the type of paint used. **Enamel repairs require the most care and dirt-control measures.** Lacquer repairs dry fast and dirt is less of a concern, because they require compounding and polishing to complete the repairs. The apprentice can do compounding and polishing as he or she becomes qualified to do so.

PAINT SHOP SAFETY RULES

Compliance with sound safety rules is essential to the success and survival of a paint shop, because one bad accident can wipe out an entire business and possibly injure or kill people working in the shop. All paint materials, especially lacquers, enamels, paint solvents, and cleaning solvents, are potential fire hazards. Paint fires burn violently and escalate quickly. Vapors from evaporating solvents can quickly spread throughout an enclosed area and be ignited by sparks from electrical motors, hot lighting fixtures, lighted cigarettes, and welding torches. Before spray painting in any given locality, it is wise to check with local fire and city inspection authorities to become familiar with existing local, state, and federal safety regulations.

The federal government has established an industrial safety regulating department as a result of the **Occupational Safety and Health Act (OSHA).** OSHA regulations protect the health and safety of all people in industry, including painters. In doing so, the safety rules also protect the owner and the paint shop. Because of OSHA regulations, many safety devices that were recommended before are now required. It is better to be safe than sorry. Paint shops and painters are urged to incorporate the paint safety rules in their everyday work operations. Because of limited space in this book, it is impossible to cover all safety rules in refinishing. According to OSHA records, the following safety rules have been compiled because they are the most frequently violated, and they happen to cause the most accidents in the refinishing industry:

1. **Wear approved rubber gloves** when handling plastic body fillers, solvents, and paint materials.
2. **Use a dust respirator** (TC-21C) and **safety glasses** when performing power sanding operations and when air dusting a car.
3. **Use a cartridge-type respirator** (TC-23C) when preparing and applying nonisocyanate paint materials.
4. **Use a mask- or hood-type, air-supplied respirator** (TC-19C) when preparing and applying refinish materials containing isocyanates. Also wear lint-free, full-length plastic coveralls.
5. **Keep all paints and solvents in approved metal or wood cabinets** or rooms with explosion-proof lights and adequate ventilation (at least six air changes per hour).
6. **Keep the amount of paint used outside the approved storage area or cabinet limited to what can be used in a day.**
7. If the spray area is used for drying with portable heaters or lamps, **remove these hot surfaces before spraying takes place.**
8. **Be sure that the use of flammable cleaning solvents and all spray painting operations are at least 20 feet from flames, sparks, nonexplosion-proof electric motors, and other sources of ignition.**
9. **Keep fire control sprinkler heads clean** of built-up paint by cleaning them periodically according to sprinkler supplier recommendations. Also, you may place and tie a plastic sandwich bag over each sprinkler head to keep it clean.
10. **Be sure that No Smoking signs are posted** in the spray area, paint room, paint booth, and paint storage area.
11. **Be sure that fire extinguishers are fully charged and mounted** in designated areas and that they are tested and serviced regularly.
12. **Be sure that the spray area and areas where cleaning solvent is used are completely ventilated** before infrared drying apparatus is turned on.
13. **Avoid fires from spontaneous combustion** by disposing of soiled rags and paper at least daily. **Use metal containers with snugly fitting lids** for rags, paper, and used paint cans.
14. **Reduce compressed air dusting guns to 30 psi** (pounds of pressure per square inch) when used for cleaning purposes. Wear a face shield or tightly fitting safety goggles when doing so.
15. **Bulk storage of large volumes of paint and solvents should be in a separate building detached or cut off from the general work area by a firewall.**
16. **Remember to keep all paint material containers closed completely when not in use.** Paint material containers include solvent cans, paint cans, waste material barrel, gun washer, and soiled rags container. See Figure 2–1.

FIGURE 2–1. Reduce VOCs by closing all containers tightly.
(Courtesy of California Air Resources Board and California EPA Compliance Division.)

17. **Label instructions always advise users of flammable solvents to keep containers tightly closed when not in use.** Never punch holes in flammable solvent containers for ease of use. Doing so is a way to start a fire or explosion.
18. **Never use halogenated hydrocarbon solvents to clean aluminum parts or equipment.** Halogenated hydrocarbons consisting of methylene chloride and 1,1,1-trichlorethane can cause a violent chemical reaction when in contact with aluminum and can lead to equipment explosion.

Material Safety Data Sheets (MSDSs)

Material safety data sheets (MSDSs) can be the most important safety tool on the job. They contain information on a product's storage, compatibility with other products, handling, combustibility, and safety and health concerns.

The material safety data sheet was created and required by law as part of the Emergency Planning and Right-to-Know Act of 1986. Check the MSDS before starting any job so you will know exactly what the risks are and how to do the job safely. (MSDSs are available from paint sources free of charge.)

If you have any questions concerning a product currently used in your shop, refer to the material safety data sheet. Each MSDS sheet covers a single product and contains information on the following topics:

- **Date of preparation.**
- **Product name and code number.**
- **Section I** includes the manufacturer's name and address, a toll-free number for product information, a toll-free number for emergency situations, and a D.O.T. hazard class designation.
- **Section II, Hazardous Ingredients,** identifies hazardous components, chemical ID, and common names. Worker exposure limits to the chemical such as the OSHA permissible exposure limit (PEL), ACGIH TLV threshold limit value set by the American Conference of Governmental Industrial Hygienists, and other recommended safe exposure limits are also included.
- **Section III, Physical Data,** covers the chemical's physical and chemical characteristics, which may include appearance, odor, boiling point, melting point, vapor pressure, vapor density, evaporation rate, solubility in water, and specific gravity.
- **Section IV, Fire and Explosion Data,** contains fire and explosion hazard data, which helps you judge the risk of these hazards. The flash point tells you the minimum temperature at which vapors become flammable or explosive. Flammability limits indicate the concentration of the substance that is needed for it to ignite.
- **Section V, Health Hazard Data,** is one of the most important parts of the MSDS. It tells you how the chemical enters the body: inhaling, swallowing, or through the skin. Then it lists possible health hazards to you if you are exposed to the chemical. Some effects, like skin burns, are acute and show up right after exposure. Others, like lung cancer, are chronic. They are the result of many years of exposure or repeated brief exposures over a long period of time.
- **Section VI, Reactivity Data,** explains what can happen if this chemical is combined with other chemicals or with water or air.
- **Section VII, Spill or Leak Procedures,** gives you the steps to follow in case material is released or spilled, as well as the waste disposal method.
- **Section VIII, Special Protection Information,** covers the protective equipment you might need, as well as the work and hygiene practices and the ventilation required to keep your chances of exposure low.

- **Section IX, Special Precautions,** explains how to handle and store the substances safely as well as any other precautions necessary for you to protect yourself. Grounding containers during transfer of flammables, for example, is essential to prevent static electricity from becoming an ignition source.

All injuries in a paint shop can be prevented. Accidents do not just happen; they have causes. It is up to each painter to become thoroughly familiar with all safety rules in and around the paint shop. **Following common-sense safety practices not only prevents injury but may save someone's life.**

PAINT IDENTIFICATION

Body Number Plate

The term **body number plate** is adopted for the purposes of this book. Its function is to familiarize painters with important information and color codes for servicing the paint on a vehicle according to factory specifications. The body number plate is known by the following names by American car factories:

1. a. **Body number plate:** 1984 GM and earlier models.
   b. **Service parts identification label:** 1984 GM and later models.
2. **Vehicle certification label:** Ford Motor Company.
3. **Body code plate:** Chrysler Corporation.

Refinish operations on all vehicles are guided by paint codes on body number plates as follows:

1. Upper body color code
2. Lower body color code
3. Middle body color code
4. Accent stripe color code
5. Interior trim color code
6. Vinyl top color code, etc.

Figure 2–2 shows the typical location of body number plates on domestic and imported vehicles. For more specific information in locating the body number plate on specific model vehicles, check with the dealer service manager of the car company involved. Each car company has its particular method of locating body number plates. No two companies are alike. Figures 2–3, 2–4 and 2–5 explain more specifically the location of body number plates on General Motors and Chrysler vehicles.

General Motors Paint and Trim Codes

Figure 2–3 describes how to locate the service parts identification label on General Motors vehicles. Figure 2–4 explains how to identify the paint

FIGURE 2–2. Typical vehicle body number plate locations. (Courtesy of DuPont Company.)

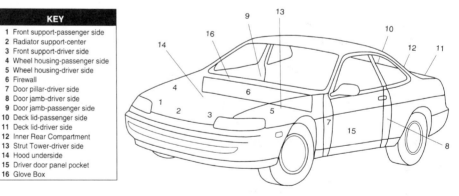

KEY	
1	Front support-passenger side
2	Radiator support-center
3	Front support-driver side
4	Wheel housing-passenger side
5	Wheel housing-driver side
6	Firewall
7	Door pillar-driver side
8	Door jamb-driver side
9	Door jamb-passenger side
10	Deck lid-passenger side
11	Deck lid-driver side
12	Inner Rear Compartment
13	Strut Tower-driver side
14	Hood underside
15	Driver door panel pocket
16	Glove Box

MODEL	POSITION	MODEL	POSITION	MODEL	POSITION
ACURA	8	FERRARI	10	MITSUBISHI	6
ALFA ROMEO	4,11	FESTIVA	7,8,9	NISSAN	6,8
AMC	7,8	FIAT	4,6,11	OPEL	1,2,3,4,5,6
AUDI	10,11	FORD	8	PEUGEOT	2,3,4,5,6
AUSTIN ROVER	15	GENERAL MOTORS	2,10,11,12,16	PORSCHE	7
BMW	4,5	GMC	16	ROVER	1,3,4,5
CAPRI	8	GEO	10,11	SAAB	5,6
CHRYSLER		HONDA	8	SATURN	11
Car / voitures / carro	2	HYUNDAI	8	STERLING	7
Imports / voitures importées /		ISUZU	6	SUBARU	13
vehículos importados	6	JAGUAR	8	SUZUKI	6
Truck & van / camions et fourgon-		KIA	8	TOYOTA	8,15
nettes / camión, camioneta	14	LAND ROVER	8	VOLKSWAGEN	2,9
DAIHATSU	1,6	LEXUS	8	VOLVO	5
DATSUN	2,6	MAZDA	8	YUGO	10
		MERCEDES BENZ	2		

HOW VEHICLES ARE PAINTED AT THE FACTORY AND IN THE REFINISH TRADE

SPARE TIRE COVER

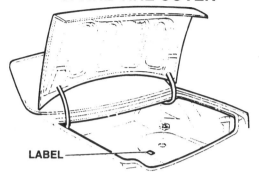

"A" - EXCEPT WAGON
"C, E, H, J, K, L, M, N, R, S, W,"
SATURNS

REAR QUARTER PANEL

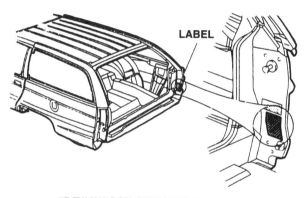

"B/D" WAGON (WITHOUT B9Q)

RIGHT WHEEL HOUSE

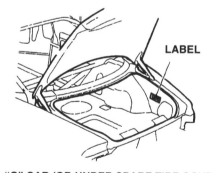

"G" CAR (OR UNDER SPARE TIRE COVER)
"A" WAGONS ONLY

REAR DECK LID

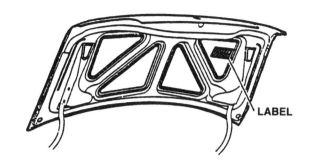

"B/D" SEDANS

I/P COMPARTMENT/ GLOVE BOX

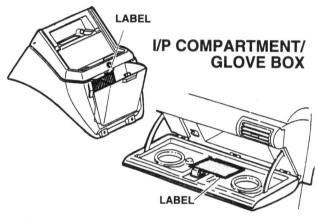

"B/D" WAGON WITH B9Q, F-CAR, VANS,
TRUCKS ("P" TRUCK MAY BE ON RH
SUN VISOR) GEO TRACKER

CONSOLE DOOR

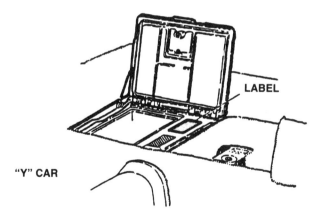

"Y" CAR

FIGURE 2–3. Service parts identification label locations on GM vehicles. (Reprinted with permission of General Motors Corporation.)

and trim codes for refinishing various components or areas of General Motors vehicles. All paint and trim codes on a service parts identification label are listed as factory options along with all the options on a vehicle. Paint and trim codes are usually found along the lower area of the label. The differences between paint and trim codes and chassis options are as follows:

1. All paint and trim codes start with a two-digit number that is followed by a single letter; an example is 93U. The 93 stands for the paint codes and the U stands for the upper body color.

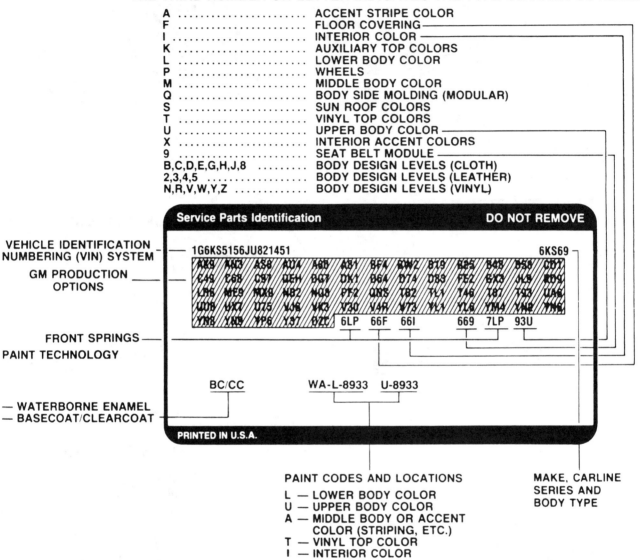

FIGURE 2-4. Location of paint and trim codes on GM vehicles. (Reprinted with permission of General Motors Corporation.)

2. All chassis codes start with a letter that is followed by a two-digit number. All chassis options or codes have nothing to do with paint. Chassis options refer to power steering, power brakes, etc.

The primary paint and trim codes are as follows:

L   lower body color
U   upper body color
A   middle body or accent color (striping, etc.)
T   vinyl top color
I   interior color

Ford Paint and Trim Codes

The paint and trim codes on Ford cars are stamped at the bottom of the vehicle certification label, as follows:

1. **Paint code:**
   a. The paint code is stamped above the words "Exterior Paint Colors."
   b. Ford paint codes consist of either
      (1) a single number followed by a single letter (example: 3P) or
      (2) a double-digit number (example: 75).

c. If a complete car is painted with a single color, the paint code appears once (example: 59).
d. If a Ford car is painted with a two-tone paint combination, a two-digit paint code appears twice. The first code stands for the lower body color. The second paint code stands for the upper body color (example: 45-9D).
2. **Trim code:** The Ford trim code is stamped on the label under the words: "Int Trim." The code consists of letters and numbers, or repeated letters (example: DD). There is no set pattern to Ford trim codes. They often change from one model year to the next. Ford trim codes are shown on paint supplier color charts.

Chrysler Paint and Trim Codes

On Chrysler vehicles, the body code plate is located in various locations on various model vehicles, as shown in Figure 2–5. Chrysler paint and trim codes consist of a combination of letters and numbers or all letters.

1. **Paint code:** The paint code is located in row 2 from the bottom and on the left side of the plate, as shown.
   a. On one-color vehicles, the paint code appears once (example: HM3).
   b. On two-tone vehicles, a second paint code appears to the right of the first code (example: DX8 – GW7).
2. **Trim code:** The trim code is located on the right side of the paint code and on the same line, as shown in Figure 2–5 (examples: FF; D5 – X9).

Whenever a painter cannot locate a body number plate, he or she is urged to call the nearest car dealer service manager or paint jobber for advice. If a body number plate is missing and the car is to be painted the same factory color, the local paint jobber can determine the color of paint for the car if the following information is known:

1. **Make and model year of car:** This information can be determined from the car owner's car registration.
2. **Sample of paint required:** Bring the car or a small exterior or interior painted part of the car for which paint is required, such as a gas tank cover door or glove compartment door, to the jobber for analysis and color matching. With this information most jobbers can determine and match the required car color with factory-approved color formulation.

COLOR CHARTS

A color chart is assembled for each car factory each model year. The color chart is the principal means by which a paint supplier communicates the ready availability of automotive refinish colors to the refinish trade. Each paint shop should have a book of color charts from each paint supplier and/or jobber in the area. Color charts should be kept in appropriate three-ring binders especially made for this purpose. Color books should cover all factory car line colors, including truck colors, for a given model year and for several years back. American and imported car color books should be kept separately for quick and easy reference. Color charts and binders can be requested from each paint supplier for paint shop use.

Exterior Colors

1. Each factory color code is listed in alphabetical or numerical order.
2. Next to each color code is a factory color control number. This number may be on the reverse side of the color chart.
3. The paint supplier color code is next to each factory color code.
4. A sample color chip next to the color code shows the painter what the color looks like.
5. At each color code is a car factory color usage designation. This information indicates that a particular color belongs to the car line shown.

Interior Colors

There is a big difference between interior and exterior colors, and the two should never be confused or interchanged. Interior colors are designed for car interiors. Each interior color is available in two **gloss levels:**

1. **Flat colors** are required in all driver-vision areas to prevent **glare** to drivers. This is a federal regulation. Most color charts list those parts that must be painted with a **flat** color. If necessary, refer to another paint supplier's color chart for this information.
2. **Semigloss colors** are usually recommended to paint all other interior trim parts.

Selection of the proper color and paint system for interior parts is covered in Chapter 15.

18 CHAPTER 2

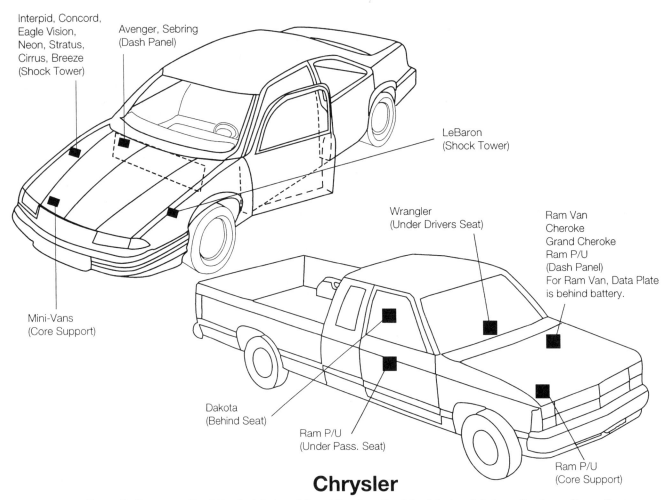

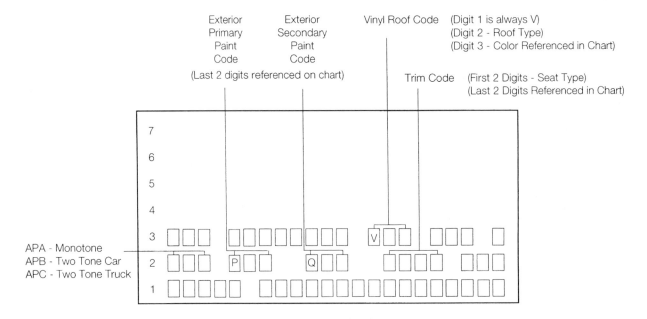

FIGURE 2–5. Location of paint and trim codes on Chrysler vehicles. (Courtesy of Chrysler Corporation.)

The trim code or interior color code is essential to ordering and painting of interior trim parts. Also, the trim code tells the following about a car interior:

1. Basic color of interior (this includes two-tone interiors)
2. Type and design of interiors:
   a. Cloth and vinyl
   b. All vinyl
   c. Leather

Color charts explain the availability of interior colors in two ways:

1. Some paint suppliers use **color chips** to aid identification and ordering of interior colors. These charts use less descriptive information and are easy to use.
2. Other paint suppliers do not use color chips to identify interior colors. They use a direct **trim code, color name,** and **stock number** system.

Additional refinish color systems are available through color charts for automotive refinishing as follows:

1. Striping colors
2. Vinyl roof colors
3. Painted vinyl roof molding colors
4. Luggage compartment interior colors

ORDERING COLORS FROM A PAINT JOBBER

Exterior Colors

> **NOTE**
>
> Repair colors are available in two basic types. **Single-phase colors** have been used for many years. **Two-phase colors** are the **basecoat/clearcoat types** used increasingly in recent years. Each system is entirely different. They are explained in detail in Chapter 9.

1. **Refer to the proper color chart (model year and make of car).**
   a. Check the paint code of the car against the color chart and color chip.
   b. Check the WA number (a paint color identification prefix) of the car with the chart (if GM). The number may be on the reverse side of the chart.
   c. If the car is a GM car, determine the type of paint on the car from the service parts identification label. When the color codes in steps (a) and (b) above match, the correct color is determined. If the car is not a GM car, determine the paint type by the color chart or by the sanding test (see Chapter 9). If problems are encountered, call the paint jobber.
2. **Decide on the repair system to be used.**
   a. Repair single-phase color systems with single-phase products. The paint jobber can help determine which system is best for the situation.
   b. If the car is basecoat/clearcoat or tricoat finish, a two- or three-phase repair system should be used. The paint jobber can help determine which basecoat system is best for the situation.

> **CAUTION**
>
> Paint systems in (a) and (b) above should never be interchanged.

3. **Recheck all paint ordering facts with the paint jobber before ordering the paint.**
   a. Make and model year of car
   b. Color codes (and WA numbers)
   c. Type of paint system and color stock numbers
   d. Amount of color and/or clear needed
   e. Amount and type of solvent, hardeners, and the like, needed

> **NOTE**
>
> **It is easy to make a mistake when reading a body number plate.** A mistake could mean that the wrong paint could be ordered. If the color is factory-packaged, this is not much of a problem. Factory-packaged paint can be exchanged without a loss. However, if the wrong paint color is ordered and formulated by the jobber (which means the jobber makes up paint with special equipment), the paint shop must pay for the loss.

Interior Colors

1. Refer to the proper color chart.
2. Locate the trim code or desired color chip on the color chart.
3. Determine if the color is to be:
   a. **Flat** color. See flat color usage on the color chart. This usage is determined by the part to be painted. Record the stock number.
   b. **Semigloss** color. All the other paintable parts on the interior are painted in this finish. Record the stock number.
4. Determine the amount of paint needed.
5. Recheck and review the color ordering facts with the paint jobber before authorizing the paint order.

REVIEW QUESTIONS

1. Describe each of the following to explain how cars are painted in a car factory:
   a. Production line
   b. Paint mixing room
   c. Metal conditioning
   d. Manual spray painting
   e. Robotic spray painting
   f. "Dip" priming
   g. Wet sanding primer–surfacer
   h. Oven baking
2. Name the basic paint color system used by the following car factories:
   a. Chrysler
   b. Ford Motor Company
   c. General Motors
3. Explain where the body number plate is found on the following cars:
   a. Chrysler
   b. Ford
   c. General Motors
4. Explain where the paint code is found on the body number plate of the following car factories:
   a. Chrysler
   b. Ford
   c. General Motors
5. Explain specifically where the trim code is found on the body number plate of the following car factories:
   a. Chrysler
   b. Ford
   c. General Motors
6. Explain how to tie in the paint code (and WA number) with the paint supplier stock number and where these stock numbers are displayed for the painter.
7. Explain how to tie in the trim code with the paint supplier stock number and where these stock numbers are displayed for the painter.
8. At what gloss level are interior trim colors available for the following?
   a. Driver's vision-area trim parts
   b. Side wall and seat trim parts
9. Explain how to order car exterior color from the paint jobber.
10. Explain how to order car interior color from the paint jobber.
11. Describe ten paint shop safety rules for fire prevention.

CHAPTER 3

Paint Spray Guns and Paint Cups

INTRODUCTION

Chapter 3 covers the basic description, construction, and operation of several popular spray guns used in the refinish trade. Many additional spray guns are available but space does not allow coverage of them all. For information on spray guns not covered in this book, check with your local paint jobber. The author does not favor any particular spray gun over any other; however, the author does mention outstanding features of spray guns discussed.

In this book, all spray guns are divided into two categories:

1. **Conventional high-pressure spray guns**
2. **High-volume, low-pressure (HVLP) spray guns**

IMPORTANT

Before spray painting, check with the local paint jobber regarding federal and local laws. Paint jobbers are always happy to advise you regarding existing laws, special techniques, or proper use of their products in any specific area of the country.

The paint spray gun is considered the heart of the spray painting operation and is the painter's most important tool. A painter should have a complete understanding of how a spray gun works. The apprentice painter learns how to understand a spray gun better by carefully taking all the parts off the gun body, studying the purpose and construction of each part, and putting the spray gun back together. A painter should clean the spray gun after every use. Also, a painter should lubricate the spray gun after each cleanup.

CAUTION

All spray gun companies recommend a degree of disassembly and assembly of spray guns for maintenance purposes when needed. However, **before disassembling any particular spray gun, be sure to read the spray gun technical manual.** Some spray guns have components that require special factory tools for proper disassembly. To be safe, follow the manufacturer's specific instructions.

CONVENTIONAL HIGH PRESSURE SPRAY GUNS

Paint Spray Gun—Definition

The **automotive paint spray gun** is a precision tool that **atomizes sprayable liquid paint materials with compressed air and applies the materials to the surface to be painted.** *To atomize* means to break up a liquid paint material into small particles. Air and liquid paint materials enter the gun through separate passages (Figure 3–3) and are ejected and mixed at the air cap in a controlled pattern (Figure 3–6). When sprayed on a surface, the paint particles, made up of paint material and solvents, stick to the surface and flow together to form a smooth, uniform coating.

Normally, all conventional high-pressure spray guns operate in the range of 30 to 70 psi. For specific air pressures to apply specific products, always follow product label directions and local regulations.

Types of Spray Guns

The spray guns used most often in the automotive refinish trade are available in three types: **suction-feed, top-feed,** and **pressure-feed.** The suction and top feed guns are the most popular.

SUCTION-FEED SPRAY GUN

The suction-feed spray gun (Figure 3–1) is easily identified by **the fluid tip, which extends slightly beyond the air cap.** This type of spray gun is operated by a stream of compressed air that passes by the fluid tip to create a partial vacuum. This vacuum causes atmospheric pressure to force paint material from an attached container through the spray head of the gun. The suction-feed spray gun is usually limited to quart-size or smaller containers. Suction-feed spray guns are used most popularly in paint repair shops because many color changes can be made quickly and small amounts of materials are generally used.

TOP-FEED SPRAY GUN

On top-feed spray guns (also known as "gravity-feed spray guns"), **the paint cup is mounted on the spray gun above the fluid tip.** As the term implies, paint material is fed through the fluid tip by gravity. Top-feed spray guns are like suction-feed guns because both must have a cup vent hole that must be kept open for proper operation of the gun. Top-feed guns are easy to use. Top-feed spray guns are adaptable for use in conventional high pressure and in HVLP paint systems.

The top-feed spray gun is very popular in automotive refinishing for the following reasons:

1. It is low in cost.
2. It is easy to operate and clean.
3. It requires no special attachments.

PRESSURE-FEED SPRAY GUN

On a pressure-feed spray gun (Figure 3–2) **the fluid tip generally is flush with the air cap.** The fluid tip and air cap on pressure-feed guns are not designed to create a vacuum. Paint material is forced to the gun by compressed air pressure (usually 8 to 11 psi) acting on the material in the cup or tank. A pressure-feed system

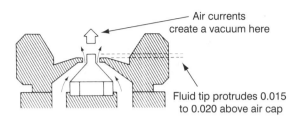

FIGURE 3–1. Suction-feed spray gun. (Courtesy of ITW Automotive Refinishing.)

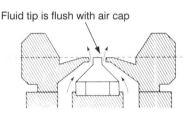

FIGURE 3–2. Pressure-feed spray gun. (Courtesy of ITW Automotive Refinishing.)

is used when large amounts of the same material are being used, when fast application is required, or when the material is too heavy for the suction-feed gun.

Parts of the Spray Gun

Figure 3–3 shows a cutaway view of a typical Binks HVLP spray gun. HVLP spray guns are designed to deliver paint with a high volume of air (HV) and a low air pressure (LP) at or below 10 psi. HVLP spray guns are the most efficient in the trade because they meet a government regulation called "transfer efficiency of 65 percent." HVLP spray guns are required by law in the most highly regulated areas, like California. HVLP spray guns have these special features:

1. A special venturi that limits the air pressure. (HVLP spray guns that are not designed with a venturi to control air pressure, like Accuspray and DeVilbiss models, require a 1 to 10 psi pressure gauge at the base of the handle.)
2. Extra large air passages.
3. High-volume air caps.
4. Specially adapted fluid nozzles.

Most other parts of HVLP guns are similar to conventional spray guns with regard to spray fan controls, fluid controls, pressure-feed controls, etc.

PAINT SPRAY GUNS AND PAINT CUPS

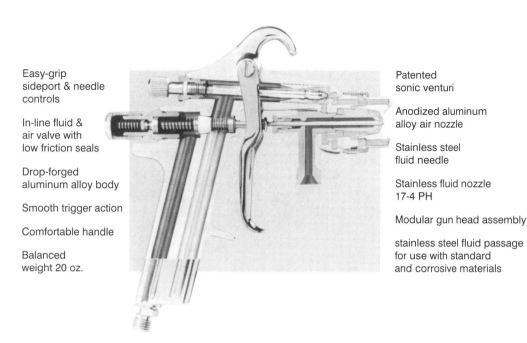

FIGURE 3–3. Cross-section of Binks HVLP spray gun. (Courtesy of ITW Automotive Refinishing.)

> **CAUTION**
>
> When available, **use special company-supplied spray gun wrenches for spray gun disassembly and assembly.** Otherwise, use a properly fitting wrench on each part to be removed. **Avoid stripping attaching parts.** Use a vise with care to hold the gun body when removing and installing the fluid tip.

The name, purpose, and description of each spray gun part follows.

AIR CAP

The purpose of the air cap (ITW, Figure 3–4, item 1) is to break up a solid stream of liquid paint into a spray fan made up of finely atomized paint particles and to direct the spray fan toward the surface to be painted. The spray fan is formed by equal and opposing forces of air coming from the horn holes (and containment and auxiliary holes, if present) of the air cap (Figures 3–5 and 3–6). The containment and auxiliary holes aid atomization. All holes in air caps must be kept clean at all times, not only after painting operations but during spraying operations. Air caps do their best work when used with matching fluid nozzles and fluid needles. For complete details regarding matching sets of fluid nozzles, needles, and air caps, check with your local equipment supplier or paint jobber.

FLUID TIP

The purpose of the fluid tip (ITW, Figure 3–4, item 3), working with the fluid needle, is to create a round, controlled stream of paint at the fluid tip when the trigger is operated. Also, the fluid tip forms a seat for the fluid needle to shut off the stream of paint when the trigger is released. Fluid tips and needles are available in various sizes and are made to fit each other in matched sets.

FLUID NEEDLE

The fluid needle (ITW, Figure 3–4, item 14) is designed to limit and control the amount of fluid leaving the fluid tip. The fluid needle is controlled by the trigger. It is fully adjustable by means of the fluid control knob.

FLUID NEEDLE PACKING NUT

The purpose of the packing nut (ITW, Figure 3–4, item 17) is to tighten the packing on the fluid needle to prevent leakage at this point. To adjust the packing more tightly, turn the packing nut clockwise with your fingers.

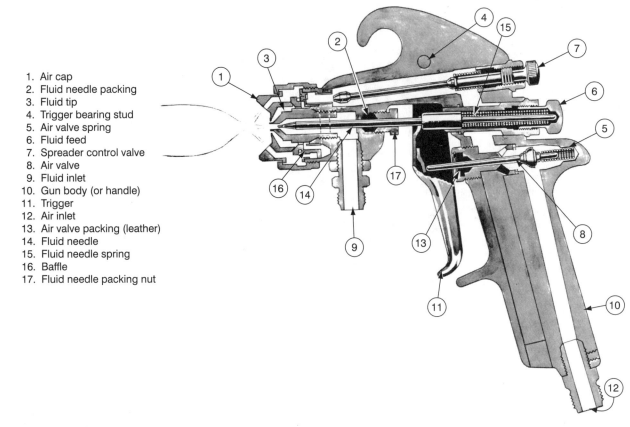

FIGURE 3–4. Cross-section of a typical ITW–DeVilbiss spray gun. (Courtesy of ITW Automotive Refinishing.)

1. Air cap
2. Fluid needle packing
3. Fluid tip
4. Trigger bearing stud
5. Air valve spring
6. Fluid feed
7. Spreader control valve
8. Air valve
9. Fluid inlet
10. Gun body (or handle)
11. Trigger
12. Air inlet
13. Air valve packing (leather)
14. Fluid needle
15. Fluid needle spring
16. Baffle
17. Fluid needle packing nut

FLUID NEEDLE PACKING

The purpose of the fluid needle packing (ITW, Figure 3–4, item 2) is to seal the fluid needle at this point during spray painting operations. The packing should be lubricated as described in the section "Spray Gun Lubrication" in Chapter 4. The packing must be tight against the fluid needle while allowing the needle to operate freely.

FLUID NEEDLE SPRING

The fluid control spring is designed to shut off the material flow when the trigger is released. The spring (ITW, Figure 3–4, item 15) should be lubricated periodically as described in "Spray Gun Lubrication" in Chapter 4.

FLUID FEED VALVE

The purpose of the fluid feed valve (ITW, Figure 3–4, item 6) is to control the amount of fluid leaving the fluid tip when the trigger is operated. When the knob is turned all the way in (clockwise), no fluid can leave the fluid tip. When the fluid feed valve is turned all the way out, counterclockwise until one thread shows, the fluid valve is wide open. Between fully open and almost closed is a full range of adjustments that the painter can use in the course of all types of refinish work.

SPREADER CONTROL VALVE

The purpose of the spreader control valve (ITW, Figure 3–4, item 7) is to control the amount of air going through the horn holes of the air cap. The amount of air going through the horn holes determines and controls the width of the spray fan. When air is cut off completely from the horn holes, the spray forms a small round pattern. This adjustment is done by turning the spreader control valve all the way in, clockwise. For spot repairs and solvent blending, a midrange setting of the valve is used. This is a medium-size fan. For full panel painting and complete refinishing, a full spray fan is used. To open the spreader control valve fully, turn the valve counterclockwise all the way. This adjustment produces a spray fan 10 to 12 inches wide.

PAINT SPRAY GUNS AND PAINT CUPS

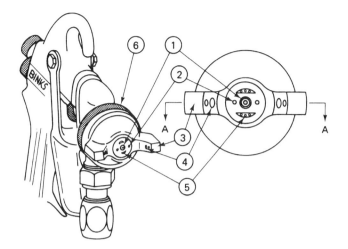

FIGURE 3–5. Air cap construction. (Courtesy of ITW Automotive Refinishing.)

1. Round opening around fluid tip
2. Containment holes
3. Horns of air cap
4. Pattern control holes
5. Auxiliary holes
6. Air cap retaining ring

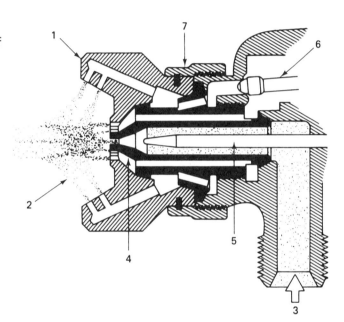

FIGURE 3–6. Atomization air and material flow. (Courtesy of ITW Automotive Refinishing.)

1. Air nozzle
2. Atomization air
3. Material flow
4. Fluid nozzle
5. Fluid needle
6. Spray pattern control stem
7. Air nozzle retaining ring

TRIGGER

The purpose of the trigger (ITW, Figure 3–4, item 11) is to actuate the air valve and the fluid feed valve. The trigger is designed to fit two fingers and the handle is designed for the other two fingers. Initial operation of the trigger allows the spray gun to work as an air gun only. Further operation of the trigger causes contact with and movement of the fluid needle. Movement of the needle from the fluid tip seat causes the gun to spray paint.

TRIGGER BEARING STUD AND SCREW

The stud and screw (ITW, Figure 3–4, item 4) retain the trigger to the gun body and allow the trigger to pivot at this point. The bearing stud should be lubricated as described in Chapter 4, "Spray Gun Lubrication."

AIR VALVE

The purpose of the air valve (ITW, Figure 3–4, item 8) is to control the flow of compressed air through the spray gun air passages when the trigger is pulled. Initial operation of the air valve allows only a small amount of air to pass through the gun. Further operation of the valve opens the fluid needle at the fluid tip, which causes paint to be sprayed.

AIR VALVE PACKING

The purpose of the air valve packing (ITW, Figure 3–4, item 13) is to seal the valve stem at this location. Packings are commonly made of leather. The packing should be lubricated as covered in Chapter 4, "Spray Gun Lubrication." To improve the seal, simply tighten the packing nut with the fingers.

AIR VALVE SPRING

The air valve spring (ITW, Figure 3–4, item 5) causes the air valve to close when the trigger is released, which shuts down the operation of the spray gun.

GUN BODY

Most gun bodies (ITW, Figure 3–4, item 10) are made of a hard, tough aluminum alloy. Thus, they are balanced, easy to handle, and very durable. The gun body is the principal unit of the spray gun to which all other parts attach. The handle of the gun has a raised flange that fits the fingers to help support the gun as it is being used. Properly cared for, gun bodies can last a lifetime.

FLUID INLET

The fluid inlet (ITW, Figure 3–4, item 9) guides paint material from the container, through the material passages, and through the fluid tip.

> **CAUTION**
>
> All fluid passages and connections must be absolutely airtight or the spray operation will be faulty. The thread on the fluid inlet is usually $\frac{3}{8}$-inch NPS.

AIR INLET

The air inlet (ITW, Figure 3–4, item 12) at the base of the gun handle has threaded provisions for an air-line adapter. A quick-detach connector or threaded connector attaches to this adapter.

BAFFLE

Some spray guns are equipped with a baffle (ITW, Figure 3–4, item 16) at the spray head of the gun. The chief purpose of a baffle is to direct a uniform amount of air to the air cap horn holes, which helps to achieve a balanced spray pattern. On some spray guns, the baffle is a separate part. On other spray guns, the baffle principle of equal air distribution is designed into the spray head of the gun.

Principles of Suction-Feed Spray Gun Operation

Refer to Figure 3–7. Initial operation of the trigger opens the air valve but does not move the fluid needle, which allows compressed air to travel past the

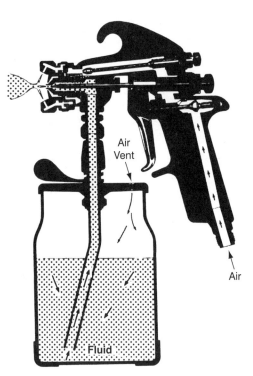

FIGURE 3–7. Principles of suction-feed spray gun operation. (Courtesy of ITW Automotive Refinishing.)

air valve, past the fluid control bypass, through the gun air passages, and out of the center opening of the air cap. At this stage the spray gun operates only as an air gun.

As the trigger is operated beyond the initial movement, it opens the fluid needle and opens the air valve more. The compressed air, traveling out of the center opening of the air cap and past the extended fluid tip, creates a partial vacuum at the fluid tip. A **vacuum** may be defined as the partial or complete absence of air. **Atmospheric pressure** (see the Glossary at the end of the book) pushing on the fluid in the cup through the vent hole forces the fluid through the fluid passages of the gun and out of the fluid tip. The ejected fluid and compressed air mix to form a spray pattern.

The air cap functions as follows. The air cap horn holes are equal to and opposite each other. The counteracting forces of air coming out of the horn holes create a flat but wide spray pattern. As the volume of air going through the horn holes increases, the spray fan becomes wider. The size of the spray pattern is controlled by the spreader control valve, which controls the amount of air passing through the air cap. The amount of fluid passing through the gun is controlled by the fluid feed valve. The range is from just barely open to full open.

Principles of Top-Feed Spray Gun Operation

On top-feed spray guns (see Figures 3–8, 3–9, and 3–10), the paint cup is mounted on the spray gun above the fluid tip. The spray gun is designed to operate while it is upright, as follows:

1. Initial operation of the trigger operates the air valve only, and the gun operates as an air gun.
2. The trigger is operated beyond the initial movement. At the same time:
    a. On gravity-feed guns, gravity initiates movement of the paint material from the paint cup, through the fluid passages, and out the fluid tip. At this stage, the air cap forms the atomization and spray fan.
    b. On pressure-feed guns, pressure in the paint cup initiates movement of the paint material from the top-load paint cup, through the fluid passages, and out the fluid tip. On pressure guns, the fluid movement is instantaneous and more positive. Pressure guns perform better when applying heavy bodied, high-solids paint materials.
    c. Because a single pressure source is used on the top-feed HVLP spray gun, the pressure on the paint in the cup is the same as the pressure at the air cap, which is similar to Accuspray's basic gun operation. Adjustment of the gun is explained in Chapter 4.

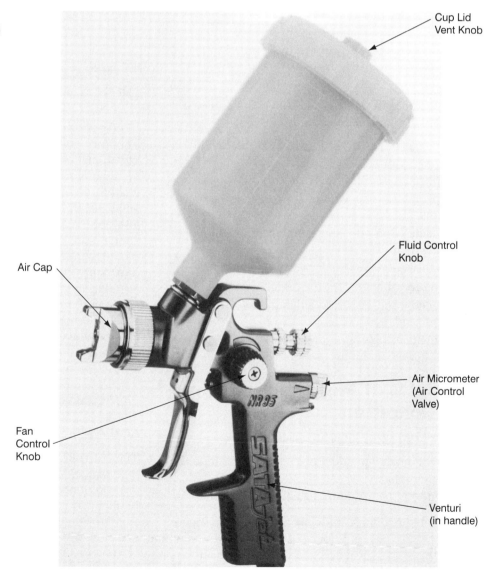

FIGURE 3–8. SATA Jet/B-NR top-feed 95/HVLP gravity-feed spray gun. Top-load spray guns are available in two forms: *gravity feed* and *pressure feed*. (Courtesy of SATA Spray Equipment Company.)

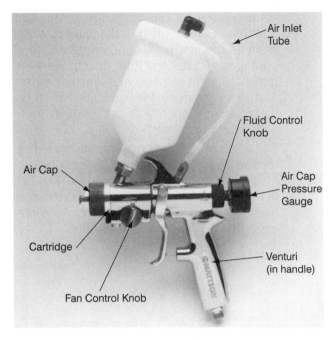

FIGURE 3–9. Mattson "IQ" top-feed HVLP pressure-feed spray gun. (Courtesy of Mattson Spray Equipment, Inc.)

Top-feed HVLP spray guns have the following additional design features:

1. Special air restrictor (venturi) in gun handle on some guns.
2. Special low-pressure gauge on some guns.
3. Special air cap with extra large holes on all guns.
4. Special fluid tip and needle combinations to match air caps.
5. Lightweight designs for ease of handling.
6. Comfortable grip for the painter.
7. Speedy and accurate adjustments for the fan and fluid feed.
8. Advanced HVLP atomization of all refinish materials, including waterborne products.

Principles of Pressure-Feed Spray Gun Operation

Pressure-feed spray guns are available:

1. With the independent pressure regulator on the paint cup.
2. Without the pressure regulator on the paint cup.

Like most spray guns, initial operation of the trigger opens the air valve but does not move the fluid needle. At this stage, the gun operates only as an air gun.

HVLP spray guns by Binks, DeVilbiss, and Mattson are examples of spray guns with independent cup regulators (see Figure 3–11). As the trigger is pulled beyond the initial movement on these guns, the following three events happen simultaneously:

1. The fluid needle opens from the fluid tip.
2. The independently regulated air pressure in the cup forces paint up the fluid tube.
3. A stream of paint leaves the fluid tip at the air cap that, in turn, forms the spray fan.

Accuspray is an example of a spray gun without a cup regulator. As the trigger is pulled beyond the initial movement for this type of gun, the following three events happen simultaneously:

1. The fluid needle opens from the fluid tip (see Figure 3–12).
2. Air line pressure in the paint cup (being the same as at the air cap) forces paint up the fluid tube.
3. A stream of paint leaves the fluid tip at the air cap that, in turn, forms a spray fan.

FLUID CONTAINERS: GENERAL DESCRIPTION

Fluid containers are made of metal, plastic, or glass. They are available in many sizes and in two basic types: suction- and pressure-feed. Very small suction-feed paint cups, 1- and 2-ounce capacity, are used in custom paint work with an air brush. Next in size, small spray guns and cups, such as the DeVilbiss Model EGA and Binks Models 15 and 26, are used for spot repairs and for gaining access to small areas where painting cannot be done with the full-size standard spray gun. The 1-quart paint cup is the most popular in the suction-feed system. Pressure-type containers range in size from 1 quart to several gallons. The most popular are the 2-quart remote cup and the several-gallon tank-type containers. These containers are used in large paint shops that do high-volume paint work.

Suction-Feed Cups

CONVENTIONAL SUCTION-FEED CUP

The conventional suction-feed cup (shown in Figure 3–7) is a small vented container ranging in size from several-ounce capacity to 1 quart. The vent hole must be clear to allow atmospheric pressure to force the fluid through the spray gun. The proper installed position of a suction cup cover is with the vent hole at the rear of the cup. This position minimizes the possibility of fluid drippage at the vent hole during paint spray operations on a horizontal surface.

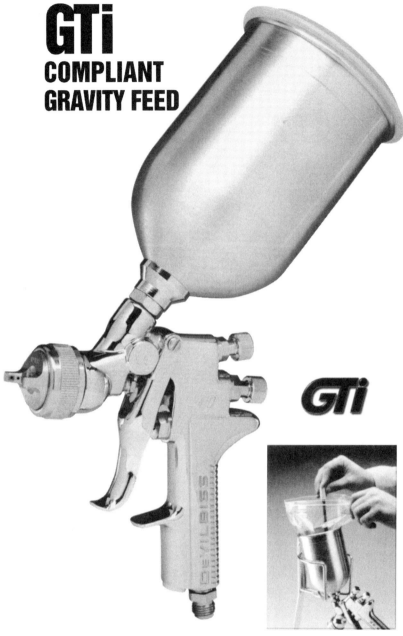

GTi COMPLIANT GRAVITY FEED

Developed using the most advanced techniques in computational fluid dynamics

Complies with regulations requiring HVLP

Waterborne compatible – stainless steel fluid tip and needle, nickel-plated gun body inside and out

Accurate, repeatable settings

Quick cleanup and upside-down spraying (with no cup lid leaks) with E-Z liner disposable cup liners

Virtually maintenance-free:
 No gaskets – self sealing
 One-piece air valve is smooth
  and efficient

Includes:
 Air adjusting valve with gauge
 32-oz. aluminum cup
 Gun hook
 9 disposable cup liners

Specifications:
 30 psi air inlet pressure delivers
 10 psi air cap pressure at 16 cfm
 air volume required

192055 (GTI-5033-100) air cap test kit

GTi Compliant Gravity Feed

Order No. Model No.	Fluid Tip (mm)	Inlet Air Pressure	Applications
170145 **GTI-600G** (includes 1.4 mm and 1.6 mm fluid tips)	1.4	20-25 psi	Base coats
		25-30 psi	High solids clear coats and single stages
		25-30 psi	Waterbornes
	1.6	20-25 psi	Base coats
		25-30 psi	Low solids clear coats and single stages

FIGURE 3–10. ITW–DeVilbiss top-feed gravity-feed HVLP spray gun. (Courtesy of ITW Automotive Refinishing.)

FIGURE 3–11. Mattson LP-2Q HVLP remote 2-quart pressure cup. (Courtesy of Mattson Spray Equipment.)

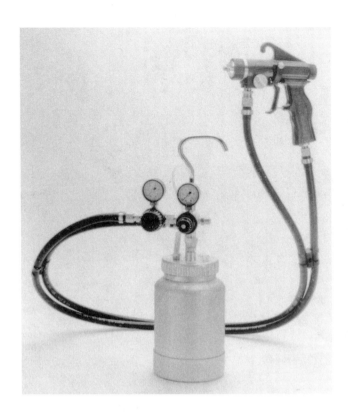

FIGURE 3–12. Binks MACH 1 Plus HVLP spray gun. (Courtesy of ITW Automotive Refinishing.)

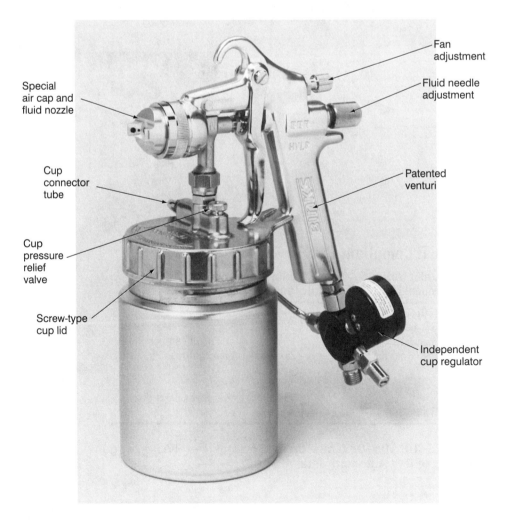

PAINT SPRAY GUNS AND PAINT CUPS

AGITATOR-TYPE SUCTION-FEED CUP

Several agitator-type cups designed for use on suction-feed spray guns are available in the refinish trade. The Binks agitator cup in Figure 3–13 works with a vibrating mechanism in standard one-quart, suction-feed cups and two-quart, pressure-feed cups. Agitation is accomplished by a large perforated metal disc near the bottom of the cup. Speed of agitation is controlled by an air-adjusting valve.

The features of the agitator cup are as follows:

1. Complete control of agitation at all times.
2. Dripless.
3. Air-pressure control valve and air gauge at gun ensure proper air pressure while spraying.
4. Available in 32-ounce (1-quart) and 40-ounce cups.

The benefits of the agitator cup are as follows:

1. Eliminates streaks in any finish.
2. Helps in color matching.
3. Great for spot repair and overall finishing.
4. Makes application of Metalflake and other metallic finishes as easy to apply as solid colors.
5. Highly recommended for custom painters.

Pressure-Feed Cups

A pressure-feed cup is a material container usually 1 or 2 quarts in size. One-quart pressure cups almost always attach directly to the spray gun. When larger containers are used, the containers are separated from the gun by hoses because of the heavy weight involved. On pressure equipment, the material is forced from the container to the gun by compressed

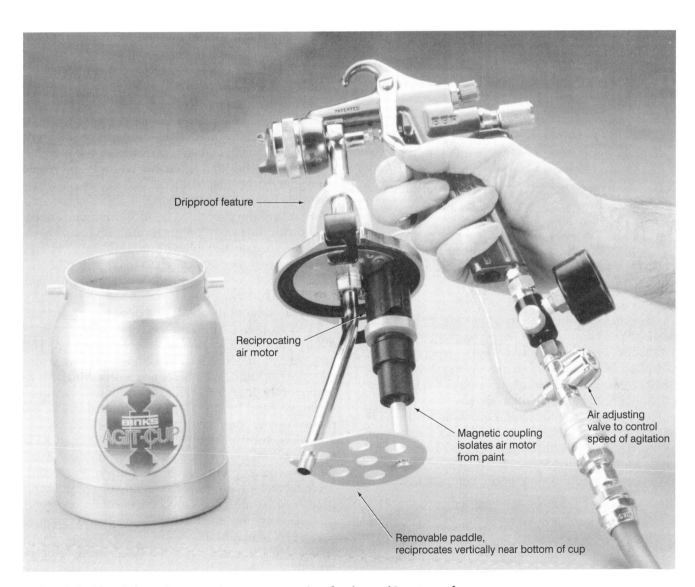

FIGURE 3–13. Binks Agit-Cup agitator-type suction-feed cup. (Courtesy of ITW Automotive Refinishing.)

air. Pressure-feed cups are used for application of fluids too heavy for suction-feed application.

PRESSURE-FEED REMOTE CUP

The remote cup (refer back to Figure 3–11) involves the use of a 2-quart pressure-feed cup with regulator-type provisions. One regulator and one gauge control the pressure on the fluid in the cup. A separate regulator controls the air pressure to the gun. As the term *remote* indicates, the cup is separated from the spray gun by means of two hoses. One hose is for fluid and the second hose is an air hose. In this spray system, the spray gun can be handled easily because it is light. The remote cup method of spraying makes possible the highest precision and fastest paint application with excellent quality control. All the things that make up the best spray patterns can be set correctly. They are:

1. **Large, full spray pattern**
2. **Controlled fluid pressure and fluid flow**
3. **Excellent atomization**

The remote cup method of painting provides the following features:

1. Improved gun handling because the cup can be held by the other hand.
2. Better and faster application of all finishes by low- and high-pressure systems.
3. Results of spraying is similar to production method.
4. Ease of cleaning considering the system used.

HIGH-VOLUME, LOW-PRESSURE (HVLP) SPRAY GUNS

Certain areas of the country affected by excessive levels of pollution, such as California, require that all refinishing be done with HVLP spray equipment. As an option, however, paint shops may use HVLP equipment anywhere in the country because it is the most efficient refinish equipment available.

Government regulations require that all HVLP spray guns be limited to a maximum air pressure of 10 psi. HVLP means that a high volume of air is delivered at low pressure. To be classified as HVLP, spray guns require the following special components:

1. A special venturi in the gun to limit air pressure leaving the gun to 10 psi maximum. No matter what the air pressure entering an HVLP gun, only 10 psi or less can leave the gun at the air cap. If guns do not have a venturi, they must have an attached air regulator that limits the air pressure to 10 psi maximum.
2. Special large hole air caps that coordinate with specific size fluid tips and needles.
3. Paint cups that may have independent pressure regulators. The regulator enables pressure on the material in the cup to be regulated. If the cup has no regulator, pressure on the material in the cup is the same as at the air cap.

All HVLP guns can be checked at any time for air pressure output with a special air pressure testing gauge and adaptable air cap. See Figure 3–14.

HVLP spray guns are available in several basic designs. Some are described in this book. Space does not allow coverage of all HVLP spray guns.

The first design we will describe is a converted, conventional, high-pressure spray gun. Similarities to conventional guns are in the gun body, handle, trigger, and fluid and air controls. Figures 3–12 and 3–15 are examples of this design.

The second design is shown in Figure 3–16. It has a specially designed gun body with extra large air passages for HVLP purposes, with an outer 0 to 10 psi, air-pressure-sensitive gauge at the handle. The air cap has extra large holes to handle the high volume of air for atomization. Since the gun does not use a venturi, it is classed as an externally regulated gun. Accuspray does not use an independent pressure regulator on the paint cup. The pressure on the material in the cup is the same as at the air cap. The Accuspray gun can be adapted to the various viscosities of materials by changing the fluid tips and air caps, and by adjusting the fluid feed control, the fan control, and gun air pressure. The Accuspray cup has a check valve with a nonstick plunger that unloads the cup pressure when the plunger is pressed down.

Figure 3–17 shows a third design, a removable cartridge–type spray gun. It was designed especially for HVLP use, but it can also be used for any and all refinish purposes. This gun has a venturi in the handle to control the air pressure from 1 to 10 psi. The fan size and material feed are adjustable with a single control knob, which speeds refinishing. Fine-tuning of the fluid feed can also be controlled with a separate fluid adjustment. The cartridge can be removed quickly. Since the cartridge contains the complete fluid passages, it can be cleaned as a bench operation or in a gun cleaner. In either case, the lubrication points of the gun are not affected. An independent cup pressure regulator controls the fluid feed to the gun. Three types of air caps and cartridges are designed to handle the broad range of viscosities of materials.

The fourth design, top-load HVLP spray gun, is available:

1. With a gravity-feed cup (see Figure 3–8)
2. With a pressure-feed cup (see Figure 3–9)

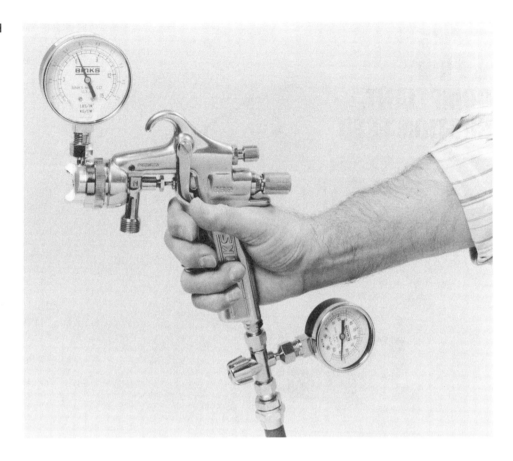

FIGURE 3–14. Binks MACH 1 HVLP spray gun showing two air pressure gauges. (Courtesy of ITW Automotive Refinishing.)

Both types of top-load spray guns are equipped with a venturi in the gun handle, which limits the air pressure to HVLP requirements. When used as a gravity-feed gun, the cup is vented to the atmosphere. When a pressure-feed cup is used, the cup lid is sealed and air pressure forces the material through the gun instantly. The air pressure in the cup is the same as in the air cap. Top-load spray guns operate with a lower volume of air (CFM) than the large spray guns. Figure 3–9 shows a typical Mattson pressure-feed, top-load spray gun.

Spray gun companies issue an operating manual with each spray gun. The manual describes the proper adjustment, use, and performance specification information. HVLP spray equipment requires more care than conventional spray equipment. The manuals include the specific fluid nozzles, needles, and air caps for application of all refinish systems, which range from low to high viscosities.

Selection of an HVLP Spray Gun

Several excellent HVLP spray guns are available on the market. The three spray gun companies (ITW Automotive Refinishing, Accuspray, and Mattson Spray Equipment, Inc.) mentioned in this book have excellent track records. Space prevents coverage of all spray guns available. **Before purchasing an HVLP spray gun, consider all factors carefully and be sure to test at your place of business on your compressor for performance capability. Do not be influenced by salestalk without a trial run.**

CONVENTIONAL HIGH-PRESSURE SPRAY PAINTING VERSUS HVLP SPRAY PAINTING

In high-pressure spray painting (40 to 70 psi), strong cross-currents of air at the air cap blast a liquid stream of paint into small particles that spread instantly in many directions at a high rate of speed while moving forward as a spray fan. Figure 3–18a shows typical bounce-back, overspray, and spread of fumes that happen instantly in high-pressure spray painting.

Transfer Efficiency

Transfer efficiency is the ratio of the amount of paint solids deposited on the surface being painted to the total amount of paint solids used. The ratio is determined by dividing the amount of paint sprayed by the amount of paint deposited. Thus, if all of the paint spray landed and remained on the surface being painted, the transfer efficiency would be 100 percent.

The transfer efficiency of high-pressure spray painting is poor because only 25 to 35 percent of the

GTi
COMPLIANT SUCTION FEED

FIGURE 3–15. ITW–DeVilbiss suction-feed HVLP spray gun. (Courtesy of ITW Automotive Refinishing.)

Developed using the most advanced techniques in computational fluid dynamics

Complies with regulations requiring HVLP

Simple, easy-to-use suction feed operation allows painters to spray with methods they are used to

Waterborne compatible – stainless steel fluid tip and needle, nickel-plated gun body inside and out

Lightweight and comfortable

Virtually maintenance-free:
　No gaskets – self-sealing
　One-piece air valve is smooth and efficient

Includes:
　Air adjusting valve and gauge
　One-quart cup

Specifications:
　30 psi air inlet pressure delivers
　10 psi air cap pressure at 16 cfm air volume required

Air Cap Test Kit 192055 (GTI-5033-100)

GTi Compliant Suction Feed

Order No. Model No.	Fluid Tip (mm)	Inlet Air Pressure	Applications
170146 GTI-600S	2.0	20-25 psi 25-30 psi	Base coats Low and high solids, clear coats, single stages

PAINT SPRAY GUNS AND PAINT CUPS

paint in a paint cup lands on the object being painted, and 65 to 75 percent of the paint in the cup is wasted through bounce-back, overspray, and vaporization.

HVLP spray guns operate normally with air pressures that range from 3 to 10 psi. Lower viscosity materials can be atomized and applied with 3 to 5 psi, while heavier materials and higher fluid deliveries require higher settings. Special air-controlling devices (like venturis), special large hole air caps, and specified fluid tips are the keys to HVLP. Each equipment company has its own patented HVLP system. The low air pressure and the high volume of air produce a quiet and soft spray fan with excellent atomization (see Figure 3–18b). Since bounce-back and overspray are reduced, more paint is deposited on the surface. The transfer efficiency of HVLP in the 7 to 10 psi range is at least 65 percent. At lower spray pressures (3 to 5 psi), the transfer efficiency is even greater. The use of HVLP equipment results in substantial savings in paint materials and in spray booth filter use. For the most efficient operation, however, HVLP equipment requires a 5-horsepower air compressor.

Performance

As the volume of air at the air cap increases, the size of the atomized particles becomes smaller. This fact is the key to HVLP. To produce acceptable atomization, HVLP spray guns require a minimum amount of cubic feet of air per minute (CFM). For average HVLP spray guns, this amount is between 15 and 25 CFM. Spray gun companies are continually searching for ways to improve HVLP equipment and gun performance. For example, the Mattson Spray Equipment Company produces the Air Miser Blue Ring Air Cap, which can achieve excellent atomization with only 9 to 12 CFM when applying low- to medium-solids paint materials at 3 to 5 psi. See Tables 3–1 and 3–2 for comparisons between HVLP and conventional paint systems.

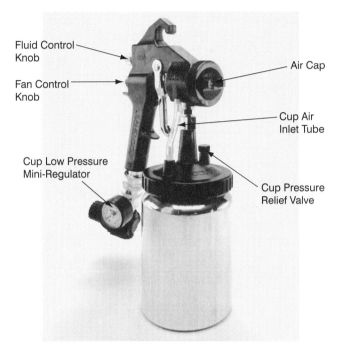

FIGURE 3–16. Accuspray HVLP spray gun. (Courtesy of Accuspray, Inc.)

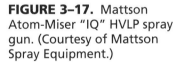

FIGURE 3–17. Mattson Atom-Miser "IQ" HVLP spray gun. (Courtesy of Mattson Spray Equipment.)

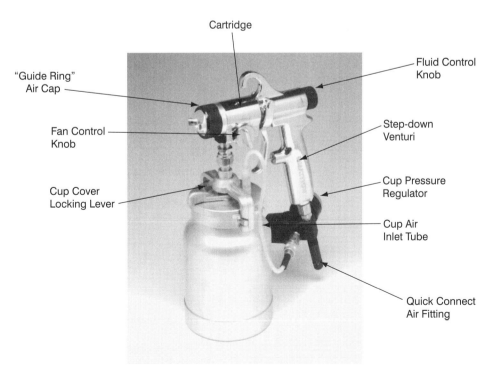

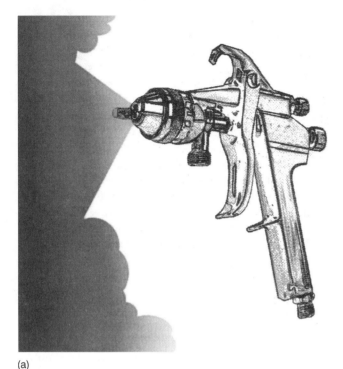

FIGURE 3–18. (a) Typical high-pressure spray gun in which overspray and spread of fumes is excessive, resulting in a maximum of paint waste, etc.; (b) typical HVLP low-pressure spray gun in which overspray is more controlled with a soft spray fan with minimal paint waste. (Courtesy of ITW Automotive Refinishing.)

TABLE 3–1 Performance of HVLP Versus Conventional Paint Systems

	CFM[a]	PSI[b]	TE[c]
Conventional high-pressure spray guns produce:	8	40	35%
	10	50	25%
	15	70	20%
HVLP spray guns produce:	10	3	85%
	14	5	75%
	19	10	65%

[a]Cubic feet of air per minute.
[b]Pressure per square inch.
[c]Transfer efficiency.

TABLE 3–2 HVLP Versus Conventional Paint Technologies

New Technology HVLP	Old, Conventional Technology
65% TE minimum	35% TE Maximum
10 PSI Maximum	40–70 PSI
11–22 CFM	12–14 CFM

REVIEW QUESTIONS

1. What is the purpose of a paint spray gun?
2. What is the purpose of an air cap?
3. Explain how the fluid tip, fluid needle, and fluid control valve work to control the amount of fluid passing through a spray gun.
4. Explain how the spreader control valve works to make a large spray fan and a small spray fan.
5. Explain how the trigger works on a spray gun.
6. What are two types of conventional high-pressure spray guns?
7. What is the primary difference between the air cap and fluid nozzle design on the suction-feed and pressure-feed spray guns?
8. Conventional high-pressure spray guns operate in the range of what air pressures (from _____ psi to _____ psi)?
9. HVLP spray guns operate at what air pressure ranges (from _____ psi to _____ psi)?
10. Describe what takes place to make paint exit the fluid nozzle on suction-feed spray guns.
11. What does HVLP stand for?
12. Which spray gun system has the highest transfer efficiency: conventional high-pressure system or HVLP system?
13. What is a venturi?
14. Name or describe three different types of HVLP spray guns.
15. Explain how atomization of paint materials is achieved on HVLP spray guns, even though the air pressure is 10 psi or less.
16. Explain how an HVLP spray gun works.

CHAPTER 4

Spraying Techniques

INTRODUCTION

Chapter 3 covered the basic construction and principles of spray gun operation for several popular spray guns. Chapter 4 covers:

1. **Adjustment of the spray guns** described in Chapter 3 **to achieve a balanced spray pattern.**
2. **Use and handling of the spray gun** in various ways.
3. **Cleaning and maintenance of a spray gun** for maximum performance.

The basic fundamentals of spraying techniques apply to both types of spray guns: conventional high-pressure and HVLP.

While the air pressure requirements of each type of spray gun are dramatically different, the technology behind each spray system produces acceptable quality atomization and surface smoothness. As apprentices and painters become familiar with the flow of the material produced by each spray gun system, they adjust spraying techniques accordingly. Painters know how to move along with the flow of paint material produced by each paint system.

A painter must be able to achieve a balanced spray pattern with the spray gun being used to achieve top quality results when applying any paint system. All spray guns are designed for this purpose. Proper techniques are easy to develop. Practicing the right way to do refinishing will end up becoming the painter's habit and he or she will be on the right road to success. Always remember that, no matter how precise or expensive a spray gun may be, its value is practically nil if it is used improperly.

Automotive refinishing is made up of two parts:

1. **Theory**
2. **Skill development**

Theory is understanding the "why" behind every refinishing step, tool, product, and procedure. Theory includes why paint finishes are good and why some paint finishes fail. Theory includes what causes a good spray pattern, how paint application is controlled by the painter, and what produces quality results. Theory is learned by reading, by study, by discussion, and by asking questions of the proper people when any doubt exists in a student's or painter's mind.

Skill development is practicing manual exercises, "doing things by hand," as outlined in this chapter. Spraying techniques are learned by practicing how to handle the spray gun under all spray painting conditions. Spraying techniques are mastered when the painter or apprentice can do them by habit: performing an operation with minimum thought. A person must learn both theory and skill development to become a qualified painter.

SPRAY GUN ADJUSTMENTS

Spray guns have several basic adjustments with which the painter must be thoroughly familiar. Before applying paint to a car, the painter decides on the type of spray gun adjustments that the job requires. The following spray gun adjustments and spray pattern checks are the most popular among the best painters.

Air Cap Adjustments (All Spray Guns)

Air caps are the simplest of spray gun adjustments. They are adjustable to create two popular spray fans:

1. The **perpendicular spray fan** is straight up and down.
   a. This fan is achieved when air cap horns are parallel to the floor (Figure 3–5).
   b. This adjustment is the most used by painters because it is the easiest to use.
2. The **horizontal spray fan** is flat and parallel to the floor.
   a. This fan is achieved when air cap horns are perpendicular to the floor (Figure 3–6).
   b. This fan is used on occasion when cross-coating or when checking a spray pattern.

To adjust the air cap:

1. Loosen the air cap retaining ring one-half to one turn counterclockwise.
2. Locate and **hold** the air cap horns in the precise position desired.
3. Tighten the air cap retaining ring.

Normal Spray Patterns

The three most popular spray gun adjustments (see Figure 4–1) and their uses are:

1. **Full-open adjustment** (full size pattern)
   a. For complete paint jobs
   b. For full panel and large-area painting
2. **Spot-repair adjustment** (medium size pattern)
   a. For spot repairs
   b. For solvent blending
3. **Small fan adjustment** (small size pattern)
   a. For special repairs during refinishing
   b. For banding and edge application

Flooding Test Method of Checking Spray Patterns (All Spray Guns)

Before doing quality spray painting on a car, the painter should **always check the spray pattern of the paint gun.** One of the best ways to check a spray pattern is with the **flooding test method.** The procedure follows:

1. **Prepare a cup of paint** according to label directions.

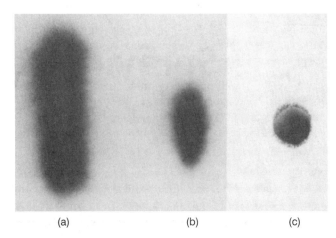

FIGURE 4–1. Normal spray patterns: (a) full-size pattern, (b) medium-size pattern, and (c) small-size pattern.

2. **Adjust the air cap on the gun** for a horizontal fan (refer back to Figure 3–6).
3. **Adjust spreader control valve** and fluid feed valve for full open position (see Figures 4–2 and 4–3).
4. **Adjust air line pressure at gun** as follows:
   a. On high-pressure suction-feed guns, set the air pressure according to label directions.
   b. On HVLP guns, set the air pressure at the air cap and at the individual cup regulator, if so equipped, according to Table 4–1, 4–2, 4–3, or 4–4. Make necessary adjustments as the flooding test progresses.
5. **Perform the flooding test** as follows:
   a. Hold the gun one hand span from the surface.
   b. Apply a flooding test pattern at one spot on a suitably masked vertical area in the spray booth for 3 to 4 full seconds so that the applied paint sags and runs.
6. **Evaluate and correct the spray pattern** as follows:
   a. If all paint runs are of approximately equal length, you have a **balanced spray pattern** (see Figure 4–4). This pattern proves that an equal amount of paint is spread uniformly throughout the pattern. Three components of a spray fan must be applied uniformly, at the same time, for a fan to be balanced:
      (1) The amount of fluid must be correct for the size of the fan.
      (2) The size of the spray fan must be correct for the amount of fluid.
      (3) The air pressure setting and volume of air must be correct to atomize the fluid and to disperse the atomized particles equally throughout the spray fan.

FIGURE 4–2. Adjusting spreader control valve. (Courtesy of DuPont Company.)

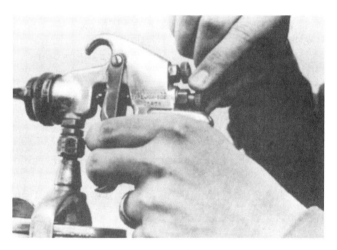

FIGURE 4–3. Adjusting fluid feed valve. (Courtesy of DuPont Company.)

FIGURE 4–4. Flooding test showing balanced pattern.

FIGURE 4–5. Flooding test showing split pattern.

FIGURE 4–6. Flooding test showing heavy-center pattern.

b. If the outer paint runs are longer and have more paint than the inner paint runs, you have a **split pattern** (see Figure 4–5). This pattern is caused by an insufficient amount of fluid that cannot be dispersed uniformly in a large fan. The fluid that is present is blown toward the outer ends by the strong criss-crossing air currents at the air cap. **The remedy for a split pattern** is to bring the three components of a proper spray fan into balance as follows:

(1) Increase the amount of fluid and follow with a flooding test.

(2) Reduce the air pressure and follow with a flooding test.

(3) Reduce the size of the fan and follow with a flooding test.

A proper combination of the above will produce a balanced spray pattern.

c. If the center area of the pattern has longer paint runs than the outer areas, you have a **heavy center pattern** (see Figure 4–6). This pattern is caused by too much material for the size of the fan and the possibility that the air pressure may be low. **The remedy for a heavy center pattern** is to bring the three components of a proper spray fan into balance as follows:

(1) Increase the size of the fan and follow with a flooding test.

(2) Reduce the amount of fluid and follow with a flooding test.

(3) Increase the gun air pressure and follow with a flooding test.

A proper combination of the above will produce a balanced spray pattern.

Description of Performance Settings Charts

Each performance settings chart lists recommended air cap, fluid nozzle, air pressure, and fluid pressure settings for a specific HVLP spray gun. These settings are recommended by each equipment company as starting points for adjusting the gun to produce the best atomization and gun performance.

> **IMPORTANT**
>
> **The procedure for preparing and adjusting each brand name of HVLP spray gun is specific and individual.** No two HVLP procedures are necessarily alike. To obtain a procedure for adjusting a spray gun not covered in this book, the author recommends that you work very closely with the company that makes the spray gun you are using.

Full-Open Spray Gun Adjustment

For Binks and DeVilbiss conventional spray guns:

1. Fill the cup with reduced paint and attach it to the gun.
2. Open the spreader control valve, counterclockwise, to full open position (refer to Figure 4–2).
3. Open the fluid feed valve, counterclockwise, to full open position (refer to Figure 4–3).
4. Adjust the air cap on the gun for a horizontal spray pattern and tighten.
5. Adjust the air line pressure according to the paint label directions.
6. Perform a Flooding Test as explained in step 5 above.
7. Evaluate and correct the spray pattern as explained in step 6 above.

Spot-Repair Spray Gun Adjustment

Spray gun adjustments for spot repairs are most often in the midrange or half-open settings. The size of the spray fan is determined by the size of the spot repair to be done. Spot-repair spray fans vary in size from just barely open to three-fourths of full open. The smaller the spot-repair fan, the greater is the saving in material and labor.

1. On the spreader control valve (Figure 4–2), determine the number of half-turns from fully closed to fully open settings. Make a note of it. The reason for the half-turns is that the control valves might be turned only a half-turn at a time with one grip. Set the valve at the midpoint range.
2. On the fluid feed valve (Figure 4–3), determine the number of half-turns from fully closed to fully open settings. Set the valve at the midpoint range.
3. Apply a flooding test pattern on a suitably masked vertical wall.

These procedures apply to:

1. Conventional high-pressure spray guns.
2. Binks and DeVilbiss HVLP spray guns.

> **NOTE**
>
> The Mattson and Accuspray HVLP guns are adjusted for spot repair by simply reducing the fan size and fluid feed. Their adjustments are different from Binks and DeVilbiss.

Before adjusting an Accuspray spray gun (see Figure 4–7), perform the following steps:

1. Adjust the fluid tip and air cap to match Table 4–1.
2. Check the mini-regulator on the gun for firm attachment.
3. Prepare reduced paint for testing and prepare a suitable area in the spray booth to test the paint spray pattern.
4. The objective is to achieve a full, open spray gun adjustment.

The adjustment procedure follows:

1. Set the mini-regulator at a low setting for material to be sprayed according to Table 4–1 while the gun is triggered with flowing air.
2. Adjust the air cap for a horizontal spray fan.
3. Fill the cup with reduced paint and install the cover tightly on the gun.
4. Open the fluid feed, counterclockwise, to full open position.
5. Open the fan control valve, counterclockwise, to full open position.
6. Perform a Flooding Test as explained in step 5 above.
7. Evaluate and correct the spray pattern as explained in step 6 above.

SPRAYING TECHNIQUES

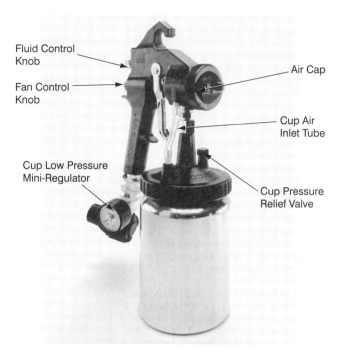

FIGURE 4–7. Accuspray HVLP spray gun. (Courtesy of Accuspray, Inc.)

For Binks and DeVilbiss HVLP spray guns with cup pressure regulators (see Figure 4–8), the air supply should be 18 CFM at 80 psi and the inside diameter hose should be a minimum of $\frac{5}{16}$-in. Check Table 4–2 for Binks equipment and gun performance settings; check Table 4–3 for DeVilbiss equipment and gun performance settings.

1. Turn the fan adjustment all the way open, counterclockwise (a).
2. Turn the fluid needle adjustment all the way open, counterclockwise (b).
3. Adjust the air cap for a horizontal spray fan.
4. Fill the cup with paint according to label directions.
5. With the trigger pulled, adjust the air pressure at the cup and at the gun for the type of material being sprayed.
6. Perform a Flooding Test as explained in step 5 above.
7. Evaluate and correct the spray pattern as explained in step 6 above.

TABLE 4–1 Performance Settings Chart for Accuspray Series 10 HVLP Spray Gun

Material Type[a] (Ready to Spray)	Fluid Tip	Air Cap Size	Regulator Settings Low-Pressure Inlet		Cup Pressure
			Wall Regulator	Mini-Regulator[b]	
UNDERCOATS					
Lacquer primer surfacer	.043	9–10	50 psi minimum	2.5–3.5	2.5–3.5
Two-component primer surfacer	.043–.051	9–10	1 psi	2.5–4.0	2.5–4.0
Primer sealer	.036–.043	9–10	pressure loss per 30′ hose	3.5–4.5	3.5–4.5
COLOR SYSTEMS					
Acrylic lacquer	.028–.036	8–9		2.0–3.5	2.0–3.5
Acrylic/synthetic enamels	.028–.036	9		4.5–6.5	4.5–6.5
Acrylic urethane	.036–.043	9		4.5–6.5	4.5–6.5
Base coat	.028–.036	8		3.0–5.0	3.0–5.0
Polyurethane	.036–.043	8–9		4.5–6.5	4.5–6.5
Low VOC	.036–.043	9–10		5.0–9.0	5.0–9.0
CLEARCOATS					
Acrylic lacquer	.028–.036	7–8		2.0–4.5	2.0–4.5
Acrylic urethane	.028–.036	7–8		3.0–5.0	3.0–5.0
High-solids urethane	.036–.043	6–8		4.0–6.0	4.0–6.0

[a] Larger nozzle sizes allow for faster rates of application and higher flow rates.
[b] Air pressures are to be set with gun triggered and flowing air.

1. Always follow paint manufacturer's ratios for reduction and catalyzation. Use a viscosity cup when possible.
2. Maximum transfer efficiencies are achieved with low-pressure inlet guns.
3. Wall mount low-pressure regulator uses 30′ length of $\frac{3}{4}''$ hose.
4. Wall mounted high-pressure regulator.

Note: To achieve maximum flowout and reduce orange peel, always use paint manufacturer's recommendations for proper thinners and reducers.

FIGURE 4–8. ITW–DeVilbiss suction-feed HVLP spray gun. (Courtesy of ITW Automotive Refinishing.)

GTi COMPLIANT SUCTION FEED

Developed using the most advanced techniques in computational fluid dynamics

Complies with regulations requiring HVLP

Simple, easy-to-use suction feed operation allows painters to spray with methods they are used to

Waterborne compatible – stainless steel fluid tip and needle, nickel-plated gun body inside and out

Lightweight and comfortable

Virtually maintenance-free:
  No gaskets – self-sealing
  One-piece air valve is smooth and efficient

Includes:
  Air adjusting valve and gauge
  One-quart cup

Specifications:
  30 psi air inlet pressure delivers
  10 psi air cap pressure at 16 cfm air volume required

Air Cap Test Kit 192055 (GTI-5033-100)

GTi Compliant Suction Feed

Order No. Model No.	Fluid Tip (mm)	Inlet Air Pressure	Applications
170146 GTI-600S	2.0	20-25 psi 25-30 psi	Base coats Low and high solids, clear coats, single stages

SPRAYING TECHNIQUES

TABLE 4–2 Performance Settings Chart for Binks Mach 1 Plus HVLP Spray Gun

MATERIAL TYPE[a] (Ready to Spray)	Air Caps and Fluid Tips		Regulator Settings		Cup (psi)	Air Cap (psi)
			$\frac{3}{8}''$ Air Hose	$\frac{5}{16}''$ Air Hose		
UNDERCOATS						
Lacquer primer surfacer	92	.046		40–46	2–3	5–6
Two-component primer surfacer	92	.046		46–50	2–4	6–7
Primer sealer	90	.030		40–46	2–3	5–6
	91	.040				
COLOR SYSTEMS						
Acrylic lacquer	92	.046		46–60	2–3	6–9
	91	.040			2	6–9
Acrylic/synthetic enamels	92	.046		46–60	3	6–9
Acrylic urethane	92	.046		46–55	3	6–8
Base coat	91	.040		46–55	2–3	6–8
	92	.046		46–55	2–3	6–8
Polyurethane	92	.046		55–64	3	8–10
Low VOC	92	.046		55–64	3	8–10
CLEAR COATS						
Acrylic lacquer	92	.046		55–60	2–3	8–9
Acrylic urethane	92	.046		55–60	2–3	8–9
High-solids urethane	94	.055		55–64	2–3	8–10
	92	.046				

[a]Regulator pressures are based on 25' of $\frac{5}{16}''$ hose in good condition without Quick-Disconnects or other restrictive fittings. Use the air nozzle test gauge accessory to confirm the atomizing/regulator pressure relationship for your actual air supply setup. These recommendations are for typical or average materials and are intended to serve as a starting point. Adjust as necessary for your specific application.

TABLE 4–3 Performance Settings Chart for DeVilbiss Super Pro-Plus 2 HVLP Spray Gun

Material Type[a] (Ready-to-Spray)	Air Caps and Fluid Tips	Regulator[b] Settings		Cup (psi)	Air Cap (psi)
		$\frac{3}{8}''$ Air Hose[c]	$\frac{5}{16}''$ Air Hose[c]		
UNDERCOATS					
Lacquer primer surfacer	FX (.042)	45–60	55–70	2–4	4–6
Two-component primer	FX (.042)	45–55	55–75	3–5	4–6
Primer sealer	GX (.034)	60–75	70–85	2–4	6–8
COLOR SYSTEMS					
Acrylic lacquer	GX (.034)	60–80	70–90	2–3	6–8
Acrylic/synthetic enamels	FX (.042)	60–80	70–90	2–4	6–8
Acrylic urethane	FX (.042)	60–80	70–90	2–4	6–8
Base coat	GX (.034)	60–70	70–80	2–3	6–7
Polyurethane	FX (.042)	70–80	80–90	3–4	7–8
Low VOC	FX (.042)	70–80	80–90	3–5	7–8
CLEAR COATS					
Acrylic lacquer	FX (.042)	60–80	70–90	3–5	6–8
Acrylic urethane	FX (.042)	70–80	80–90	4–6	7–8
High-solids urethane	FX (.042)	80	90	5–7	10

[a]Always follow label directions for reduction and catalyzation.
[b]Wall-mounted shop high-pressure regulator.
[c]These figures are effective with 30 feet of air hose.

Note: To achieve maximum flowout and reduce orange peel, use the next slower paint solvent and/or a touch of retarder, especially in high temperatures.

> **NOTE**
>
> If too much paint is being sprayed, turn the cup regulator (D) knob off all the way, counterclockwise, and remove all cup pressure (on the DeVilbiss gun, depress the poppet; on the Binks gun, loosen the cup cover). Then repeat steps 6 and 7 at a lower pressure.

For the Mattson LP-DC HVLP spray gun shown in Figure 4–9, note that it is completely different from any other HVLP spray gun on the market. You must understand the construction and adjustments of Mattson's guns before using them. The following procedure outlines how to adjust the LP-DC HVLP spray gun.

1. Turn the fan control knob all the way, clockwise, to "MAX." The total flow fan control knob has a four-position adjustment, from "MIN" to "MAX." This control knob adjusts the size of the spray fan and the amount of material being sprayed simultaneously.
2. Turn the fluid control knob all the way, clockwise, until it bottoms. Then back it off, counterclockwise, until the metal ball marker is at the 12 o'clock position. This position is good for normal use.
3. Adjust the air cap for a horizontal spray fan.
4. Fill the cup with paint according to label directions.
5. With the trigger pulled, adjust the air pressure at the gun and at the paint cup for the type of material being sprayed. Start at a low figure as recommended in Table 4–4.
6. Perform a Flooding Test as explained in step 5 above.
7. Evaluate and correct the spray pattern as explained in step 6 above. Always start at a low figure and increase gradually.

Fine tuning can be speeded up by adjusting the fluid control knob as follows:

1. Turning the fluid control knob counterclockwise from the 12 o'clock position increases the amount of fluid being sprayed. Do not go past the 6 o'clock position.
2. Turning the fluid control knob clockwise from the 12 o'clock position decreases the amount of fluid being sprayed.

> **NOTE**
>
> You may increase the cup pressure at any time by unlocking the regulator adjusting knob and by turning to a higher setting. Then you may continue painting. However, if reducing the cup pressure is required, it must be bled off by releasing the cup cover locking lever and by starting again at a lower cup pressure.

Evaluate and correct the spray fan to achieve a balanced spray pattern with the least movement of the controls.

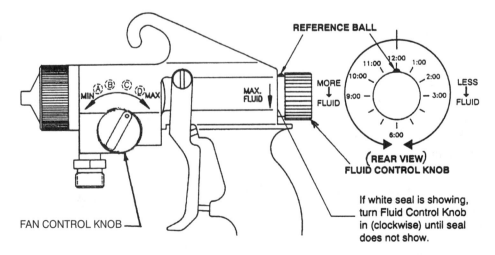

FIGURE 4–9. Fan and fluid adjustment on Mattson LP-DC HVLP spray gun. (Courtesy of Mattson Spray Equipment.)

SPRAYING TECHNIQUES

TABLE 4-4 Performance Settings Chart for Mattson HVLP[a]

Material Type[b] (Ready to Spray)	Cartridge Size	Regulator[c] Settings		Cup (psi)	Air Cap (psi)
		3/8" air hose[d]	5/16" air hose[d]		
UNDERCOATS					
Lacquer primer surfacer	.040—gold locking ring[e]	34–42	36–44	5–6	3–4
Two-component primer surfacer	.040—gold locking ring[e]	42–48	44–50	5–9	4–5
Primer sealer	.032—black locking ring	34–42	36–44	5–6	3–4
	.040—gold locking ring[e]	42–48	44–50	5–6	4–5
COLOR SYSTEMS					
Acrylic lacquer	.032—black locking ring	34–42	36–44	5	3–4
	.040—gold locking ring[e]	34–42	36–44	5	3–4
Acrylic/synthetic enamels	.032—black locking ring	42–48	44–50	5–8	6–9
	.040—gold locking ring[e]	56–62	58–64	5–8	6–9
Acrylic urethane	.032—black locking ring	34–42	36–44	5	3–4
	.040—gold locking ring[e]	34–48	36–50	6–7	3–5
Base coat	.032—black locking ring	34–42	36–44	5	3–4
Polyurethane	.032—black locking ring	34–48	36–50	5–6	3–5
	.040—gold locking ring[e]	34–48	36–50	6–7	4–6
Low VOC	.032—black locking ring	48–74	50–76	6–8	6–9
	.040—gold locking ring[e]	48–74	50–76	6–8	6–9
CLEAR COATS					
Acrylic lacquer	.032—black locking ring	34–42	36–44	5	3–4
	.040—gold locking ring[e]	34–42	36–44	5	3–4
Acrylic urethane	.032—black locking ring	34–42	36–44	5	4–5
	.040—gold locking ring[e]	34–42	36–44	5–7	4–5
High-solids urethane	.032—black locking ring	42–56	44–58	6–8	6–9
	.040—gold locking ring[e]	42–56	44–58	6–8	6–9

[a]Regulator settings based on a system with 30' hose length.
[b]Always follow paint manufacturer's ratios for reduction and catalyzation. Use a viscosity cup when possible.
[c]Wall-mounted shop high-pressure regulator.
[d]These figures are effective with 30 feet of air hose.
[e]Select .040 gold locking ring to spray more material faster and when using high-solid material.

Note: To achieve maximum flowout and reduce orange peel, use the next slower solvent temperature, especially in high-volume paint booths and in high temperatures. If orange peel persists, decrease cup pressure and/or increase air cap pressure.

Calibrating Shop Air System for HVLP Use

Any qualified paint shop, high-pressure, wall regulator can be calibrated for use in doing HVLP spray painting without the necessity of using a low-pressure gauge all the time. You can calibrate your shop's high-pressure regulator for HVLP use if you have the following items on hand:

1. HVLP spray gun.
2. 1 to 15 psi test gauge (check with gun supplier for gauge).
3. A blank HVLP calibration chart as shown in Figure 4–10.

> **NOTE**
> You can make several copies of this chart at the size you need at your local print shop. Also, check your paint jobber for chart availability.

1. Install a low-pressure gauge on the HVLP gun (see Figure 4–11).
2. Use the same hose that will be used in spray painting. Connect the gun and hose to the wall regulator.
3. With the wall regulator set at about 30 psi, pull the gun trigger to read the actual air cap pressure on the calibration gauge.

FIGURE 4–10. Chart for calibrating wall air regulator for HVLP use. (Courtesy of Mattson Spray Equipment.)

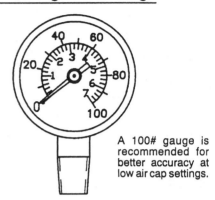

A 100# gauge is recommended for better accuracy at low air cap settings.

Static settings—make all final readings with the trigger off.

ACTUAL AIR CAP	WALL REGULATOR SETTING	ACTUAL AIR CAP	WALL REGULATOR SETTING
2		6.5*	
2.5		7*	
3		7.5*	
3.5		8*	
4		8.5*	
4.5		9*	
5		9.5*	
5.5		10*	
6			

*Settings for higher solid paints.

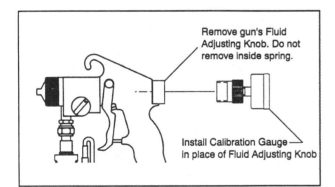

FIGURE 4–11. Installing calibration gauge (gauge part no. 700138) on Mattson HVLP gun. (Courtesy of Mattson Spray Equipment.)

4. Adjust the wall regulator until the calibration gauge on the gun reads 3 psi. Mark this regulator setting on the chart opposite "3." Trigger the gun on and off to ensure that the setting is correct. Adjust the wall regulator for each additional low-pressure setting and fill in the chart with regulator settings for each low pressure shown.
5. When you are finished, remove the calibration gauge and store it in a safe place.

SPRAY PAINTING STROKE
The Grip

Most painters use the conventional grip shown in Figure 4–12. In this grip the spray gun is supported by the lower two fingers, the hand, and the thumb as

SPRAYING TECHNIQUES

FIGURE 4–12. The conventional grip. (Courtesy of DuPont Company.)

shown. The first two fingers are used for triggering. Some painters use an alternative grip in which the gun is supported by only the small finger at the bottom, the index finger (or forefinger), the hand, and the thumb. The middle two fingers are used for triggering. Selecting a grip is a matter of "feel" to the painter and quickly becomes a habit.

Position of the Gun to the Surface

The spray gun should be held perpendicular to the surface or as close to perpendicular as possible (see Figure 4–13). Tilting the gun excessively results in flooding the surface with one side of the spray pattern and starving the opposite side.

Stroke Movement

The conventional spray painting stroke is made by moving the gun parallel to the surface while maintaining the correct gun distance and perpendicular position (see Figure 4–13). This is easier said than done. To maintain a parallel and perpendicular position of the spray gun to the surface, the painter must make a special effort at the beginning and end of each stroke to change the arm and wrist position as the painter remains still. As shown in Figure 4–13, the spray gun is the important thing as it travels back and forth in spray painting. **The painter's job is to adapt to the spray gun movement.**

Distance

Paint spray guns do their best work at a distance of about 6 to 8 inches from the surface (refer to Figure 4–14). Holding the gun too close to the surface can cause paint runs, mottling, and/or excessive orange peel. Holding the gun more than 8 inches from the surface causes excessive dry spray, poor flowout,

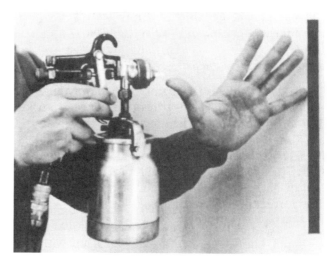

FIGURE 4–13. Distance of spray gun from surface. (Courtesy of DuPont Company.)

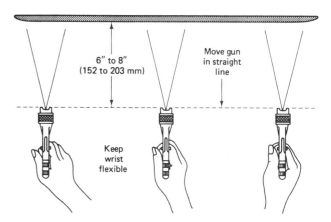

FIGURE 4–14. Keep spray gun parallel and perpendicular to surface. (Courtesy of ITW Automotive Refinishing.)

and overspray. Any of these conditions may require rework. **It is up to each painter to find the distance that works best for a smooth, even coat.**

Speed

The speed of a paint spraying stroke is vitally important to the end result. If the gun is moved too fast, the surface is starved, resulting in poor flowout, rough surface, and insufficient material. If the gun is moved too slowly, the material is piled on and may sag. There is a proper speed for every spray pattern and fluid feed. The best speed is a steady deliberate pass that leaves a full, wet coat just short of running. In actual practice, the painter must be able to use several speeds to meet changing conditions and equipment. **Whatever the equipment used, the painter must be able to vary** the speed and be able to apply thin-wet coats, medium-wet coats, and full-wet coats.

Triggering

The trigger controls the action of the spray gun and the painter should use the trigger during each stroke. The farther the trigger is drawn back, the greater is the flow of material. During conventional strokes, the trigger should be pulled back completely, not partially. To avoid excessive buildup of material at the end of each stroke, the trigger should be released. The correct procedure for triggering is:

1. First, begin the (arm movement) stroke; then pull the trigger as the gun lines up with the starting work edge. This can be:
   a. Two inches before the starting work edge, or
   b. In the middle of a vertical banding stroke.
2. Release the trigger as the stroke lines up with the finishing work edge. This can be:
   a. Two inches past the finishing edge, or
   b. In the middle of a vertical banding stroke.

To the expert painter, the art of triggering becomes automatic. The triggering action is performed with no concentrated thought. Thus, triggering is the heart of the spray painting technique. The objective of the painter is to hit the edge of the work and to maintain full coverage with minimum overspray.

Proper triggering makes possible uniform coverage of the surface, minimum waste of material, and a reduction of overspray. Also, proper triggering cuts down muscle fatigue because a muscle that alternately flexes and relaxes does not get as tired as one that is flexed all the time.

Feathering

Feathering is a type of control a painter uses during spot-repair and color-blending operations. **Feathering** is the application of a spray material starting with very little material at the start of a stroke (fluid feed barely open) to a gradually thickening material (fluid feed opens gradually with trigger pull) at the end of the stroke. At this point, the trigger is released suddenly. The correct procedure for feathering follows:

1. Start the spray gun motion smoothly and start pulling the trigger gradually as the gun approaches the target area and apply color at the correct wetness. Then release trigger suddenly but smoothly. Spray gun motion continues. This is considered spraying "from the outside in."
2. Start the spray gun motion smoothly and pull the trigger fully over the target area; begin releasing the trigger as the spray gun nears the blend area. For smoothness of operation, spray gun motion continues. This is considered spraying "from the inside out."

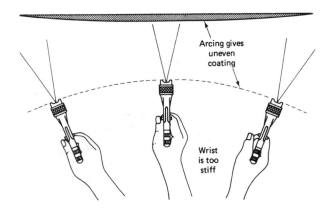

FIGURE 4–15. Arcing the spray gun. (Courtesy of ITW Automotive Refinishing.)

Arcing

Arcing is not recommended when spraying full panels, as shown in Figure 4–15. **Arcing** is a common fault among apprentices and beginners when conventional spraying techniques are required on full panels. Arcing in this case causes an uneven application of paint at the beginning and end of each pass as shown. Also, this poor technique causes excessive overspray and dry spray at the outer ends. But arcing is used in spot repairs and blending. Arcing is used to good advantage because the painter can control the arcing and feathering at the same time. The best painters have excellent control of and great confidence in a combination of good arcing and feathering techniques. Arcing can be done by pivoting the gun at the wrist or by using an arm movement pivoting at the elbow. To maintain uniformity of distance and gun position, the painter's arm and wrist must be conveniently flexible.

Spraying a Panel with a Single Coat

When spraying panels on cars, the most popular technique is to spray left to right and right to left alternately, as shown in Figure 4–16a and b. While the trigger starts and stops fluid flow at the end of each stroke, the spray gun continues moving for positioning into the next stroke. Thus, the movements are smooth and coordinated.

1. In the first stroke (Figure 4–16a), the center of the spray pattern is aimed at the top edge of the panel.

SPRAYING TECHNIQUES

FIGURE 4–16. Panel spraying technique (single coat): (a) directions for application. (Courtesy of ITW Automotive Refinishing); (b) what the technique looks like. (Courtesy of DuPont Company.)

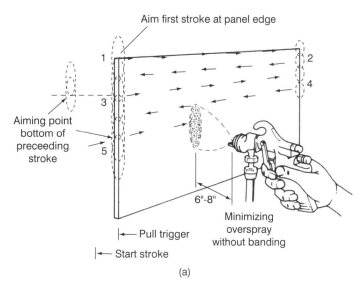

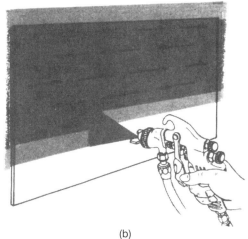

2. On all other strokes, the center of the spray fan is aimed at the bottom of each previous pass.
3. The top and bottom edges are sprayed twice for proper coverage.
4. Actually, triggering takes place 1 to 2 inches before and after each panel edge to be sure of complete and uniform panel coverage.
5. The 50 percent overlap described in this procedure and depicted in Figure 4–16b is known as a **single coat** in the refinish trade.

Banding

To reduce overspray in a shop, many painters use a **banding** technique, as shown in Figure 4–17. The narrow vertical stroke at each end of the panel is made with a small spray fan about 4 inches wide as

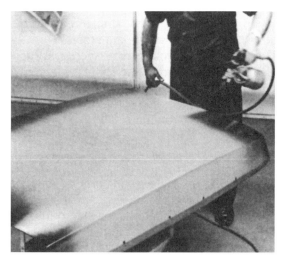

FIGURE 4–17. Banding technique. (Courtesy of DuPont Company.)

shown. Banding ensures complete coverage of a panel and reduces the waste of paint that results from trying to spray right up to the vertical edge with the usual horizontal strokes. When using the banding technique, start and stop triggering action on the sprayed band, preferably at the center.

Spraying a Panel with a Double Coat

A **double coat** is one single coat followed immediately by a second single coat. A double coat most often applies to the application of fast-drying paint systems such as lacquers. Sometimes paint suppliers recommend that the second coat be sprayed in the opposite direction to the first coat, one horizontal and one vertical. Enamel system products are almost always applied in single coats with a waiting period between coats.

The painter should study and determine in advance a way of spraying every paint job with a definite plan of strokes and panel sequence. In this manner, all paint jobs are easier to do, missing panels or areas is avoided, and the results are always better.

Once mastered, the spray painting stroke is done with little effort and with extreme smoothness. Each of the components that make up the spray painting stroke is explained in the following sections.

TYPES OF SURFACES TO BE SPRAYED

Spraying a Long Panel

A long panel can be sprayed with vertical strokes, but most painters have better control with natural horizontal strokes. Spray a long panel in short sections, 18 to 36 inches long, with similar triggering and motions as used on a small panel (see Figure 4–18). Banding or vertical strokes are not necessary at other than edge areas. However, use about a 4-inch overlap for each coat. When applying succeeding coats, it is wise to change the location of the overlap area to prevent excessive color buildup.

Spraying Edges and Corners

When spraying panel edges and corners, aim the center of the spray pattern at the edge so that 50 percent of the spray is deposited on each side of the corner (Figure 4–19). In these cases, hold the gun an inch or two closer than the normal distance. One stroke along each corner coats the edges and bands the face of the panel on each side at the same time.

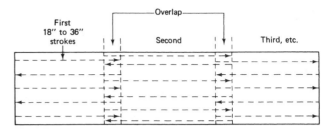

FIGURE 4–18. Spraying long work. Long panels are sprayed in sections of convenient length. Each section overlaps the previous section by about 4 inches.

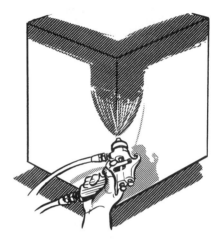

FIGURE 4–19. Spraying edges and corners. (Courtesy of ITW Automotive Refinishing.)

Spraying Large Horizontal Surfaces

When spraying a large flat surface such as a roof, hood, or the rear deck of a car, always start at the near edge and work to the center. Then, working on the opposite side, continue spraying from the center to the near edge. This process results in a fully wet sprayed panel while minimizing overspray and dry spray.

Spraying Curved Surfaces

When spraying a large-diameter curved surface, keep the gun at the proper distance by following the curves of the surface. When spraying slender curved surfaces of a smaller diameter, it is best to spray with the length of the part.

SPRAY GUN MAINTENANCE

The spray gun is the painter's most important tool. It should be cleaned thoroughly after each use. Every painter appears to clean a spray gun in a different manner. Also, not all painters lubricate spray guns as they should be lubricated. Neglect and carelessness cause most of a spray gun's problems. Proper care of a spray gun takes very little time and effort and it is well worth it.

Two basic methods of cleaning spray guns are used in the refinish trade. **Method 1 is used in the least regulated areas of the country; method 2 is required for use in all the most highly regulated areas of the country.**

Method 1 is used in those areas of the country least affected by air pollution. These areas are classed as *least regulated* because they have minimal pollution problems. In least regulated areas, spray guns may be cleaned by spraying solvent through the gun into the air to clean out the fluid passages, as in the past; containers of solvent may be left uncovered during the cleaning of parts as a bench operation; etc.

Method 2 (for compliant areas) must be used in areas most affected by clean air regulations. These areas are known as *compliant areas* because all refinishing is controlled by specific clean air regulations. For example, spraying solvent through a gun into the air to clean fluid passages is prohibited. All cleaning of spray guns must be done in enclosed containers to prevent escape of solvent into the air. This law caused the development of enclosed automatic spray gun washers and equivalent washers that are not completely enclosed. (Safety Kleen models are examples.)

To determine which method of spray gun cleaning is used in your area, check with your local paint jobber or your local clean air district. They can also tell you which cleaning solvent is recommended for your area.

Cleaning the Air Cap

For least regulated areas, the air cap should be cleaned by removing it from the gun, soaking it in clean solvent, brush cleaning all holes with solvent, and then keeping it clean. Some painters like to blow the air cap dry with compressed air after cleaning. This is a good idea. Other painters like to keep their air caps soaking in clean solvent at the bottom of their cleaned paint cups. This, too, is a good idea. If cleaning small holes in the air cap becomes necessary, do it in a recommended manner: (1) soak the air cap in lacquer thinner, (2) clean out holes with soft items such as round-type toothpicks or suitable plastic bristles, and (3) finish by brush cleaning holes and blowing out the air cap with compressed air. Then test the air cap for proper spray fan. Resoaking and recleaning may be necessary. If a problem with the air cap persists, contact the equipment supplier for factory help. They can fix it if the air cap is not damaged.

Cleaning a Suction-Feed Gun and Cup

Clean a suction-feed gun and cup in least regulated areas as follows. Loosen the cup from the gun and, while the fluid tube is still in the cup, unscrew the air cap (two or three turns). Hold a folded cloth over the air cap and pull the trigger. Air, diverted into fluid passages, forces material back into the container. Empty the cup of paint material and clean the cup and the cup cover with solvent and a suitable brush and then with a clean rag dampened with solvent. Then fill the cup about one-third full with clean solvent and spray solvent through the gun to clean out the fluid passages. The painter can usually see when the fluid passages are clean. Wipe off the gun with a solvent-dampened rag. The quality of cleanup is proportional to the attention and detail of a painter's good housekeeping habits.

Cleaning a Pressure-Feed System

In a pressure-feed system with a tank or 2-quart cup (Figure 4–20), the fluid is fed to the spray gun under pressure usually from a separate container through a hose. The pressure-feed system is speedy because the spray gun responds immediately when the trigger is pulled. The material hose and the container have paint material in them and they are under pressure. This is where the cleanup begins.

The cleaning procedure for pressure-feed systems in least regulated areas follows:

1. Turn the air pressure regulating adjusting screw on the fluid container counterclockwise to cut off air pressure to the container.
2. Release pressure from the container by means of the relief valve or safety valve.
3. Loosen the spray gun air cap three turns.
4. Hold a folded cloth over the air cap and pull the trigger to force paint fluid from the hose back into the container.
5. Clean out the material container as required and partially fill it with solvent.
6. Reassemble the fluid container; turn on all air pressure controls and run the solvent through the fluid hose by pulling the trigger to clean out the hose.
7. Dry the fluid hose by running compressed air through it for 10 to 15 seconds.

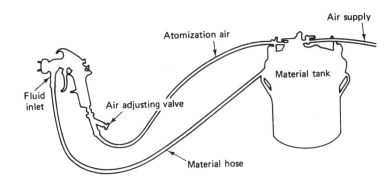

FIGURE 4–20. Typical pressure-feed hookup (tank type). (Courtesy of ITW Automotive Refinishing.)

8. Clean the spray gun and air cap.
9. Clean out the container and reassemble it for future use.

Plastic Liners for Paint Containers

Plastic liners of polyethylene (or equivalent construction) that fit snugly in material containers are available from some equipment suppliers. These liners save the shop considerable amounts of solvent because they leave the containers nearly clean when they are removed after a paint job. For availability of plastic liners, check with your local paint jobber.

Spray Gun Lubrication

Parts of the spray gun that require lubrication with a drop of light machine oil at the end of each day the spray gun has been used are indicated by letters A through E in Figure 4–21, as follows:

Location	Description
A	Oil trigger bearing stud
B	Oil fluid needle and packing (apply oil to needle)
C	Oil air valve stem and packing (apply oil to stem)
D	Fluid control screw threads (apply oil to threads)
E	Spreader control screw threads (apply oil to threads)

If the spray gun is used every day, apply a light grease to the springs, items F and G, twice per year. Disassembly of the fluid needle spring and air valve spring is necessary to perform this operation. If the spray gun is used two or three times per week, apply light grease to the springs once per year.

SPRAY GUN CLEANERS

Several different types of spray gun cleaners are available in the refinish trade. Many gun cleaners are completely enclosed during cleaning cycles and operate much like automatic dishwashers. Vertical pegs and special construction are provided to hold spray guns, paint cups, and other components while the cleaner runs through its cycles. The primary cleaning ingredients are permissible cleaning solvents that contain VOCs. It is easy to understand how evaporation of solvents into the air is prevented because, with a sealed cover, the system is completely enclosed.

Because of the differences from one spray gun cleaner to the next, no single procedure suits all cleaners. The best advice is to follow manufacturers' instructions. The most common precaution among enclosed gun cleaners is to remove the air gauge, if one is present at the base of the handle, before proceeding with cleaning instructions.

Painters will experience lubrication problems when using a completely enclosed gun washer to clean a Binks, DeVilbiss, Sharp, or equivalent construction spray gun. The system removes the lubrication at the following points:

1. Fluid needle packing
2. Air valve stem packing

The above problem may require gun disassembly and service parts replacement to keep the spray gun in top shape.

A second type of popular spray gun cleaner is called a spray gun cleaning station. The Safety-Kleen gun washer shown in Figure 4–22 is an example. To meet government regulations, a spray gun cleaning station must have the following capabilities:

1. All cleaning solvents (VOCs) must be collected in a closed container immediately upon cleaning.
2. Hydraulic, vacuum, or mechanical (non-atomized) means may be used to force cleaning fluid

FIGURE 4–21. Spray gun lubrication. (Courtesy of ITW Automotive Refinishing.)

A. Oil trigger bearing stud
B. Oil fluid needle and packing
C. Oil air valve stem and packing
D. Fluid control screw threads
E. Spreader control screw threads
F. Springs
G. Springs

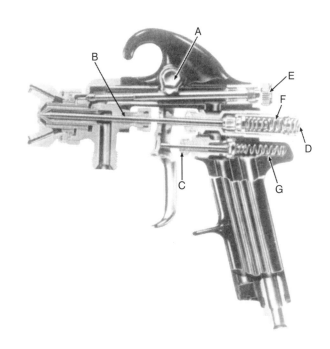

through a spray gun or over components to be cleaned. Then the solvents collect immediately in the closed container.

3. The spray gun cleaner must be equipped with a filter system so that solvents may be reused.
4. The company providing the spray gun cleaner must be connected with a waste disposal service company that meets EPA waste minimization regulations.

Whenever cleaning a spray gun in any part of the country, it is permissible to wipe the gun and its parts with a cloth dampened with solvent to complete the cleanup. The cloth must be disposed of in a closed waste container.

There is no need to soak a complete spray gun in solvent for the purpose of cleaning only the fluid passageway. That is what happens, however, when a conventional spray gun is cleaned in a conventional spray gun cleaner. Usually the air valve and fluid feed valve packings dry out with this cleaning method and need replacement.

The following is a list of advantages to using the Safety-Kleen gun cleaner system for cleaning any spray gun on the market safely and efficiently:

1. All fluid passages of the gun are cleaned thoroughly.
2. The air and fluid feed valve packings are not affected.
3. All air passages of the gun remain clean and are not affected.
4. This spray gun cleaner is a great time saver.

CLEANING A MATTSON SPRAY GUN

The Mattson Quick Switch cartridge contains the entire fluid passageway. The cartridge can be removed and reinstalled on the gun quickly, which makes the cleaning of a cartridge a simple matter that can be done in one of two ways:

1. As a bench operation, the cartridge can be cleaned with a solvent cleaning bottle furnished by Mattson when the spray gun is purchased. A bottle can also be purchased through a paint jobber handling their equipment under part number 400161 (see page xxx).
2. The Quick Switch cartridge can be washed in any gun cleaner or in an OSHA approved spray gun cleaning station.

The purpose of the cleaning clip shown in Figure 4–23 is to keep the fluid needle extended so cleaning fluid can be flushed through the fluid passages while they are cleaned.

IMPORTANT

When cleaning a gun as a bench operation in a compliant area, all flushed solvent must be captured in a sealed container according to clean air regulations.

FIGURE 4–22. Safety-Kleen's three easy steps to clean a spray gun (manual method). (Courtesy of Safety-Kleen Corporation.)

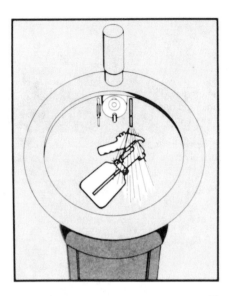

Step 1: Wash outside of gun under running solvent, which drains off immediately into a sealed container.

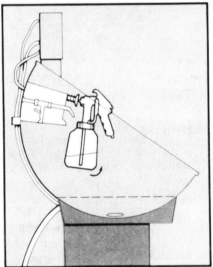

Step 2: Fill cup 1/3 full with solvent. Simply position gun tip into canister as shown. On suction feed guns, pull trigger and suction draws solvent through fluid passages to clean gun. On pressure feed guns, keep cup cover loose and pull trigger to clean gun.

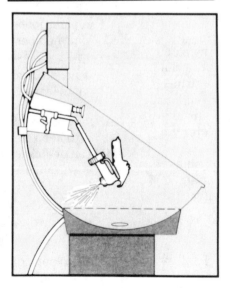

Step 3: Remove cup, place fluid feed tube of gun under solvent spigot, pull trigger, and pump clean solvent through gun. The clean solvent drains off immediately into a sealed container. The gun is cleaned in a very short time.

SPRAYING TECHNIQUES

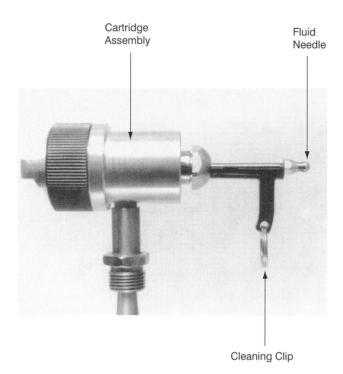

FIGURE 4–23. Cartridge fluid needle extended by installation of cleaning slip. (Courtesy of Mattson Spray Equipment.)

Description of Mattson HVLP Quick Switch Cartridge System

The Mattson Atom-Miser HVLP spray gun uses a design feature unique to today's painting industry, the Quick Switch Cartridge System. This system allows the professional painter to change quickly from applying one type of material to another of varying viscosity—from undercoats to basecoats to clearcoats and to single stage finishes—while using a single gun handle assembly. Before attempting a Quick Switch change, make sure all pressure is relieved from the system.

WARNING! The cup stays pressurized even after the shop air line has been disconnected from the gun. **Do not pull the trigger after disconnecting the air hose.** Release cup pressure by slowly releasing the lock lever.

Removal of Mattson Quick Switch Cartridge

Removal of the cartridge is necessary when changing to a different size of cartridge or when performing necessary cleanup operations.

1. Loosen the set screw securing the cartridge into its slide holder. Loosen the upper knurled black thumbscrew on the right side of the gun at least two full turns (see Figure 4–24).
2. As shown in Figure 4–25, hold the gun with your right hand and turn the knob clockwise until the slide holder moves forward enough to remove the cartridge.
3. While holding the cartridge stem with your left hand as shown in Figure 4–26, rotate the gun handle to the 3 o'clock position. This position frees the cartridge stem from the U slot of the slide so the cartridge can be removed.
4. Pull the cartridge straight forward and out of the gun (see Figure 4–27).

The cartridge can be reassembled in the gun by reversing the above steps. However, it is often faster to assemble the cartridge to the cup and remove or install this as a unit into the gun handle. The next four steps assume you are using that combined cup/cartridge method.

Installation of Mattson Quick Switch Cartridge

1. Position the cup/cover assembly in front of you so the clamping bridge is perpendicular to you and the locking lever is away from you toward the front. The air inlet tube will be located on the left side of the cup closest to you. With the cartridge nozzle pointing straight forward, connect the swivel nut on the cover to the cartridge stem fitting. Hold the cartridge body with one hand, and tighten the swivel nut with a $\frac{3}{4}$-inch open-end wrench (see Figure 4–28).
2. With the U slot of the slide extended to its maximum position as in Fig. 4–28, hold the gun handle at the 3 o'clock position and insert the cartridge into the gun body as shown (Figure 4–29).
3. Rotate the gun handle down to the 6 o'clock position to engage the cartridge stem in the U slot. Turn the total flow knob counterclockwise until the cartridge shoulder is flush against the gun body (Figure 4–30).
4. Keep a slight pressure on the total flow knob to hold the cartridge flush against the gun body. Tighten the thumb screw to set the cartridge into the slide (Figure 4–31).

Once the thumb screw is set, the total flow knob no longer moves the cartridge but becomes a size adjustment control for the spray pattern. In preparation for painting, turn the knob clockwise to the "MAX" position.

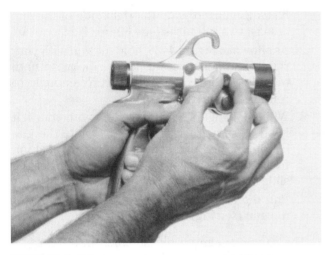

FIGURE 4–24. Loosening set screw on LP-DC gun cartridge. (Courtesy of Mattson Spray Equipment.)

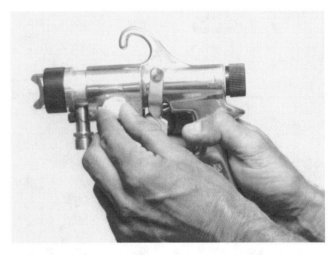

FIGURE 4–25. Turning control knob clockwise. (Courtesy of Mattson Spray Equipment.)

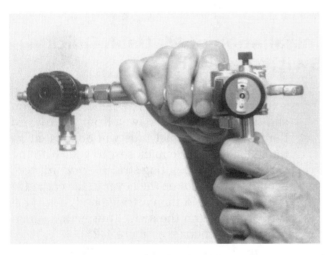

FIGURE 4–26. Rotate gun handle to 3 o'clock position. (Courtesy of Mattson Spray Equipment.)

FIGURE 4–27. Pull the cartridge straight and forward out of the gun. (Courtesy of Mattson Spray Equipment.)

Mattson Spray Gun Quick Clean Procedure

The following procedure applies to both compliant and noncompliant areas:

1. Disconnect the gun from the air line and move it to the designated area of the shop for cleaning guns.
2. Open the cup *slowly* to release pressure, hold the feed tube over the cup, and pull the trigger to allow any remaining material to drain back into the cup.
3. Transfer any remaining material in the cup into an appropriate container for storage or disposal. Clean the cup with a cloth dampened with solvent. Do *not* use Tri Chlor 3 solvent because it is corrosive to aluminum.
4. Remove the spray gun handle from the cartridge/cover assembly. (See page 55.)
5. Remove the Catch-All filter from the paint tube and place it in a jar of solvent to soak. Install cover on jar.

FIGURE 4–28. Attaching cartridge to cup/cover assembly. (Courtesy of Mattson Spray Equipment.)

FIGURE 4–29. Engage gun handle and cartridge at 3 o'clock position. (Courtesy of Mattson Spray Equipment.)

FIGURE 4–30. Turn control knob counterclockwise to engage gun. (Courtesy of Mattson Spray Equipment.)

FIGURE 4–31. Tighten thumb screw on the right side of the gun. (Courtesy of Mattson Spray Equipment.)

> **NOTE**
>
> The Mattson Spray Gun Company supplies a specially designed solvent bottle, part no. 400161, for flush cleaning the gun. The pointed tip of the solvent bottle is designed to fit all of the tube openings and small hose connections used in the Mattson Spray Gun. Do *not* cut or enlarge the tip of the solvent bottle.

6. Disconnect the air inlet tube from the quick-connect on the built-in regulator.
7. Insert the tip of the solvent bottle into the opening of the air inlet tube and run the solvent into the tube. Starting at this point will push the solvent through the check valve and LiquidLock (see Figure 4–32).
8. After flushing, remove the LiquidLock assembly from the cover and place it in clean solvent to soak. Install the cover on the jar.
9. Using the clip supplied, retract the fluid needle by pulling on the round ball stud at the rear of the needle assembly and attach the clip on the needle stem (see Fig. 4–23). This adjustment will hold the needle in the open position. The cartridge/cover unit can now be placed on the spigot of the gun washer, or it can be cleaned by flushing with solvent from solvent bottle.
10. If using the solvent bottle, invert the gun unit, insert the tip of the solvent bottle into the paint feed tube, and flush the system, allowing solvent to flow through the fluid tip into the catch container.
11. Remove the LiquidLock assembly from the container of solvent, insert the tip of the solvent bottle into the LiquidLock housing and flush thoroughly. Do the same with the catch-all filter.

> **CAUTION**
>
> **Always wear the proper safety equipment when painting and cleaning your spray equipment,** including:
> 1. Rubber gloves
> 2. Eye protection
> 3. Respiratory mask (TC-23C)
> 4. Proper clothing

> **IMPORTANT**
>
> After cleaning the spray gun, do the three Mattson daily performance checks before reassembling the gun.

Store the solvent bottle and all equipment cleaning tools in an OSHA approved metal cabinet. The Quick Clean procedure usually takes about five minutes. Consult the troubleshooting guide in Figure 4–33 when adjusting Mattson spray guns and equipment during use and cleaning.

> **IMPORTANT**
>
> Each spray gun company issues an operating manual for each spray gun it makes. These manuals explain the design, construction, operating procedures, and, most important, a troubleshooting procedure for each spray gun. All painters, apprentices, and others using spray equipment must pay close attention to all directives in these manuals to keep each spray gun operating at its peak efficiency. Always read the MSDSs on the products with which you work.

Space in this book does not allow complete coverage of all spray guns on the market or even those spray guns discussed in this book. The best way to get training on a particular paint spray gun or equipment line is to contact a representative of the company whose spray equipment you prefer. Your local paint jobber can help you contact the company.

SPRAY GUN EXERCISES

The purpose of the exercises in this section is to guide a person through the movements involved in a spray painting stroke so that, with practice, the person can become an efficient spray painter. The best advice is to master each component of the spray painting stroke and then to combine all the components together as explained in the following sections. Then it is only a matter of practice (see Figure 4–34). To practice the spray gun exercises described in this section, several basic items are needed:
1. One or more, depending on the size of the class, paint spray stands on which to do the spray gun exercises (see Figure 4–34).

After thoroughly cleaning the Mattson Atom-miser, always do these 3 simple checks.

Is the LiquidLock™ Assembly Clean?

Squeeze the solvent bottle to send clean solvent into pressure cup's air inlet tube. The solvent should flow freely through the LiquidLock.

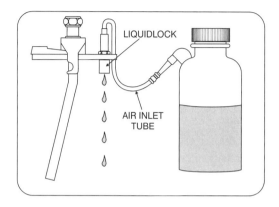

If Not:

1. Remove LiquidLock and soak in solvent until clean. If still obstructed after soaking, discard and replace.
2. Blow low-pressure air into the air inlet tube to clear out solvent. Be certain all traces of solvent are gone from the tube before reattaching tube to the regulator.
3. Reinstall the LiquidLock. Reassemble the Atom-miser.

Does the Needle Snap Back Freely?

When you pull back on the needle stem and release it, the needle should snap forward until it seals off the fluid tip nozzle.

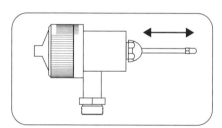

If Not:

1. Remove needle assembly from cartridge.
2. Be sure needle and wiper seals are clean.
3. Apply a small amount of petroleum jelly to needle on both sides of the retaining clip.
4. Replace needle assembly into cartridge. Tighten packing nut (#4 at right) only until snug. Needle should snap back freely when released.

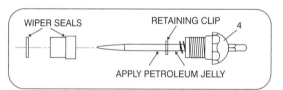

Does the Cartridge Neck Move Freely?

You should have about 1/8" travel in either direction when moving fluid pick-up neck along cartridge body.

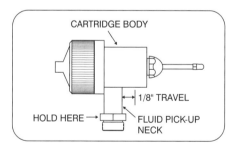

If Not:

1. Be sure the white fluid nozzle gasket (#1 right) is not crushed or distorted.
2. Make sure needle seals (#2 and #3 right) are not interchanged. (Seal #2 is thicker and has a larger inside diameter.)
3. If clear or color paint is trapped in cartridge, soak in high quality solvent until clean, then blow dry.

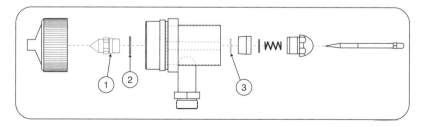

FIGURE 4–32. Daily performance check. (Courtesy of Mattson Spray Equipment.)

PROBLEM: Paint will not spray from fluid nozzle when trigger is fully depressed.

Check:

- Is the main air supply system for the shop functioning properly?
- Is the high pressure regulator in the paint booth set between 30 psi and 90 psi? Can you maintain this pressure?
- Have all hose connections been hooked up properly to the Atom-miser® System?
- Is the fluid control knob on the spray gun set correctly? See "Spray Gun Adjustment."
- Is the LiquidLock™ assembly obstructed? Clean or replace.
- Is the paint tube or "Catch-All" filter obstructed? Clean tube, and clean or replace filter.
- Is the cup regulator diaphragm swollen shut? Replace.
- Does the needle retract (open) when spray gun trigger is pulled? If you hear a clicking each time the gun is triggered, the fluid needle is stuck in the fluid tip. Clean fluid tip and needle.
- Piston and O-ring in regulator stuck in the closed position. Disassemble regulator and put Vaseline on piston and O-ring sliding surfaces.

PROBLEM: Spray gun drips paint from the area of the air cap/fluid tip after releasing trigger.

Check:

- Has the cartridge locking screw loosened up, allowing the cartridge to move out or forward from the spray gun body? Reset cartridge flush into gun handle; then tighten locking screw.
- Fluid needle is not closing completely—dirt or material buildup has occurred from improper cleaning. Remove and clean needle and fluid tip. If problem still exists, fluid needle or fluid tip must be replaced.
- Cartridge assembly may be worn or missing seals. Inspect and replace.
- Cartridge paint flow fitting has been damaged and must be replaced.
- Loose fluid tip. Tighten.

PROBLEM: Erratic spray pattern.

Check:

- Is the air cap clean? Clean in fresh solvent if necessary.
- Has air cap been damaged? If so, it must be replaced.
- Is fluid tip or needle dirty or damaged? Clean or replace.
- Is the nozzle seal (part # 81021) missing or distorted? Replace.
- Is the cartridge frozen up without 1/8" stem free play? Clean cartridge in solvent.
- Are main fluid control adjustment knob and fan control knob adjusted properly?
- Is shop air pressure regulator adjusted correctly for the material being sprayed?
- Is the paint overreduced? Incorrect reduction may cause an erratic spray problem.

PROBLEM: Spray gun squirts material before spraying.

Check:

- Is there sufficient air cap pressure? Check that shop air inlet pressure is at least 30 psi.

FIGURE 4-33. Troubleshooting guide. (Courtesy of Mattson Spray Equipment.)

- Cartridge has loosened and moved forward out of spray gun body. When spray gun is triggered, material is flowing before air is activated. **Air must always flow before material.** Reset cartridge into gun handle, then secure with the locking screw.
- Needle seal retainer is sticking. Apply petroleum jelly lightly to needle at least twice weekly and after each cleaning.

PROBLEM: Paint leakage from cup at cover.
Check:
- Is lid seated and locked firmly on cup? Reseat lugs and lock with lever.
- Defective or worn lid gasket—replace gasket.
- Severely damaged lid—replace cover.
- Material cup damaged, nicked, or bent out-of-round or locking lugs are loose. See below.

If the cup mouth has received minor nicks, it can be resurfaced by simply placing a full sheet of #400 sandpaper on a flat surface and sanding the cup mouth lightly. Keep steady, even pressure on the cup when performing this operation. **This resurfacing procedure should be done only on minor nicks. Replace cup if damage is severe. Too much removal of the cup neck will result in the cup leaking because the locking assembly does not have sufficient cup length in the neck area to seat the cover. If the cup is bent or out-of-round, it must be replaced.**

PROBLEM: Paint/air spray mixture not concentrated enough (sprays too "dry").
Check:
- Set fluid control knob to maximum (6 o'clock) position to increase fluid flow.
- Reduce wall-mounted regulator setting to lower air cap pressure.
- Switch from .032 fluid tip cartridge to .040 fluid tip cartridge (optional accessory).
- Increase fluid pressure by turning up cup pressure regulator.
- Achieve balanced spray pattern. See "Spray Gun Adjustment."

PROBLEM: Paint/air spray mixture too concentrated (sprays too "wet").
Check:
- Reduce cup pressure at cup pressure regulator.
- Improper spraying distance. Spray at 8" from surface.
- Atomizing air pressure too low. Increase air cap pressure by turning up wall regulator.
- Reduce setting on fluid control knob at rear of gun.
- Piston and O-ring in regulator stuck in maximum open position. Disassemble regulator and put petroleum jelly on piston and O-ring sliding surfaces.
- Achieve balanced spray pattern. See "Spray Gun Adjustment."

PROBLEM: Total Flow™ fan control knob binding and hard to turn.
Check:
- Thumbscrew on the right side of the gun and that locks in the cartridge is turned in too hard and is preventing free movement of the cartridge. Loosen thumbscrew until just snug.

FIGURE 4-33. *continued*

FIGURE 4–34. Use of paint spray stand in BASF Refinish Training Center spray booth. Note how students are prepared to take turns at practicing correct spraying techniques under the instructor's supervision. (Courtesy of BASF Corporation.)

2. A supply of paper on which to do the spray gun exercises. Choose from among the following:
   a. A special pad of chart paper to fit the spray stand.
   b. A supply of used newspapers that can be tack taped to the chart stand one sheet at a time.
3. A practice spray painting location:
   a. A fully equipped and approved paint spray booth is best.
   b. As an alternate, use a properly ventilated and lighted work area. This work area should include an adequate air supply, hose, air filter, and regulator.
   c. Spray painting equipment as required:
      (1) In the most highly regulated areas, only HVLP equipment can be used.
      (2) In all least regulated areas, conventional high-pressure or HVLP spray guns may be used.
4. An adequate supply of paint material and solvent as required. As an option to using paint material for beginners, you can use plain tap water.
5. Masking tape and a supply of used newspapers to cover paint stand for practice purposes.
6. Solvent cleanup materials as required.
7. Container for trash paper, cans, and shop debris.
8. Container for hazardous waste materials with sealed cover.
9. Proper safety equipment for all students:
   a. Respiratory equipment (mask or hood)
   b. Rubber gloves
   c. Eye protection
   d. Required clothing

Aids for Developing a Spray Technique

1. Some apprentices may want to start by going through the motions of spray painting without pulling the trigger. Allow the wrist to bend at the end of each stroke to maintain the perpendicular position of the gun to the surface. **Avoid arcing when painting complete panels.**
2. Strive for a uniform and wet application of color. Also strive for accuracy in hitting the target (banding) with proper triggering and smoothness of stroke movement.
3. When applying full spray painting strokes, check the following fundamentals of each stroke to see that it is done properly:
   a. Spray gun grip (Figure 4–12): Be sure that your grip is comfortable and fully controlled.
   b. Distance of the gun from the surface (Figure 4–13): Follow the contour of the surface. Always maintain the same distance from the surface.
   c. Position of the gun to the surface (Figure 4–14): Keep the gun position as perpendicular as possible.
   d. Speed of gun travel: Guide yourself by the sight of visual application. Never spray when you cannot see the application.
   e. Stroke movement: Keep your application smooth and relaxed. Use continuous motions at the beginning and end of each stroke.
   f. Triggering: Always keep the initial air on. Keep the application smooth and accurate. Trigger the material at target areas only.
   g. Avoid arcing on full panels (Figure 4–15): You must make an extra effort at the beginning and end of each stroke to bend the wrist and thus keep your arm in the perpendicular position of the gun to the surface.

SPRAYING TECHNIQUES

When learning spray painting, concentrated thought must be given to each fundamental. As practice continues and your stroke becomes more of a habit, you'll give less thought to each fundamental. You'll then have the opportunity to watch for the most important items:

1. Wetness of application.
2. Desired overlap.
3. Target areas where triggering takes place.
4. Review of complete panel paint application.

EXERCISE 1
Adjusting a Spray Gun

All spray guns are designed to achieve a balanced spray pattern with a full open spray gun adjustment. Balanced spray patterns can also be achieved with a reduced spot repair adjustment and small fan adjustment. Each student will perform and practice each of the adjustments, with instructor guidance, as follows:

1. Prepare the paint material for practice.
2. Adjust the air pressure at the wall regulator and at the paint cup (if provided).
3. Adjust the gun for full open spray gun adjustment and perform the flooding test described in Figure 4–4.
4. Make final adjustments and perform the final flooding test.
5. Adjust the gun for spot repair adjustment and perform the flooding test. Use a 50-percent fan (refer to Figure 4–1).
6. Make final spot repair adjustment and perform the final flooding test.
7. Adjust the gun for small fan adjustment and perform the flooding test.
8. Make the final small fan adjustment and perform the final flooding test.

EXERCISE 2
Triggering Practice (The Big X)

Refer to Figure 4–35. The purpose of this exercise is to practice triggering while the spray gun is in motion according to the following procedure:

1. Position fresh paper on the paint stand with masking tape.
2. With a banding stroke (and adjustment) make a large **X** on the spray stand from the upper left corner to the lower right corner, and from the upper right corner to the lower left corner (see Figure 4–35).

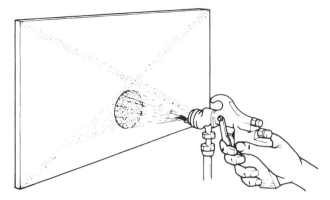

FIGURE 4–35. Triggering practice: The big "X."

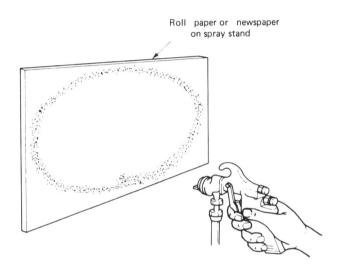

FIGURE 4–36. Overlap practice: The big "O."

3. Adjust the spray gun for spot-repair or midrange adjustment. Using only horizontal strokes and starting at the top or bottom, fill in one section of the **X** without crossing the lines. Use 50 percent overlap. Keep the arm and spray gun moving continually until the complete segment of the triangle is filled. Use good spray technique. Practice until arm motions are smooth and free flowing. Achieve full coverage.
4. Fill in a second section of the **X** without crossing the lines.
5. Fill in each additional section without crossing the lines. Strive for accuracy, uniformity, and wetness of application.
6. When done, the **X** should not be readily visible.

EXERCISE 3
Overlap Practice (The Big O)

Refer to Figure 4–36. The purpose of this exercise is to practice overlapping in a way that is unique and at the same time gives the sprayer an opportunity to develop eye-to-muscle coordination and control. For this exercise, adjust the spray gun for the smallest spray fan that can be achieved. Turn off the spreader and fluid feed controls completely. Then open the fluid feed just slightly and do the same with the spreader control to achieve the smallest balanced spray pattern.

1. Position fresh paper on the paint stand with masking tape.
2. With a banding stroke, make as large a circle as possible on the paint stand.
3. Fill in the circle with one continuous spiral pass. Spray from the outside in or from the inside out. Maintain overlap at 50 percent. Avoid streaks or sags by moving at the proper speed. Strive for accuracy, uniformity of coverage, and wetness of application.
4. When done, the circle and all overlapping passes should not be readily visible.

EXERCISE 4
Spraying a Rectangular Panel

The purpose of this exercise is to practice the technique of spraying a rectangular panel (see Figure 4–37). The instructor will explain and demonstrate the techniques on the practice paint stand. Each student will be given an opportunity to practice the techniques on the practice paint stand under the guidance of the instructor. Each student should practice the technique repeatedly, until all elements of the technique are performed properly.

FIGURE 4–37. Spraying a rectangular panel.

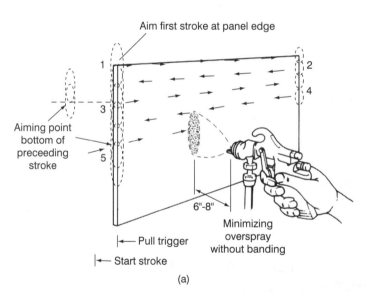

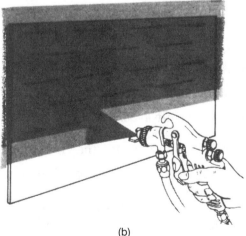

EXERCISE 5
Spot Repair Technique

The purpose of this exercise is to practice basic spray techniques used in making successful spot repairs on single-stage paint systems. (Spot repairs on basecoat and clearcoat systems are more involved and are covered in Chapter 13.) Spot repair techniques require special gun adjustments, triggering, and gun movement (see Fig. 4–38). Also, the techniques require greater color control by the painter. All automotive colors are spot-repairable. Each painter must learn how each color behaves and learn the latest factory recommendations on color shading to achieve the best color matches.

1. Prepare paint material for practice. (Black paint is preferred.)
   a. Set up one cup with reduced color. Mark the cup "Color."
   b. Set up a second cup with a blender. Mark the cup "Blender."
2. Set up a suitable paint stand for spray practice.
3. Adjust the spray gun to the midrange adjustment for spot repairing. (See "Spray Gun Adjustments," page 37, for details.)
4. Using a suitable marking pencil (black felt tip), draw about a 6-inch diameter circle at the center of the practice panel (see Figure 4–36). Then draw five additional circles about 2 inches larger than each previous one.
5. Apply color as follows while trying to stay within each target. Concentrate on grip, distance, arm motion, triggering, and feathering.
   a. Spray the first coat while spraying from the inside-out.
   b. Spray the second coat while spraying from the outside-in.
   c. Spray the third and fourth coats beyond the previous coats while concentrating on wrist arcing and feathering. Continue the inside-out and outside-in feathering technique.
   d. Apply the fifth and sixth coats by combining all the above techniques. At the conclusion of color application, adjust the gun for the blender coat by leaving the fan adjustment alone but by reducing the material feed just slightly. Switch the paint cups to the blender coat material. Apply the blender coat material at all the edges. Concentrate on proper technique, especially triggering, while uniformly wetting down all blending areas.

REVIEW QUESTIONS

1. Spray painting involves theory and skill development. Explain each.
2. Explain how to make and test the full-open spray gun adjustment on a conventional high-pressure spray gun.
3. Explain how to make and test the full-open spray gun adjustment on an HVLP low-pressure spray gun.
4. Explain how to make and test a spot-repair spray gun adjustment on a high- or low-pressure spray gun.
5. Describe the normal panel spraying technique.
6. Describe how the blender coat is applied when making spot repairs.
7. Describe the conventional single-coat method of spray painting.
8. Describe the banding spray pattern technique.
9. Describe the technique used for spraying large, horizontal surfaces such as a roof, hood, and/or rear deck of a car.
10. Describe the conventional spot-repair technique with regard to final color application.
11. Describe the flooding test method of checking spray patterns.
12. What is the air pressure regulation for an HVLP spray gun in a most highly regulated refinish area like California?
13. Describe how spray guns are cleaned in a most highly regulated area.
14. Describe an HVLP performance settings chart.
15. What does transfer efficiency mean in HVLP spray paint application?

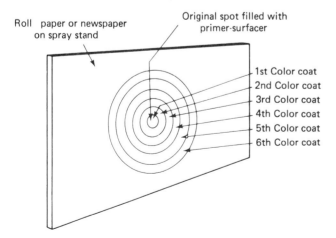

FIGURE 4–38. Spot repair blending technique.

16. What minimum percentage of paint material in a cup is applied to a surface when HVLP refinish equipment is used?
17. Explain how to calibrate a conventional paint shop air system for HVLP use.
18. Describe the Mattson HVLP Quick Switch Cartridge and explain how it is cleaned after painting.
19. Explain how to lubricate a spray gun after cleaning.
20. Explain the procedure for cleaning a hose on a 2-quart HVLP spray outfit.

CHAPTER 5

Spray-Painting Equipment and Facilities

INTRODUCTION

Automotive refinishing can be divided into four areas (Figure 5–1):

1. Paint materials
2. Paint equipment
3. Painter skill and know-how
4. Safety and health regulations

This chapter deals with the general description of paint equipment and the role equipment plays in the overall refinish picture. How to use each piece of equipment is covered in applicable chapters of the book.

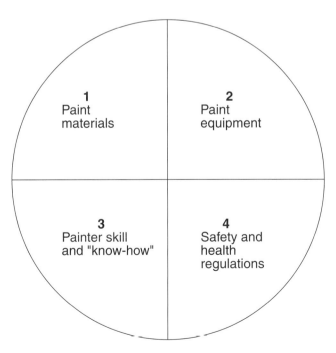

FIGURE 5–1. The four parts of automotive refinishing.

ROLE OF OSHA AND NIOSH

In 1970, Congress passed the **Occupational Safety and Health Act.** The purpose of the act is "to assure as far as possible every working man and woman in the nation safe and healthful working conditions and to preserve our human resources." To administer the act, the Labor Department's Occupational Safety and Health Administration (OSHA) issues standards and rules, called regulations, for safe and/or healthful working conditions, tools, equipment, facilities, and processes, and conducts inspections to ensure that they are followed.

In the automotive refinish trade, OSHA regulations protect the health and safety of the painter and protect the investment of the owner. OSHA does not test, approve, or certify products or devices in industry. OSHA establishes safety and health regulations and enforces them. OSHA encourages states to develop plans for their own safety and health programs. OSHA grants states up to 50 percent of the cost of operating an OSHA-approved plan that "is at least as effective as" OSHA's program. The act and the standards issued by OSHA apply to every employer with one or more employees.

Under the act, employers have the general duty of providing both conditions of employment and a place of employment free from recognized hazards to safety and health and the specific duty of complying with OSHA standards. Employees must comply with standards and with job safety and health rules and regulations that apply to their own conduct.

Prior to 1970, the federal government had established the **National Institute for Occupational Safety and Health,** or **NIOSH.** This agency now serves in cooperation with OSHA. NIOSH has many

67

responsibilities. The two most important responsibilities that affect the refinishing trade are:

1. **Recommendations for particular safety regulations.**
2. **Testing and certifying the safety of tools and equipment designed and built for industry.** Tools and equipment certified by NIOSH are so marked, and approval limitations are stated on each box or container.

Paint shop owners are urged to cooperate with both OSHA and NIOSH. To do business, a body and paint shop must be licensed each year by the city and state. To become licensed, each shop must be inspected. Compliance with local, state, and federal regulations is observed closely by authorities when licenses are renewed each year.

ENVIRONMENTAL PROTECTION AGENCY

The federal government's clean air regulations are implemented by the Environmental Protection Agency (EPA) and may be divided into two parts:

1. Part 1 of the plan involves setting up a national rule on VOCs for the entire country.
2. Part 2 of the plan consists of clean air rules and regulations that must be enacted by each state based on local pollution problems. (Clean air laws are explained in more detail in Chapter 1.)

RESPIRATORS

Smart, qualified painters always use the proper respiratory protection for the job. Many factors are involved in selecting the proper respirator. Most important are the job being performed and the atmosphere surrounding the worker for determining the type of respirator needed. Each respirator on the market is marked with a NIOSH rating, which lets you know the level of protection it offers.

Three types of respirators are needed by the painter to do automotive refinishing:

1. **General spray painting respirator** (NIOSH specification number TC-23C; see Figure 5–2) is recommended when spray painting and when mixing nonisocyanate finishes like acrylic lacquers, enamels, and conventional undercoats. Basically, all these finishes are nonisocyanate finishes.
2. **Air supplied mask or hood respirator** (NIOSH specification number TC-19C) is recommended whenever spray painting and mixing any and all finishes that contain isocyanates.
3. **Dust particulate respirator** (NIOSH specification number TC-21C) is recommended whenever sanding and/or grinding operations are done. (TC-23C may also be used for dusty work conditions.)

Painters should be instructed in the following:

1. How to select the proper respirator.
2. How to fit and test a respirator before starting to paint.
3. The health hazards involved.
4. How to clean and sanitize the respirator.
5. How to store the respirator for the next use.

General Spray Painting Respirator (TC-23C)

The general spray painting respirator is a chemical and mechanical (charcoal filter) respirator designed to protect the painter against atomized paint particles, dirt, and toxic vapors when spray painting nonisocyanate paint finishes. These respirators are available from leading paint equipment suppliers and from paint jobbers.

Before using your respirator:

1. Inspect it before every use to make sure it is in proper working order.
2. Check to see that each rubber exhalation valve flap is secured to the valve seat, and that both the valve and seat are clean and free from damage.
3. Check that the filter cartridges and prefilters are attached properly to the respirator.

To wear your respirator, adjust the position of the facepiece for comfort and fit. If the straps are too tight, remove the mask and adjust each of the straps the same amount so the mask will be centered on your face. Tighten them just enough to hold the mask securely to your nose to prevent leakage. Never overtighten. Straps that are too tight might distort the mask and actually cause leakage.

NOTE

Do not try to wear a half mask respirator if you have facial hair because it will not seal properly to your face. You may have to wear a full face-covering mask for protection.

SPRAY-PAINTING EQUIPMENT AND FACILITIES

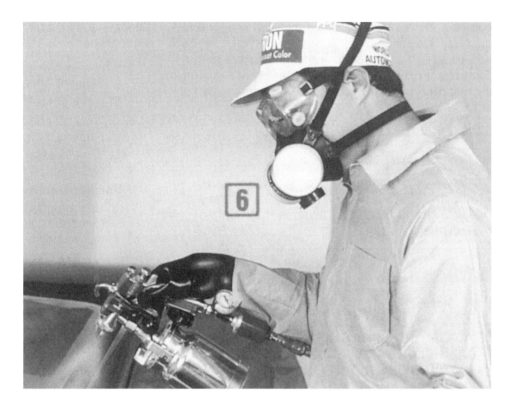

FIGURE 5–2. General spray painting respirator (cartridge type TC-23C) for preparing and applying single-stage undercoats and topcoats. (Courtesy of PPG Industries.)

Qualitative Fit Test

Before entering an area containing toxic materials, the mask wearer must verify the proper fit and working order of the respirator with a qualitative fit test. Every time the respirator is worn, verify proper fit by performing one or both of the following tests immediately before entering a contaminated atmosphere. If you detect any leakage, readjust the mask as needed and repeat the test until there is no detectable leakage.

> **NOTE**
> Do *not* move the facepiece while performing these tests.

Negative Pressure Test

1. Fit the mask to your face as described. Remove all of the filters on your respirator.
2. Close off the air inlet openings by covering them with the palms of your hands or temporarily seal them using tape or a plastic film like plastic wrap (remove it after the test).
3. Inhale and hold your breath so the facepiece collapses inward slightly, indicating there is negative pressure in the facepiece. The fit is considered satisfactory if the facepiece remains in this slightly collapsed condition for the duration of the test and no inward leakage of air is detected.
4. Remove hands, tape, or plastic wrap from the blocking inlet openings.
5. Reinstall filter cartridges.

Positive Pressure Test

1. Fit the mask to your face as described.
2. Close off the air outlet opening to the exhalation valve by temporarily covering it with your hand or sealing it with tape or plastic film (remove it after the test).
3. Exhale so the facepiece is slightly enlarged and hold your breath for about 10 seconds. The fit is considered satisfactory if the facepiece remains in this slightly enlarged condition for the duration of the test and no outward leakage of air is detected.
4. Remove hands, tape, or plastic film from the blocking exhalation valve cover.

You should clean, inspect, and sanitize your respirator at the end of each day. After cleaning your respirator, it is important to store it in a clean, dry, cool place. You can keep it in a tightly sealed plastic bag or a clean container, such as a leftover coffee can with a resealable lid. Always replace the filters according to the directions of the respirator manufacturer. Filters need to be replaced when you detect increased breathing resistance, odor, or taste.

> ### WARNING
> Always read a label like the one below carefully and slowly. This warning applies to spray painting and mixing all finishes that contain isocyanates.

> ### WARNING!
> **VAPOR AND SPRAY MIST HARMFUL. MAY CAUSE LUNG IRRITATION AND ALLERGIC RESPIRATORY REACTION. MAY IRRITATE SKIN AND EYES. COMBUSTIBLE.**
>
> **Gives off harmful vapor of solvents and isocyanates (a hazardous material). DO NOT USE IF YOU HAVE CHRONIC (LONG-TERM) LUNG OR BREATHING PROBLEMS, OR IF YOU HAVE EVER HAD A REACTION TO ISOCYANATES. USE ONLY WITH ADEQUATE VENTILATION. WHERE OVERSPRAY IS PRESENT, A POSITIVE PRESSURE, AIR-SUPPLIED RESPIRATOR (TC19C NIOSH/MESA) IS RECOMMENDED. Follow directions for respirator use. Wear the respirator for the whole time of spraying and until all vapors and mists are gone.**
>
> **Avoid breathing of vapor or spray mist. Avoid contact with eyes and skin. Keep away from heat and open flame. Keep closures tight and upright to prevent leakage. Keep container closed when not in use. In case of spillage, absorb and then dispose of the material in accordance with local applicable regulations.**
>
> **FIRST AID: If affected by inhalation of vapor or spray mist, move to fresh air. If breathing difficulty persists or occurs later, consult a physician and have label information available. In case of eye contact, flush immediately with plenty of water for 15 minutes and CALL A PHYSICIAN. In case of skin contact, wash thoroughly with soap and water.**
>
> **KEEP OUT OF THE REACH OF CHILDREN.**

Recommended safety guidelines for painters are listed below:

- **When in the paint booth,** there must be no smoking or open flame.
- **For eye protection:**
  a. Wear a hood or goggles when mixing or painting.
  b. Wear safety glasses when sanding and cleaning.
- **For lung and respiratory protection:**
  a. Use supplied air when painting or mixing (NIOSH TC-19C; hood or eye goggles).
  b. Wear a dust mask or face respirator when sanding.
- **Protective clothing must be worn:**
  a. Wear a Tyvex suit while painting or mixing.
  b. Always wear safety toe shoes and a work uniform when working in the shop area.
- **For hand protection:**
  a. Wear rubber gloves while cleaning equipment.
  b. Wear latex gloves while mixing paint.
  c. Wear surgical gloves while spraying paint.
  d. Wear cotton gloves while sanding.

Air-Supplied Mask or Hood Respirator

Air-supplied mask or hood respirators, as approved by NIOSH specification number TC-19C, are suitable for spray painting urethane paint finishes and other paint finishes even when spraying in confined areas with poor air circulation, when the painter would be forced to breathe high concentrations of toxic vapors, chemicals, and overspray. Air-supplied mask or hood respirators are approved by NIOSH as a "package" consisting of several components. Violation of any single component by painters or paint shop owners violates the complete NIOSH approval. An air-supplied mask respirator is shown in Figure 5–3; an air-supplied hood respirator is shown in Figure 5–4.

No fit testing is required with the hood. Beards, eyeglasses, and long hair pose no problem. The hood is made of a tough plastic material and weighs just 4.5 ounces, making it easy to use. It has a large window that allows maximum peripheral vision. It features a built-in twin lens system, which doubles hood life and makes it easy to remove overspray. Also, the hood does not inhibit verbal communication. Painters can be heard when speaking in normal conversation.

The oil-less air pump for the hood respirator (see Figure 5–4) is available through paint jobbers. The

SPRAY-PAINTING EQUIPMENT AND FACILITIES

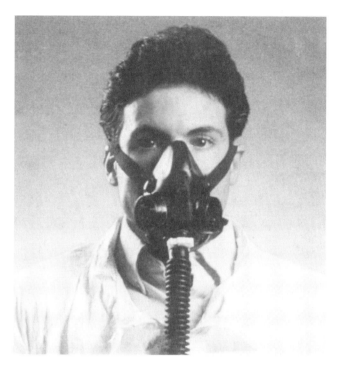

FIGURE 5–3. Air-supplied mask respirator (TC-19C). (Courtesy of DuPont Company.)

FIGURE 5–4. Air-supplied hood respirator (TC-19C) for preparing and spraying two-pack isocyanate primers and top coats. A typical air pump is shown below. (Courtesy of ITW Automotive Refinishing.)

pump draws and passes ambient air through an efficient filter, providing clean air to painters without any fancy air compressors. The pump is powered by an oil-less electrical engine with a low power requirement. Therefore, it does not generate any contaminants such as carbon monoxide, oil vapor, or mist. There is no need for expensive carbon monoxide monitors or high temperature alarms. The pump is easy to maintain. For more information on all air supplied respirators, contact your local paint jobber.

CAUTION

Painters should never use conventional compressors that are oil lubricated to supply air to air-supplied respirators because carbon monoxide, a poisonous gas, can be produced by these compressors, and warning signals can always fail.

Dust Respirator

Dust respirators like the TC-21C shown in Figure 5–5 are mechanical-type respirators designed with filters that remove dust and solid particles from the air a painter breathes. Dust respirators cover the nose and mouth. Most dust respirators are of the throw-away variety: after they become saturated with dust or dirt, they are discarded. Dust respirators are used mostly in the paint shop when dry-sanding paint and plastic fillers, particularly with power equipment. Dust respirators should be used whenever you are working in a very dusty situation.

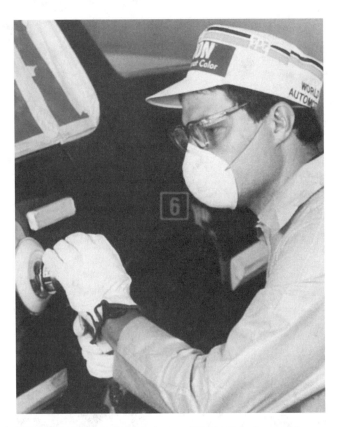

FIGURE 5–5. Dust respirator (TC-21C) to be used when sanding or filing. A general-purpose respirator (TC-23C) may be used as an option. (Courtesy of PPG Industries.)

FIGURE 5–6. Dusting guns (Binks models 190 and 152). (Courtesy of ITW Automotive Refinishing.)

WARNING

Simple face masks or loosely fitting respirators should not be used. For best results, use dust respirators that have been approved by NIOSH, the Bureau of Mines, or MESA. They fit snugly.

Respirators should be stored in a clean box. Cleaning by light brushing, tapping on a hard surface to free dirt, and vacuuming can make dust respirators last longer. When heavily clogged with plastic dust such as fiberglass residue, it is better to discard the respirator.

DUSTING GUN

A dusting gun (see Figure 5–6) is a small air gun that attaches directly to an air-line hose by means of a quick attach and detach connector. A dusting gun is a very popular tool in a body and paint shop because it is used to clean cars with compressed air before painting.

Before OSHA came into existence, full line pressures were often used to clean cars. These pressures often exceeded 100 psi. At times, painters or helpers accidently blew dirt or foreign matter into their own or someone else's eyes, which resulted in serious problems. Good eyesight is one of a painter's most important assets.

OSHA Regulation Regarding Dusting Guns

"Compressed air shall not be used for cleaning purposes except where reduced to less than 30 psi, and then only with effective chip guarding and personal protective equipment." To meet this OSHA regulation:

1. Paint equipment suppliers have specially designed dusting guns available that develop less than 30 psi at the nozzle even when high air pressure is used.
2. Also, conversion nozzles, which meet the same OSHA requirements, are available.

To meet the second portion of the OSHA regulation, the painter must observe the following safety rules:

1. Before dusting, the painter should wear suitable eye protection (such as a face shield or goggles).
2. No other person may be close to the person with the dusting gun unless he or she is wearing suitable eye protection.

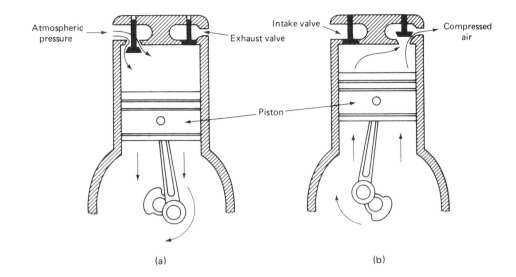

FIGURE 5–7. Piston-type air compressor (single-stage): (a) downstroke: intake. (b) upstroke: compression. (Courtesy of ITW Automotive Refinishing.)

AIR COMPRESSORS

Compressor Capacity

For air-operated equipment to function properly, it is necessary to have a certain amount of continuously supplied air available at a required pressure. The equipment that performs this operation is the air compressor. The most important part of this air supply is the amount or volume of air that is measured in cubic feet of air per minute (CFM). Air pressure is measured by pounds per square inch (psi).

The compressor used most often in paint shops is the 5-horsepower compressor with an 80-gallon air tank. This unit will deliver 12 to 20 cubic feet of air per minute, which is adequate for spray painting needs. Most automotive paints require 9 to 16 CFM. Low VOC and waterborne finishes (which have been developed in conjunction with the Clean Air Act) require 11 to 20 CFM. The Mattson Spray Equipment Company supplies an Air Miser Blue Ring Air Cap that can achieve excellent atomization with only 9 to 12 CFM when applying low- to medium-solids paint materials at 3 to 5 psi.

For more information regarding types of compressors available or to answer any questions regarding compressors, contact your local paint jobber or spray equipment company.

Types of Compressors

Compressors can be divided into many categories. Two of the most popular types are the piston and diaphragm types.

FIGURE 5–8. Single-stage compressor. (Courtesy of ITW Automotive Refinishing.)

PISTON-TYPE COMPRESSORS

This type is available as stationary units, which are anchored in place, and as portable units, which are on wheels and can be moved to the job. A piston-type compressor develops pressure through the action of a reciprocating piston much like that of an automobile engine. Air enters the compressor through an intake valve on the **downstroke** (Figure 5–7a). The air is compressed and expelled through an exhaust valve on the **upstroke** (Figure 5–7b) to an air tank or an air line.

When air is drawn from the atmosphere and compressed to a given pressure in a single stroke, the compressor is classified as a single-stage unit (Figure 5–8). **The efficiency of single-stage units is good up to 100 psi but poor over 100 psi.**

FIGURE 5–9. Two-stage compressor. (Courtesy of ITW Automotive Refinishing.)

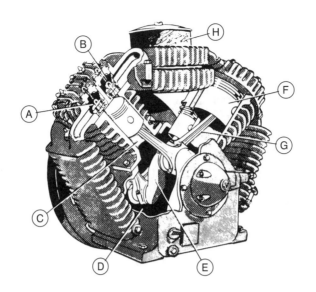

A. Intake valve assembly
B. Exhaust valve assembly
C. Cylinder
D. Crankcase
E. Crankshaft
F. Piston
G. Connecting rod assembly
H. Air-intake filter

When air is drawn from the atmosphere and first compressed to an intermediate pressure (approximately 40 psi), then passed through an intercooler to a high-pressure cylinder for recompression, the air compressor is a two-stage unit (Figure 5–9). The high-pressure cylinder is approximately one-half the diameter of the low-pressure cylinder. **Two-stage air compressors are used when air pressure in excess of 100 psi but less than 200 psi is required.**

DIAPHRAGM-TYPE COMPRESSORS

These compressors are usually small and portable, and are designed for low-air-volume use. The diaphragm compressor develops pressure through the reciprocating (pushing and pulling) action of a flexible diaphragm. An electric motor causes the pushing and pulling action on the diaphragm. On the downstroke action, a valve opens to fill the chamber above the diaphragm with air (Figure 5–10b). On the upstroke, the valve closes and air is pushed out of the compressor by the diaphragm (Figure 5–10a).

Installation of Compressor and Air Regulator

Compressors and air regulators should be installed as follows: (see Figure 5–11):

1. Install the compressor or air regulator in a cool, clean area with plenty of access for maintenance.
2. If floor-mounted, be sure the compressor is level, with all four feet resting firmly on a solid floor or foundation.

Automatic Pressure Switch

An automatic pressure switch is an air-operated electric switch for starting and stopping electric motors at predetermined minimum and maximum pressures. Switches with various cut-in and cut-out pressures are available for different requirements. They are used when it is convenient and economical to start and stop the motor. Any outfit that runs intermittently and less than 60 percent of the time is best controlled with a pressure switch.

Maintenance

A compressor crankcase should be filled with a good grade of oil to the proper level, SAE 10 for ordinary conditions and SAE 20 for temperatures above 100°F. The oil should be changed every 2 to 3 months and the level should be checked every week. The bearings on the electric motor should be oiled weekly unless they are life-lubricated bearings.

The belt should be checked for proper tension and alignment so that the proper power transmission is achieved. All dust should be blown away from the cooling fins, including the intercooler and aftercooler. The air intake strainer should be cleaned once a week. The safety valve handle on the tank should be lifted at least once a week to check if it is functioning properly. The flywheel should be checked for tightness on the crankshaft. The tank should be drained of moisture every day, especially in high-humidity areas (Figure 5–11).

A compressor, if properly cared for, will last a long time. But if trouble develops, consult the manufacturer's manual supplied to the purchaser.

SPRAY-PAINTING EQUIPMENT AND FACILITIES

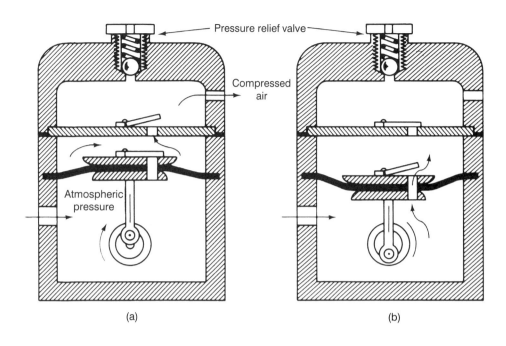

FIGURE 5–10.
Diaphragm-type air compressor: (a) upstroke: compression. (b) downstroke: intake. (Courtesy of ITW Automotive Refinishing.)

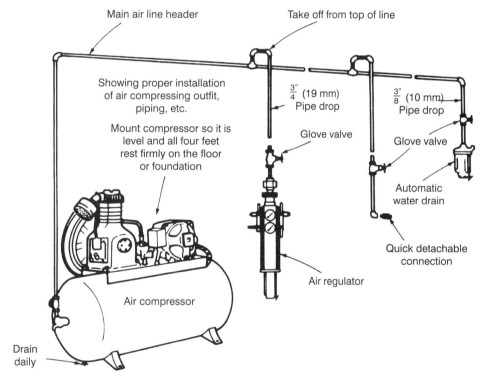

FIGURE 5–11.
Installation of compressor and air regulator. (Courtesy of ITW Automotive Refinishing.)

COMPRESSED AIR FILTER AND REGULATOR

To do top-quality paint repair work, painters need a supply of compressed air that is clean and regulated, which is the purpose of an air cleaner and regulating device. Figure 5–12 shows the Mattson 5050 Integral Filter Regulator. This advanced air-filtration system almost completely eliminates harmful water vapor and airborne dirt particles for outstanding spraying performance and finish-quality consistency.

The 5050 delivers the complete range of CFM requirements for conventional and HVLP spray equipment. An extremely sensitive pressure regulator offers precise setup without guesswork. Specialized

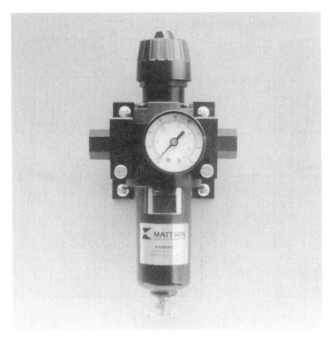

FIGURE 5–12. Mattson 5050 Integral Filter Regulator. (Courtesy of Mattson Spray Equipment.)

filter elements allow incredible filtering properties without significant pressure loss, while a unique modular design allows for filter replacement in under a minute. For more information, contact your local paint jobber.

Installation

The best location for an air regulator in a spray booth is at the side or at the center of the spray booth. Figure 5–11 shows how a typical regulator is installed. Regulators are usually hooked up off the main line at least 25 feet from the compressor. Notice that the takeoff line for the regulator goes up first; then the line is routed down to the regulator. The purpose of this design is to control any water that may be in the main line. From the point of regulator takeoff, the main line slopes downward toward the compressor. Also, from the regulator takeoff, the main line slopes downward to the rear. The reason for this arrangement is to control water in the line. Water should be drained at the regulator, at the rear drain, and at the compressor periodically. When highly humid air is compressed for long periods of time, a good amount of water collects in the lines as a result of condensation. If not removed, this water could be squirted out of the air line.

The air regulator should be bolted securely to the spray booth wall or similar sturdy object near the painter for convenience in reading the gauges and operating the valves. Use piping of sufficient size for the volume of air passed and the length of pipe used (see Table 5–1). The pipes must always be of the recommended size or larger. Otherwise, excessive pressure drop will occur.

TABLE 5–1 Minimum Pipe Size Recommendations

Compressing Outfit		Main Air Line	
Size (hp)	Capacity (cfm)	Length (ft)	Size (in.)
$1\frac{1}{2}$ and 2	6–9	Over 50	$\frac{3}{4}$
3 and 5	12–20	Up to 200	$\frac{3}{4}$
		Over 200	1
5–10	20–40	Up to 100	$\frac{3}{4}$
		Over 100 to 200	1
		Over 200	$1\frac{1}{4}$
10–15	40–60	Up to 100	1
		Over 100 to 200	$1\frac{1}{4}$
		Over 200	$1\frac{1}{2}$

Courtesy of the DeVilbiss Company.

Description of Air Filter and Regulator

A typical working air regulator does the following for the painter:

1. Condensers and filters remove water, dirt, and oil.
2. The regulator knob operates the regulated air line.
3. The regulated air gauge tells the painter the setting for proper air pressure.
4. The main line air gauge tells the painter the amount of pressure in the main air line.
5. Regulated air outlets are provided for attaching the spray gun hose.
6. Main line air outlets are provided for operating the air tools.
7. The drain valve allows the painter to drain the filter and regulator.

REFINISH HOSES

The two most popular sizes of hose used in the refinish trade are $\frac{5}{16}$- and $\frac{3}{8}$-inch inside diameter hoses. Hoses most commonly used in the refinish trade are 25 to 50 feet in length. The 2-quart pressure remote

SPRAY-PAINTING EQUIPMENT AND FACILITIES

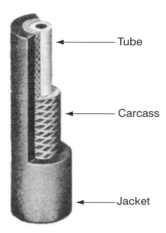

FIGURE 5–13. Typical refinishing hose construction. (Courtesy of ITW Automotive Refinishing.)

cups use a short length of hose. Red hoses are generally used for air, and black hoses are used for fluid feed lines.

Two basic types of hoses are used in the automotive refinishing trade: **air hose** and **fluid hose.** Hoses are made up of four basic components, depending on design. Three of the components shown in Figure 5–13 are vulcanized together; then connections are added. Typical hose construction is as follows:

1. **Tube:** The tube is the inner part of the hose and is carefully selected to resist solvent action (fluid hoses), pressure, and temperature.
2. **Carcass:** This section is made up of one or more layers of strong (high-tensile-strength) fabric braid that is bonded to the tube and jacket. The carcass must provide satisfactory high work pressures and maximum flexibility.
3. **Jacket:** The jacket is the protective outer cover of the hose. It is chosen to resist damage from abrasion and water, contact with oils and chemicals, and general exposure.

Pressure Drop

Pressure drop is the loss of air pressure in the hose between the air regulator and the spray gun (see Figure 5–14). The regulated air gauge on the wall may show 46 psi at the transformer but, because there is a 6-pound air pressure loss in a 25-foot hose, the actual delivered air pressure at the spray gun is 40 psi. Table 5–2 shows the amount of pressure drop for varying lengths of hose. The estimated pressures shown in Table 5–2 already have the correct amount of pressure drop subtracted for the various lengths of hose. This calculation is a help to the painter.

FIGURE 5–14. Checking drop of air pressure between air regulator and spray gun (see also Table 5–2). (Courtesy of DuPont Company.)

Pressure drops are caused by friction between the flowing air and the walls of the hose, constricted fittings, and the passages the air travels through. The smaller the inside diameter of a hose and the longer the hose, the greater is the pressure drop. The painter must understand the following about pressure drop:

1. Pressure drop exists.
2. It has an effect on the quality of paint repair work.
3. How to determine correct atomization air pressure at the gun.

Table 5–2 shows how pressure drops for $\frac{1}{4}$- and $\frac{5}{16}$-inch hoses of varying lengths. As illustrated in Figure 5–14, the difference between the regulated air pressure at the source and at the gun is determined by the size and the length of the air hose.

Hose Care and Maintenance

1. The hose should never be subjected to any form of abuse, and it should always receive reasonable care.
2. The hose should never be dragged over sharp objects, tools, car parts, or abrasive surfaces.
3. Care should be taken never to inflict severe "end pull," for which the hose and coupling were not designed.
4. The hose should always be used at or below its working pressure. Changes in pressure should be made gradually without subjecting the hose to excessive surge pressures.
5. The hose should never be kinked severely, purposely or accidentally.

TABLE 5–2 Estimated Air Pressures at the Gun

Hose Diameter (in.)	Pressure Reading at the Transformer (lb)	Pressure at the Gun for Various Hose Lengths[a] (ft)					
		5	10	15	20	25	50
$\frac{5}{16}$-inch hose	30	29	$28\frac{1}{2}$	28	$27\frac{1}{2}$	27	23
	40	38	37	37	37	36	32
	50	47	48	46	46	45	40
	60	57	56	55	55	54	49
	70	66	65	64	63	63	57
	80	75	74	73	72	71	66
	90	84	83	82	81	80	74

[a]Snap hose connectors will further reduce the pressure that reaches the gun by 1 lb. Pressures given are approximate and apply only to new hoses.
Courtesy of the DuPont Company.

6. The hose should never be run over by a car or heavy equipment.
7. If the hose is accidentally smeared with body fillers, paint materials, or grease, it should be cleaned as soon as practicable.
8. At the end of each day, the hose should be rolled up in a 2- or 3-foot loop and hung on a designated hose hanger.

OSHA Regulation Regarding Pressure Hoses

"All pressure hose and couplings shall be inspected at regular intervals appropriate to this service. The hose and couplings shall be tested with the hose extended, and using the in-service maximum operating pressures. Any hose showing material deteriorations, signs of leakage, or weakness in its carcass or at the couplings, shall be withdrawn from service and repaired or discarded." Air and fluid hoses are classified as "pressure hoses."

SPRAY BOOTHS

The spray booths of leading paint equipment suppliers are engineered and installed to conform to the safety regulations of **OSHA** and the **National Fire Protection Association (NFPA)** and local code requirements. After installation, compliance with all requirements become the responsibility of the spray booth owner.

Paint booths are available in two basic models: solid back and drive-thru. Each is designed to meet various requirements. Both models accommodate full size pickups, vans, and two small cars at one time. Today's booths remain exceptionally clean, eliminating the need to wash down between paint jobs.

A paint job that meets the top standard of the trade is not always an easy accomplishment, in spite of a competent painter, an efficient spray gun, and the best materials available. The right kind of spray booth is often the difference between a second-rate paint job and the gleaming paint finish that car owners have come to expect.

There is no question that a common problem in automotive refinishing is control of dirt in paint. There is no better solution to the problem than to do paint work in a top-quality spray booth. The purpose of a spray booth is to:

1. Provide the best available conditions for top-quality spray painting.
2. Provide maximum safety to the painter and paint shop.
3. Protect the painter's health by removing atomized paint particles and solvent fumes.
4. Guard against fire hazards.
5. Prevent dust and dirt from entering the booth.

Two additional advantages to a spray booth are:

1. **Lower insurance rates:** Some states even provide a tax advantage to paint shops.

SPRAY-PAINTING EQUIPMENT AND FACILITIES

FIGURE 5–15. Typical down-draft spray booth showing how air surrounds the car on the way to the floor exhaust. (Courtesy of ITW Automotive Refinishing.)

2. **Better community relations:** Proper filtration of exhausted air minimizes air pollution in the surrounding neighborhood.

Three types of spray booths are available in the refinish trade:

1. **Cross-draft spray booths:** A cross-draft spray booth is so named because power-ventilated air comes into the booth from the side or end through filters. The air moves across the booth, then exits the opposite side or end of the booth through filters that lead to a roof exhaust duct. Over the past few years, cross-draft spray booths have been the most popular in the refinish trade.
2. **Down-draft spray booths:** A down-draft booth is so named because power-ventilated air comes into the booth through filters in the ceiling and travels straight down, taking all overspray and air movement through filters in a grated floor. Air is exhausted through a floor pit that leads to the outside of the building.
3. **Waterwash spray booths:** A waterwash spray booth is equipped with a water-washing system designed to minimize dust or residue entering exhaust ducts and to permit the recovery of atomized paint particles. The waterwash spray booth reduces fire hazards and is most popular in car factories and in high-volume parts paint shops. Waterwash spray booths are more costly to operate and require more space.

There are several paint spray booths available from equally competitive equipment companies. Because of limited space, it is impossible to describe in detail all the spray booths that are available. The author has described three types of the available spray booths that are used frequently by many painters. Figures 5–15 and 5–16 are representative of just two spray booths that are popular because of their dirt control advantages.

When selecting a down-draft spray booth, be sure it offers the following:

1. **A clean, safe working environment:** The booth should provide a continuous flow of clean filtered air from overhead and along the sides of the booth during spray operations and the curing cycle.
2. **Low turbulence and higher transfer efficiency:** The booth should have no inherent turbulence. It should provide an envelope of fresh air that moves with the painter as it creates air stability and pressurizes all areas of the booth except the floor. A low uniform air velocity will enhance paint transfer efficiency with significant reductions in overspray.

FIGURE 5–16. Features of down-draft spray booth. (Courtesy of ITW Automotive Refinishing.)

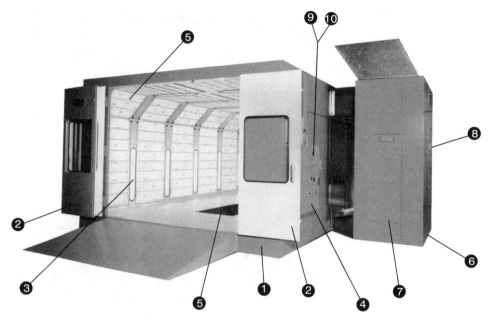

3. **Minimal operating noise:** The booth's interior environment should be quiet to minimize painter stress, hearing loss, and fatigue. Reduced noise increases productivity, job satisfaction, and quality.
4. **Shadow-free booth with color-corrected lighting:** The vehicle and painter should be surrounded with bright, uniform, and shadow-free lights. To provide such an environment, color-corrected lights should be installed in the sides, gables, and ceiling of the spray booth.

Spray Cycle (Paint Application)

When set to the spray cycle, the down-draft spray booth functions like a standard automotive down-draft spray booth. An envelope of fresh, temperature-controlled air surrounds the vehicle and painter in a stable, turbulent-free environment. The air flows from the insulated plenums through the ceiling and side filters down to the floor exhaust pit. The air is then filtered through a unique spun wool, fiberglass filter medium and exhausted.

Cure Cycle (Paint Drying)

After the car is sprayed, a painter can simply set the cure time and temperature switch on the control panel to the cure cycle; the paint force drying takes over. First, the booth is purged of any solvent vapors or volatiles for six minutes as fresh filtered air is drawn through the booth and exhausted. Then the cure cycle begins.

The direct-fired burner raises the temperature to its preset level. The return air damper opens and heated air is circulated through the burner and over the vehicle until the preset timer times out. The booth exhaust fan turns slowly to bleed off excess air during the cure cycle to compensate for the intake of fresh air.

The following is a list of features of the down-draft spray booth:

1. Rigid construction.
2. Large front vehicle access doors.
3. Unique shadow-free illumination system.
4. Central control panel system.
5. State-of-the-art filter systems.
6. Reliable air make-up unit.
7. High output heater unit.
8. Variable frequency exhaust fan.
9. Automatic digital temperature control.
10. Programmable temperature limits.
11. Compressed air line/exhaust fan interlock.
12. Ease of cleaning and maintenance.
13. Compliance with all OSHA and local fire and safety regulations.

Down-draft spray booths are designed to meet or exceed all OSHA and local fire and safety regulations for proper use and ventilation. Be sure to check with your local authorities regarding codes. Study all instructions carefully and observe all safety precautions during operation.

SPRAY-PAINTING EQUIPMENT AND FACILITIES

Description of Lower Cost Spray Booths

Lower cost spray booths are constructed of 18-gauge galvanized unpainted panels that are readily bolted together. Special floor channels ensure alignment to all side panels.

As shown in Figures 5–17 and 5–18, the booth employs the down-draft principle of air flow. Air is introduced through inlet filters in the ceiling of the booth by an air makeup system. The air is drawn over the vehicle and through the floor grate, and is filtered and exhausted via the floor pit. Air velocity over the vehicle is increased to exhaust overspray quickly.

Interior lighting in the down-draft booth is designed to provide shadow-free illumination. Booths are equipped with ten light windows, each with four tube fluorescent fixtures. Light brackets provide quick changing of bulbs and easy window cleaning.

The filter medium used in the down-draft booth is made of a special nonflammable paper and is formed into double accordion folds with staggered holes to provide a highly efficient filter (see Figure 5–18). The filters last five or more times longer than pad filters. The filter is collapsible for easy storage and lower shipping costs. It takes only minutes to remove and replace the filter and you will do it less often.

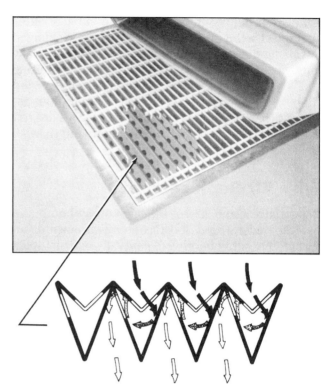

FIGURE 5–18. View of floor grating in down-draft spray booth showing cross-section of exhaust filter (located under floor grating). (Courtesy of ITW Automotive Refinishing.)

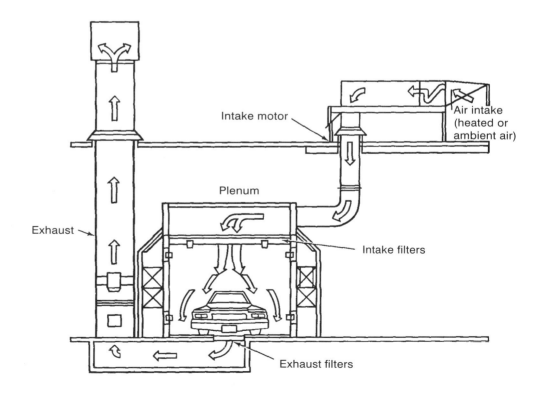

FIGURE 5–17. Cross-section view of down-draft spray booth showing path of ventilation air travel. (Courtesy of ITW Automotive Refinishing.)

Located below the grating is the exhaust filter medium. The highly efficient filter is easily changed by raising the grate and collapsing the filter for easy removal. The filters are installed by clipping in one end, stretching the folded filter across the pit, and clipping the opposite side in place: an operation taking less than one minute. If any short pieces are left over, they can be attached to a new length with masking tape or staples.

Why Air Replacement?

Depending on the size of the paint spray booth and specific federal and local code requirements, an automotive spray booth will generally exhaust from 9000 to 12,000 cubic feet of air per minute. There is no practical way of reusing the same air because it is filled with volatile fumes and must be exhausted. OSHA regulations state that the air expelled from a spray booth must be replaced. Furthermore, the replacement air must be heated whenever the outside temperature remains below 55°F for appreciable periods of time and the general building heating system is not capable of maintaining 65°F in all parts of the building area whenever the booth exhaust is in operation.

PRIMARY SOLUTION

The ideal solution to the problem is an air replacement unit (Figure 5–19) that takes in fresh air from the outside, filters it, and, if necessary, heats and dries the air. The replacement air is then delivered either into the spray booth or into the general shop area. The unit operates only when the spray booth fan is on and heats the air only when required.

ALTERNATIVE SOLUTION

If the heating plant for the entire shop is large enough to replenish the heat loss within the shop area when the booth exhaust fan is operating, the paint shop owner may or may not want to select an air replacement package assembly. However, if the owner chooses the alternate solution, care should be taken to provide a way to have adequate air enter the building without strong drafts.

Spray Booth Maintenance

1. Sweep the spray booth floor before the start of any paint job. Also, wet down the spray booth floor to minimize dust.
2. Keep the spray booth free from haphazard storage of boxes, junk, miscellaneous parts, water buckets, trash cans, brooms, and dust pans.

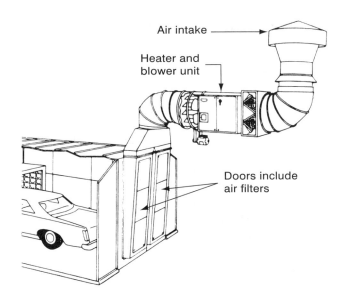

FIGURE 5–19. Paint spray booth air replacement unit. (Courtesy of ITW Automotive Refinishing.)

3. Change the spray booth intake and exhaust filters periodically as required.
4. Check and repair all door seals to ensure that they provide good dust seals for the booth.
5. Clean the glass panels over the lamps periodically for better illumination.
6. Wash down the spray booth walls periodically to remove excess dust accumulation.

DRYING EQUIPMENT

With the increased use of waterbase materials, drying equipment is very important to the success of a paint shop. After cars are painted, the next objective is to help speed the drying of the paint finishes. The success of any paint shop is determined by the type and amount of drying equipment on hand. Basically, two types of drying are used in the refinishing trade.

1. **Air drying:** The temperature range for air drying is from 60° to 100°F. In this process cars are usually left where they are painted until they are dry enough to be moved. All refinish materials are designed to air-dry in a reasonable period of time.
2. **Force drying:** The temperature range for force drying is from 100° to 180°F. Infrared drying equipment is best suited for the refinish trade. Force drying reduces drying time from 1 day to 30 or 45 minutes. Infrared drying equipment provides ideal drying conditions for the paint

SPRAY-PAINTING EQUIPMENT AND FACILITIES

repair trade and is available in various sizes to fit any equipment need:

a. Very small, one- or two-bulb units on an adjustable stand are ideal for small spot repairs.

b. Medium-size units with four bulbs or quartz tubes on portable and adjustable stands are ideal for panel repairs.

c. Multibanked units (24, 36, or more bulbs) are ideal for a complete side of a car (see Figure 5–20).

One might ask, What is an infrared heat unit, and how does it work? The answer follows:

1. Infrared heat lamps radiate or transfer energy in the form of heat waves that travel in a straight line.
2. The heat waves are invisible to the human eye.
3. The heat waves go from the lamp right through the paint finish and contact the metal or plastic substrate.
4. The substrate becomes hotter as it absorbs greater amounts of infrared heat waves.
5. The hot substrate drives the solvents from the paint finish and lacquers. At this point the paint is essentially dry (but hot). In enamels, after initial solvent evaporation, an additional chemical change takes place (polymerization) and the enamel dries and hardens. In urethane finishes, there is less solvent evaporation, the chemical change takes place faster than in enamels, and the paint dries and is hot.
6. Thus, under infrared lamps, paint dries from the inside out, or from the substrate to the outer paint surface.

For best results in using infrared heat units, follow the manufacturer's directions and maintenance instructions. Every modernly equipped shop has a drying area where force drying is done.

Use of an infrared heat lamp involves certain safety precautions. One of these precautions is that the area should be properly ventilated. A spray booth could be used for drying purposes if it is not needed at the time for spray painting.

Painters should know what specific temperatures work best for each specific paint job. A painter can determine how hot any given paint surface gets when positioning a heat lamp at the surface by employing the following procedure:

1. Secure a **candy-type** thermometer from a kitchenware store, department store, or instrument equipment firm.
2. Remove the thermometer from the container and secure it to a dry section of the car paint finish with a piece of masking tape.
3. Position the heat lamp 18 to 20 inches from the car panel with the heat lamp parallel to the car panel. The thermometer should be centered under the heat lamp.
4. Turn on the heat lamp and let it heat the surface for 7 to 10 minutes.
5. Check the temperature on the thermometer, which should be resting flush against the car panel. The heat lamp should never allow the thermometer to hit or go over 140°F. If the temperature goes over 140°F, move the heat lamp away from the panel the required distance and recheck the temperature. It is much better to know the proper distance rather than to guess at it.

PAINT MIXING ROOM AND EQUIPMENT

Every paint shop should be equipped with its own complete set of color mixing materials and equipment so that it can formulate any color, at any time, on demand. Thus equipped, a shop can paint any

FIGURE 5–20. Multibanked infrared force-drying unit.

car, in any color, without waiting for a delivery or experiencing any other delay. It also saves time and increases business because a shop can make any color immediately, to match any car. This system provides a shop with a complete set of tinting colors for year-round use to match all automotive colors. A color mixing system consists of:

1. A color mixing room with paint agitator.
2. Special shelving with a full range of base colors, with positive agitation.
3. A microfiche reader and complete sets of formulas.
4. An accurate and easy-to-use color-mixing scale.

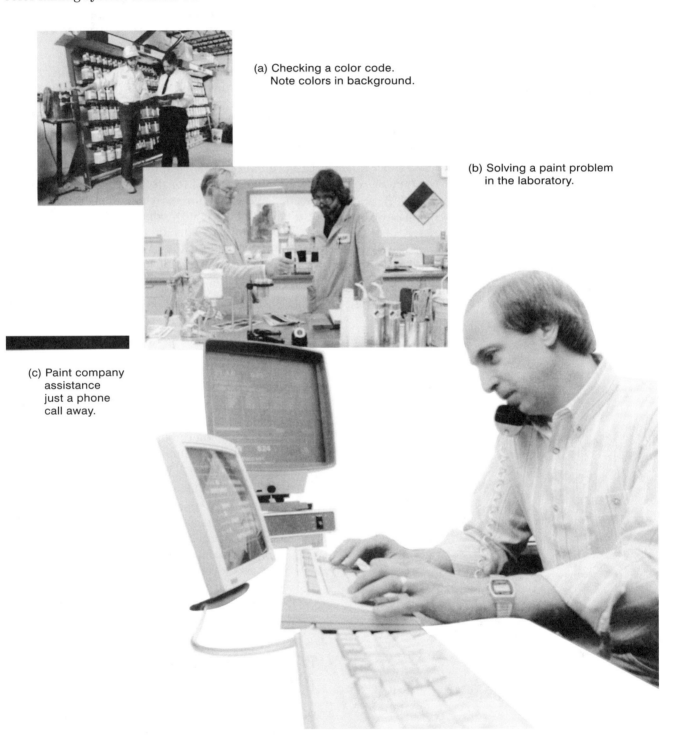

(a) Checking a color code. Note colors in background.

(b) Solving a paint problem in the laboratory.

(c) Paint company assistance just a phone call away.

FIGURE 5–21. Technical support for shops with color-mixing system. (Courtesy of BASF Corporation.)

A complete training program on how to use the mixing system comes with the system.

As shown in Figure 5–21, continual communication with the color-service department of the paint company is always available to the paint shop. With all this assistance, no color problem can go unsolved.

Automatic Color-Matching Tool— The Spectrophotometer

Major paint companies now have an automatic color-matching tool, which is actually a **spectrophotometer** (see Fig. 5–22). Matching automotive color even with an experienced eye can often be a challenge and may involve a certain amount of guesswork. The spectrophotometer is a tool that takes the guesswork out of color matching. For availability of a spectrophotometer and a color-management system, check with your local paint jobber.

To use a spectrophotometer, follow this procedure:

1. Place the tool over a sample of color to be analyzed. Rock the unit forward and wait for the beep that signals a scan has been completed. Multiple scans should be taken from areas around the repair. The software within the tool unit will calculate the number of measurements required. The unit software also identifies erroneous measurements and discards them, requiring additional measurements. Once the tool software accepts the measurements, the unit will beep twice, indicating a complete measurement.

2. Connect the tool unit with a single cable to your computer running the color-management system. With a couple of key strokes, the scanned color is downloaded automatically to your computer for analysis.

3. Enter the search criteria into the color-management system. The system will search for the closest match and display the formula for mixing. It will also adjust an existing formula to bring it closer to the scanned color.

You are now ready to mix the correct color and make the repair.

SURFACE PREPARATION WORK STATION

Several equipment companies have developed surface preparation work stations (see Figure 5–23 for an example) that can be used anywhere in the shop while keeping the shop very clean. A surface preparation work station is designed as follows:

1. The floor is raised and has a drive-on ramp.
2. The unit has a strong down-draft exhaust system.
3. The unit can be installed anywhere in the shop.
4. All surrounding areas are kept clean.

FIGURE 5–22. The Spectrophotometer, an automatic color-matching tool. (Chroma Vision Unit by DuPont; courtesy of DuPont Company.)

FIGURE 5–23. Prep-Master Work Station (down-draft design construction). (Courtesy of ITW Automotive Refinishing.)

The following surface preparation operations can be performed:

1. All surface cleaning and hand and power sanding.
2. Initial spot priming and primer–surfacer application.

For the availability of this unit, check with your local paint jobber.

When the sanding mode switch of the surface preparation work station is turned on:

1. A strong down-draft of air around the car moves the air downward and into the floor exhaust, where the air is filtered as it passes through special floor filters.
2. All sanding dust caused by hand or power sanding is carried by this strong envelope of air downward and into the filters. None of the dust can go into the shop area, so the shop stays perfectly clear of sanding dust, which is exactly what the surface preparation unit is designed to do.
3. The recirculating air moves from the floor grate area to the overhead (plenum) air supply. The air goes through a final filter as it starts another cycle.

When the prime mode is turned on, an important change takes place: the prime mode switches the exhaust system from recirculating to outside exhaust. The system works by venting all priming exhaust air and vapors out and away from the building. Together with the use of HVLP spray equipment, the result is an extremely clean, safe, and productive working environment.

> **IMPORTANT**
>
> Your local and/or state fire and building codes may include very specific requirements and codes for priming operations. Be sure to check with the proper authorities prior to installation. Study all instructions carefully and observe all precautions during operation.

SPRAY GUN CLEANER AND SOLVENT RECYCLER

Several different types of gun cleaners are available in the refinish trade. The most popular gun washers are those that can save the paint shop money by

FIGURE 5–24. Safety-Kleen spray gun cleaners: (a) automatic model, (b) manual model. (Courtesy of Safety-Kleen Corporation.)

using recycled solvents and by being connected with a reliable waste paint disposal service. According to clean air laws, paint and body shops can face stiff penalties for improper disposal of hazardous wastes, which include leftover paints, dirty spray booth filters, paint sludge, and other hazardous materials. The Safety-Kleen Corporation (and other equivalent companies) provide an excellent gun cleaner and complete waste paint disposal service. By working with a government-approved waste disposal service, paint shops are free of any government waste-paint regulations.

As shown in Figure 5–24a and b, the Safety-Kleen Corporation has two types of spray gun cleaners available for conventional paint shops. The gun cleaner on the right (Model 1107) is classed as a spray gun cleaning station and is used for cleaning spray guns manually, as described in Fig. 4–23.

The gun cleaner on the left (Model 1111) is classed as an automatic gun cleaner. When the cover is closed, the unit is sealed and works automatically to clean a spray gun according to Safety-Kleen's instructions.

By selecting the two cleaner service options, the shop has the ability to clean guns automatically or manually in any part of the country. (Use of eye protection and rubber gloves is highly recommended when cleaning a spray gun.)

The Safety-Kleen waste paint disposal service works by:

1. Thoroughly cleaning your gun cleaner on each service visit.
2. Providing you with a solvent designed specifically for cleaning spray gun equipment.
3. Removing and properly labeling all your waste solvent and helping you with all documentation to meet EPA and Department of Transportation (DOT) regulations.
4. Providing a certificate of assurance that promises the proper recycling and disposal of your gun cleaner waste—along with a pledge to pay for any costs should a spill occur while your waste is in their control.
5. Working on a schedule you have agreed to and based on your needs.
6. Providing free use of a Safety-Kleen spray gun cleaner for up to four weeks should your machine break down for any reason. This provision includes servicing a non–Safety-Kleen gun cleaner if the paint shop has a written agreement with the Safety-Kleen Corporation.

To be competitive, a paint shop should be equipped with the following:

1. An updated spray booth with force drying facilities.
2. A paint mixing room with the following items:
   a. Rack with a complete set of mixing colors.
   b. Electronic paint-mixing scale.
   c. Suitable mixing bench.
   d. Special mixing colors and popular factory pack colors.
   e. Required solvents.
   f. Required hardeners and other additives.
3. A computer system with the following capabilities:
   a. Automatic cost estimating program.
   b. System for preformulated formulas.
   c. A spectrophotometer and color-match program.
   d. Inventory of products and general recordkeeping.
4. Portable force drying equipment.
5. Instant access to product bulletins and MSDS information.
6. Spray gun cleaner service and hazardous material storage and waste removal program.
7. Continual communication with a primary paint supplier for color information and product assistance.
8. Automatic paint agitating machine.
9. Complete shop towel service.

REVIEW QUESTIONS

1. What is the purpose of OSHA?
2. What is the purpose of NIOSH?
3. Name three types of respirators needed by a painter to do automotive refinishing.
4. NIOSH-approved dusting guns operate at what maximum air pressure?
5. How often should an air compressor be drained of moisture?
6. What is the purpose of an air cleaner and air regulator?
7. What is the typical average length of hose in a paint shop?
8. What is pressure drop?
9. What is the meaning of psi and CFM?
10. What is the purpose of a paint spray booth?
11. What is an air replacement package assembly for a paint shop? How does it work?
12. What is the procedure for cleaning and preparing a paint spray booth before applying an enamel paint job?

13. Automotive refinishing can be divided into what four areas?
14. Explain how a down-draft spray booth air exhaust system differs from a cross-draft system.
15. What sizes of hose (inside diameter) are used most frequently in the refinish trade?
16. What size of air compressor (in horsepower) is used most often in paint shops?
17. What is the range, in cubic feet of air per minute, required by paint shops to apply all automotive finishes?
18. Explain how an air-supplied respirator works. When is it required to be used?
19. Dusting guns operate at what maximum air pressure while staying within OSHA regulations?
20. What is the maximum temperature beyond which heat lamps should not be used on a car?
21. Explain when a dust respirator is required to be used. How does it work?
22. Explain how a general spray painting respirator works. When is it required to be used?
23. What are two of the most popular types of air compressors?

CHAPTER 6

Surface Preparation

INTRODUCTION

Automotive refinishing, like any type of high-quality paint work, requires that a surface be properly prepared before repair painting or finishing can be done. Surface preparation consists of many forms and types of repairs, ranging from simple cleaning and/or compounding of a surface to complete removal and rebuilding of a paint finish. Paint repairs range from small brush touch-up and small spot repairs to large panel repairs and complete refinishing. In this range lie many thousands of paint repair categories. Most of these categories are summarized and presented in Chapter 17.

Considering the refinishing materials and repair methods available, for every **paint condition,** that is, every automotive paint problem, there is a specific way that each problem surface should be prepared before high-quality painting can be done. For example, take the case of a rust spot on a panel. All rust must be removed completely, and the metal must be conditioned properly if a paint repair is to be guaranteeable. Every paint repair should be done with the intent that it is guaranteed consistent with the price paid for the work.

CAR WASHING

After a repair order is written up and before the car is brought into the paint shop for repairs, the first operation, generally, is a car wash. Every paint shop should have a car wash stall equipped with the following minimum equipment:

1. Cement or equivalent floor with adequate water drainage.
2. Adequate lighting and water supply.
3. Suitably long hose with "squeeze-handle" nozzle, which shuts off the water when the handle is released. These kinds of hoses are available in hardware and department stores.
4. Water buckets, suitable sponges, a chamois, car wash detergent, window cleaning solvent, cloth towels, and paper towels.
5. Air line, air hose, dusting gun, safety goggles and/or face shield, rubber boots, and waterproof apron.

The inside and outside of the car should be cleaned as required. Having an owner take home a car with a high-quality paint finish that is clean outside and inside builds good customer relations.

Procedure: Indoors

1. Remove the floor mats (wash these separately, rinse, and hang to dry). Empty and reinstall ashtrays and litter bag or bucket. Vacuum the floors, seats, and trim, as required.
2. Close all the windows tightly and close the doors.
3. Soak or wet down the entire car before starting the washing operation. This step reduces the possibility of scratching the car finish.
4. Prepare the car wash detergent (according to label directions) or prepare a bucket of water with half a cup of laundry soap. Using a suitable soft sponge, wash the car in 2-by-2-foot sections progressively around the car.
5. Rinse the car thoroughly.
6. With safety goggles or face shield in place, use compressed air and blow off the car in the same sequence as in rinsing. Use a handy clean towel to wipe off any splatters.

> **NOTE**
>
> If compressed air is not available, use a chamois to wipe off the car as required:
>
> a. Soak the chamois in clear water until it is saturated.
> b. Fold the chamois in half several times and wring out the chamois thoroughly.
> c. Unfold the chamois and fold it in half several times until it is a suitable wiping pad.
> d. With the pad, wipe off all remaining water from the car surface working a 2-by-2-foot area at a time.

> **CAUTION**
>
> Never wipe a car surface with a chamois right after wiping the wheels or wheel openings because severe scratching could result. Soak and rinse the chamois in clean, clear water to remove all sand and dirt before wiping the car.

7. Wash each car window, first outside and then inside, with a water-dampened cloth and an application of ammoniated window solvent. Wipe each window dry with a clean cloth.
8. Wash the hinge and lock pillars of the door and body, and wash the rocker panel on each side of the car with a sponge. Use care not to contact the door lock and/or lock pillar striker to prevent smearing with dirty oil or grease.

Procedure: Outdoors

> **CAUTION**
>
> Never wash a car on a hot day in direct sun if the temperature is over 85°F, especially between 10 A.M. and 3 P.M., when the sun is hottest. Allowing soapy water to evaporate from a paint finish in bright sun on a very hot day may stain the paint finish. See "Water Spotting" in Chapter 17.

MATERIALS NEEDED FOR SURFACE PREPARATION

Paint Finish Cleaning Solvent

The purpose of a paint finish cleaning solvent is to dissolve and remove road oils, road tars, grease, pollution contaminants, wax and silicone polishes, rubbing compound oils, and fingerprints. If not removed, these contaminants may affect the adhesion or appearance of the repair finish. All cleaning solvents are not compatible with all finishes. Follow label recommendations. Refer to Table 6–1.

> **CAUTION**
>
> Be sure to wear solvent resistant gloves and safety glasses while using paint finish cleaning solvents during surface cleaning and/or wax and grease removal.

PROCEDURE FOR GENERAL CLEANING

1. Using a clean white cloth, apply paint finish cleaning solvent freely to the painted surface work area and adjacent surfaces.
2. Clean the surface up to 2 square feet at a time with as much pressure as is required. Overlap the previously cleaned surface by 2 to 4 inches.
3. For maximum results, wipe the surface dry with clean white cloths while the surface is still wet. Change cloths frequently when doing large areas or a complete car.

PROCEDURE FOR REMOVING SILICONES

1. Using a clean white cloth, apply cleaning solvent freely to the painted surface, wash as required, and wipe dry with a clean white cloth.
2. Apply cleaning solvent to the surface freely with a clean white cloth and sand the surface thoroughly with No. 500, No. 600, or Ultra Fine sandpaper.
3. Reclean the surface with solvent and then wipe dry with a clean white cloth. **CAUTION: Do not rewipe the surface with used or contaminated cloths.**

SURFACE PREPARATION

TABLE 6–1 Paint Finish Surface Cleaners

	Code	Mix Ratio	Reducer
BASF TRADE NAME			
Prep-Cleaner	905	1:1	Water
Final Wipe	909	1:1.5	Water
Universal Sanding Liquid	960	RFU[a]	not needed
Pre-Kleano	900	RFU	N/N[b]
Pre-Paint Cleaner	901	RFU	N/N
DUPONT TRADE NAME			
LOW VOC Final Clean	3909S	RFU	N/N
Kwik Clean	3949S	RFU	N/N
Prep Sol	3919S	RFU	N/N
Plastic Prep	2319S	RFU	N/N
Final Clean	3901S	RFU	N/N
PPG TRADE NAME			
LOW VOC Cleaner	DX380	RFU	N/N
LOW VOC Cleaner	DX390	RFU	N/N
Acryli-Clean	DX330	RFU	N/N
Ditz-O-Wax and Grease Remover	DX440	RFU	N/N
Plastic Cleaner	DX103	RFU	N/N

[a]RFU: Ready for use.
[b]N/N: Not needed.

> **CAUTION**
>
> Never clean the surface of a plastic-filled repair with paint finish cleaning solvent. Plastic fillers absorb and retain solvents, which later results in blistering.

Cleaning solvents should always be used before and after sanding operations.

> **CAUTION**
>
> Before using a cleaning solvent on acrylic finishes, read the label instructions. Some cleaning solvents are not safe for use on all acrylic finishes. Be sure that the label instructions say specifically: "Designed for safe use on acrylic finishes." The solvents listed in Table 6–1 are the best available for acrylic finishes. Cleaning solvents that are incompatible with acrylic finishes should be so labeled.

Metal Conditioner

Metal conditioners (see Table 6–2) for refinishing are available in a two-part system:

- **Part 1:** The purpose of part 1 is to clean the metal of any oil, dirt, grease, fingerprints, casual corrosion, and other foreign contaminants through a process known as acid cleaning and chemical etching.
- **Part 2:** The purpose of part 2 is to impart a heavy chemical coating on the surface, which promotes finish adhesion and durability.

Painters encounter three types of metal when doing paint repair work:

1. **Steel:** Most exterior car panels are made of steel. (*Note:* Most front end panels are plastic.)
2. **Galvanized steel:** Most car rocker panels have been made of galvanized steel since the early 1960s.
3. **Aluminum:** Use of aluminum is increasing for car panels to reduce overall weight. Early major uses were in hoods and rear compartment lids.

TABLE 6–2 Metal Conditioners (By Paint Supplier Identity Number)

Use	DuPont	PPG	Rinshed-Mason
For steel			
Part 1: acid cleaner	5717-S	DX-579	801
Part 2: phosphate coating	224-S	DX-520	
For galvanized steel			
Part 1: acid cleaner	5717-S	DX-579	801
Part 2: phosphate coating	227-S	DX-520	802
For aluminum			
Part 1: acid cleaner	225-S	DX-533	801
Part 2: phosphate coating	226-S	DX-503	

To determine if metal is aluminum or steel, use a magnet as follows:

1. If the metal panel shows a magnetic pull toward a magnet, the metal is steel.
2. If the panel shows no magnetic pull, the panel is aluminum or plastic.

The two-part metal conditioning system provides the best-known methods of conditioning metal in the refinish trade. The metal conditioning, priming, and color coating must be used together properly to provide the designed paint finish durability.

Metal conditioners are acids and are made available in bottled, concentrated form. In this form, the metal conditioner is inactive. Therefore, reduction with water according to the manufacturer's directions is necessary.

PROCEDURE FOR USE

- **Part 1:** Clean bare metal with a prescribed metal conditioner by reducing and applying the conditioner as described on the label directions.
- **Part 2:** Treat bare metal with the second metal conditioner according to the label directions. An important part of the application is the time factor. Allow the proper amount of time for the chemical reaction to take place. Then wash the bare metal area and adjacent surfaces with a cloth and water and wipe dry. Simple application and immediate wipe-off does not do the job.

Shop Towels

In the general operation of a paint shop, three types of towels are most popular (check with a local paint jobber for the best type available):

1. Shop towels are about 12-by-12-inches square and can be rented from a local towel supplier. Shop towels are available in a given color, which helps to locate and control them. Clean towels should always be kept in a definite location. A special closed container should be provided for soiled shop towels.
2. Disposable white cheesecloth and miscellaneous white rags are available by the pound from special towel suppliers. These rags are ideal for any general purpose, after which they can be thrown away in sealed containers.
3. Disposable paper towels of industrial weight are very popular in paint shops for equipment and general cleanup operations. These towels are available in many forms: the most popular are 12-by-12-inch folded sheets or 12-inch rolls. These towels are available through paint jobbers.

Tack Rags

Tack rags, also called tack cloths, are made of cheesecloth treated with a nondrying, nonsmearing, sticky varnish. They are about 12 by 12 inches in size and are available through paint jobbers. When folded, tack rags are about $4\frac{1}{2}$ by $8\frac{1}{2}$ inches in size and are convenient for hand use. Tack rags safely remove loose dirt, dust, sanding particles, and overspray from metal, plastic, or painted surfaces prior to painting. Use of an air gun in conjunction with a tack rag is an efficient and fast way of tack-wiping and cleaning a surface prior to painting. Avoid too much pressure on a fresh tack rag because it may leave an imprint of varnish on the surface.

Squeegees

Squeegees (see Figure 6–1) are made of flexible rubber and are available through paint jobbers in several sizes. The most popular size is 2 by 3 inches and is used primarily to remove slush and to check the progress of wet sanding. The $2\frac{3}{4}$-by-$4\frac{1}{4}$-inch size is

SURFACE PREPARATION

used most often to apply plastic fillers and glazing putty. However, these two sizes of squeegee are interchangeable. Also, both squeegees can be used as small sanding blocks by wrapping cut sections of wet or dry sandpaper around them.

Sanding Blocks

The purpose of a sanding block (see Figure 6–2) is to spread the cutting action of sandpaper uniformly over the surface of the sanding block. This uniformity automatically eliminates highlights in work surfaces caused by finger pressure. Sanding blocks are available through paint jobbers. A sanding block is made of plastic, which holds sandpaper securely during wet or dry sanding operations.

Sponge Pads

Small sponge rubber pads, about the size of rubber squeegees (see Figures 6–3a and b) and conventional sanding blocks are available through jobbers. A sponge sanding pad helps spread the sanding effort uniformly on concave and convex surfaces while sanding. Avoid cut-throughs at creaselines and edges by avoiding them and/or by sanding with due care.

Sandpaper

> **WARNING**
>
> **Always think safety.** Safety, like a work skill, improves with practice until it becomes a strong automatic habit. To make safety become an unconscious habit takes a certain amount of conscious, deliberate practice.

FIGURE 6–1. Small and large squeegees. (Courtesy of 3M Automotive Aftermarket Division.)

FIGURE 6–2. Firm hand sanding block. (Courtesy of 3M Automotive Aftermarket Division.)

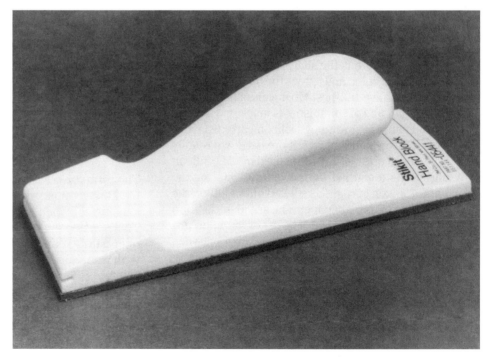

FIGURE 6–3. (a) Sponge sanding pad. (Courtesy of 3M Automotive Aftermarket Division.) (b) Sanding concave surfaces with sandpaper backed with a sponge pad. (Courtesy of 3M Automotive Aftermarket Division.)

(a)

(b)

Before using power sanding equipment (and in some cases when dry sanding by hand):

1. **Wear an approved dust respirator.**
2. **Wear an approved face shield or properly fitted safety goggles.**

The purpose of sandpaper is to do required sanding operations on a surface before paint repairs can be made, such as:

1. To remove rust.
2. To remove paint.
3. To featheredge.
4. To sand undercoats.
5. To make surfaces smooth.
6. To make a smooth surface coarse or rough.

The two most popular abrasives in the refinish trade are silicon carbide and aluminum oxide. The abrasive does the actual cutting or sanding operation. Table 6–3 is a summary of abrasives used in the automotive refinishing and metal repair trades.

1. **Silicon carbide** is the sharpest abrasive known, it is black in color, and it does the best job of sanding automotive finishes. Most sandpaper used to do fine sanding of car finishes is made with silicon carbide abrasive. Although silicon carbide lasts a long time on paint finishes, it breaks down readily when used on steel and other hard metals. Silicon carbide is identified by the letters "SiC" or the name "silicon carbide" is spelled out in full on the back side of the sandpaper. The grit identification number is also shown. The most popular grit usage of silicon carbide is in the range between No. 220 and No. 2500 grit (see Table 6–3).
2. **Aluminum oxide** is the best abrasive for metal repair because it is the hardest and most durable abrasive known. All discs for portable power sanders are made with aluminum oxide abrasive. Normally, aluminum oxide is reddish brown in color and that is how discs appear in metal repair. However, to assist in paint removal, open-coat discs are treated with a white chemical and appear white in color. Aluminum oxide abrasives are identified with the letters "AlO" or the name "aluminum oxide" is written out in full on the reverse side. For the grit sizes available, see Table 6–3.

A recent development in abrasives made by 3M, Norton, and other companies is combining ceramic with aluminum oxide to form a ceramic–aluminum oxide abrasive. The new ceramic–aluminum oxide abrasive is sharper, tougher, and stronger. These characteristics result in an abrasive that does more cutting, cuts faster, and lasts much longer. The REGALITE line of abrasives by 3M are examples of ceramic–aluminum oxides. Ceramic–aluminum oxide abrasives are available in grit sizes that range from 24 to 100.

CLOSED COAT

Closed coat abrasives are designed so that all possible space on the sanding surface is covered with the maximum amount of abrasive. Abrasives are applied and bonded to sandpaper backing by an electrostatic process that enables the sharpest cutting edges of the abrasives to be exposed.

OPEN COAT

The term *open coat* means that the grit particles on a backing are spaced a short distance from each other, causing an open space between the grit particles. This spacing is done with a screen. Additionally, the open surface and the abrasive particles are covered with a chemical treatment called zinc stearate, which is white. The chemical prevents anything from sticking to it. This combination of construction makes possible the sanding of paint finishes for long periods of time without paint accumulating on the disc or sandpaper.

SURFACE PREPARATION

TABLE 6–3 Commonly Used Abrasives in U.S. Automotive Refinishing and Metal Repair Trades

General Category	Sheet Sandpaper: Silicon Carbide (Closed Coat)	Sheet Sandpaper: Aluminum Oxide (Closed Coat)	5-to-8-inch Sandpaper Discs: Aluminum Oxide (Open Coat)	Portable Sander Discs: Aluminum Oxide Open Coat	Closed Coat
PAINT REPAIR ABRASIVES					
Micro Fine[a]	2500				
Micro Fine[b]	2000				
Micro Fine[c]	1500				
Micro Fine[d]	1200				
Ultra Fine[e]	800				
Fine	600	600			
	500	500			
Medium	400	400	400		
	360	360			
Coarse	320	320	320		
	280	280			
	240	240	240		
	220	220	220		
METAL AND PLASTIC REPAIR ABRASIVES					
Fine	180	180	180		
	150	150	150		
	120	120	120		120
Medium	100	100	100		100
	80	80	80D		80
Coarse	60	60	60D	Fine	60
	50	50			50
	40	40	40D	Medium	
	36	36	36D	36	36
	30	30		Coarse	
Extra coarse	24	24		24	24
	16	16		16	16

[a] and [b]Micro Fine 2000 and 2500 is used for the removal of minor imperfections like dirt, in BC/CC finishes.
[c]Micro Fine 1500 is used for the removal of minor imperfections like dirt in BC/CC finishes. 1500, 2000, and 2500 sanding followed by light compounding produces the best scratch-free "show car" finish.
[d]Micro Fine 1200 is used to level orange peel and to remove minor imperfections in BC/CC finishes.
[e]Ultra Fine is used for leveling orange peel in standard acrylic lacquer topcoats. Also reduces compounding time.

WATERPROOF SANDPAPER

Waterproof sandpaper, available in 9-by-1 1-inch sheets, is the most popular abrasive in the refinish trade. This sandpaper can be used wet with water or mineral spirits and dry for sanding old finishes, fine featheredging, and surface sanding. Waterproof sandpaper cuts fast and stays flexible. The terms *wet* or *dry* means that the sandpaper can be used for wet or dry sanding, respectively.

DRY-TYPE SANDPAPER

This terminology is given to the sandpaper because it is always used dry. To promote durability and to prevent loading, dry-type sandpaper is made with three features:

1. Aluminum oxide abrasive, which remains durable
2. Use of open-coat design
3. Nonloading chemical treatment (zinc stearate)

THE ART OF SANDING

Sanding operations range from sanding the color coat only to removal of the complete paint finish. How fast and efficient sanding operations are depends on the painter's ability to select the proper abrasive in the proper form to fit the needs of the repair job at hand. This ability is best learned through experience. Sanding is divided into two general categories: **hand** and **power sanding.** For greatest efficiency in hand and power sanding, sanding is done in two or more stages. Coarse abrasives speed up sanding in the early stages and fine abrasives finish the work.

Sanding and the Sander's Health

Wherever there is considerable dry sanding taking place, the air in the area is filled with particles of pollutants. The pollutants are inhaled by people in the area unless they are protected with a dust respirator. Wearing a respirator is a commonsense way of protecting one's lungs. Most types of pollutants (steel, certain dusts, plastics, etc.) are inhaled into the lungs and stay there. In time, they contribute to emphysema or something similar.

HAND SANDING

1. Select the proper grade of sandpaper for the work to be done (see Table 6–3).
2. Cut sandpaper to the suitable size according to one of three popular methods:
   a. Reducing sheet sandpaper to thirds (for small-area work; see Figure 6–4).
   b. Reducing sheet sandpaper to quarters (for large area work; see Figure 6–5).
   c. Reducing sheet sandpaper for standard sanding block (see Figure 6–6).

How to Hold and Use Sandpaper

Grip is important to sandpaper control. Three basic grips are most popular among painters:

1. The most natural grip is holding the sandpaper between the thumb and the hand as the hand is laid flat on the surface (Figure 6–7).
2. An optional grip is holding the sandpaper between the little finger and the third finger as the hand is laid flat on the surface.
3. Many painters like to combine both grips, holding the sandpaper by the thumb and little finger.

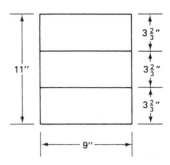

FIGURE 6–4. Reducing sheet sandpaper to thirds.

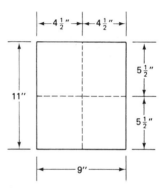

FIGURE 6–5. Reducing sheet sandpaper to quarters.

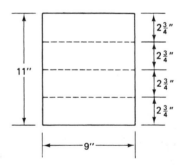

FIGURE 6–6. Reducing sheet sandpaper for standard sanding block.

POSITION OF SANDING HAND ON SURFACE

1. The position that is the most comfortable is the best.
2. A straight fore-and-aft finger position is best for large panels (Figure 6–7).
3. The hand must adapt to the special conditions of surface construction.

SURFACE PREPARATION

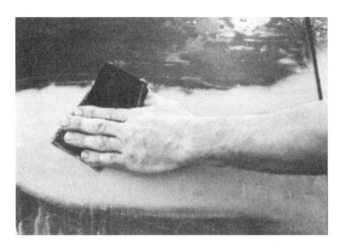

FIGURE 6–7. How to hold and use sandpaper. (Courtesy of PPG Industries.)

PRESSURE ON SANDING HAND (VARIES WITH NEEDS OF SANDING JOB)

1. Keep pressure on sandpaper to a minimum.
2. Do not press hard.
3. Use about the same effort as used in washing a car.
4. Let the weight of the entire hand and sandpaper do the sanding.
5. Distribute weight uniformly over the entire hand.
6. Use just enough "feel" and hand pressure to do the job. At times, a little extra pressure is needed. This factor is variable.

FINGER SANDING TECHNIQUE

1. Raising the palm of the hand slightly shifts the weight of the hand to the fingers. At times, this is where most sanding or cutting action needs to be done.
2. Raising the hand still more and stiffening the fingers shifts the weight to the fingertips. At times, sanding with the fingertips is required.

All three techniques of sanding are required from time to time. The choice depends on what the surface requires. This decision is known as "sanding control."

CIRCLE SANDING TECHNIQUE

Circle sanding with the fingers means **moving the sandpaper in a circle over a small area.** This technique results in a faster sanding action with good-quality results if certain cautions are observed. What the sander actually does is simulate an orbital sander.

> **CAUTION**
>
> **Circle sanding by hand should never be done on a large panel with large circular motions. Circle sanding should never be done with a sanding block.**

CROSS-CUTTING TECHNIQUE

The best sanding strokes for large-area sanding are in straight lines. This form of sanding blends best with adjacent surfaces. The best leveling action is achieved when a sander changes directions frequently while sanding. A change of direction causes a cross-cutting action that levels any surface much more quickly. If cross-sanding at a 90° angle cannot be done because panel construction does not allow it, even a 30° or 45° direction change is helpful.

Selection of a Backup Agent

Before sanding, the sander has the option of selecting the most efficient method of sanding. In addition to selecting the proper sandpaper, it is important to use a suitable backup block or pad as required. Any of the following can be used to back up the sandpaper and can be used for any form of sanding.

1. Use of conventional sanding block: best for large flat areas.
2. Use of squeegee as small sanding block: works well with threefold wrap of sandpaper.
3. Use of suitable sponge pad: excellent for sanding concave areas or surfaces; many types are available.
4. Painter's homemade sanding block. Examples include thick felt pad like a school blackboard eraser, small rectangular wooden block, small section of wooden paint paddle.

Featheredging Technique

Featheredging, as the word implies, means tapering the thickness of a paint film at the broken edges from full paint film thickness gradually to a fine feather point (Figure 6–8a). When a break occurs in a paint film, tapering the edge is necessary to make flawless paint repairs. As shown in Figure 6–8a, a properly made

FIGURE 6–8. Featheredge preparation and priming. (a) Featheredge cross-section. (Courtesy of General Motors Corporation.) (b) Featheredge technique. (Courtesy of General Motors Corporation.) (c) Featheredge priming. (Courtesy of PPG Industries.)

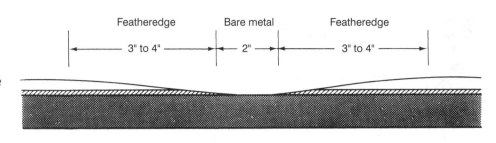

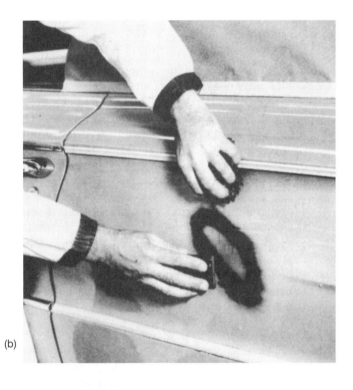

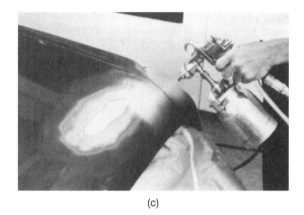

featheredge extends several inches from the bare metal area to the top of the paint film. There are many ways to featheredge paint. Each painter develops special sanding techniques based on the type of paint and experience. The following is a suggested procedure:

1. Select the proper sandpaper (see Table 6–3). A finer sandpaper grit is used when sanding by hand. However, whether sanding by hand or with power, the two-sandpaper system is fastest and provides the highest quality results (see Table 6–4).
2. If the repair area is small, 6 to 8 inches in diameter, use a squeegee as a sanding block. Otherwise, use a large sanding block (Figure 6–8b).
3. Apply water and sand from the outside in or from the inside out. On a small area, circle sanding speeds the operation. On a large area, use straight-line sanding motions. Soak the sandpaper frequently. Soak the sponge and half wring out occasionally.
4. After a featheredge is formed at the center (bare metal area), concentrate the sanding effort at the perimeter to extend the edge of the sanding area 1 to 2 inches from bare metal in all directions. Use plenty of water and continue the cross-cutting action with the sandpaper by changing directions frequently.
5. Change to fine grit sandpaper and continue sanding to remove all initial sand scratches, and extend the featheredge 2 to 3 inches from bare metal in all directions (Figure 6–8b).

TABLE 6–4 Two-Step Systems for Featheredging

Type of Sanding	Acrylic Lacquers		Acrylic Enamel and All Enamels	
	Start	Finish	Start	Finish
Hand	320	400	220	360
Power	220	320	180	280

6. Change the water in the sanding area frequently by rinsing the sponge and sandpaper as sanding continues to completion.

> **NOTE**
>
> Sanding is finished when the featheredge extends several inches from bare metal in all directions (Figure 6–8c). A featheredge is ideal when checking the area with extended fingers and hand reveals that there is no marked or noticeable depression in the sanded area.

> **NOTE**
>
> For better water control in an area with no floor drain, spread several thicknesses of used newspaper on the floor below the wet sanding area. At the conclusion of sanding and car cleanup, simply roll up the wet newspapers and place them in a trash container.

TABLE 6–5 STIKIT Products

STIKIT SHEET-ROLL SANDPAPER

Part Number	Grade	Size
02556	320A	$2\frac{3}{4}'' \times 50$ yds
02558	240A	$2\frac{3}{4}'' \times 50$ yds
02559	220A	$2\frac{3}{4}'' \times 50$ yds
02560	180A	$2\frac{3}{4}'' \times 50$ yds
02561	150A	$2\frac{3}{4}'' \times 50$ yds
02562	120A	$2\frac{3}{4}'' \times 35$ yds
02563	100A	$2\frac{3}{4}'' \times 35$ yds
02564	80A	$2\frac{3}{4}'' \times 30$ yds

STIKIT HAND-SANDING TOOLS AND DISPENSER

Part Number	Description
05452	Double roll dispenser
05330	Soft hand pad
05442	Soft hand block
05440	Firm hand block, 5″
05441	Firm hand block, $7\frac{3}{4}''$
05444	Hand file board

The STIKIT Sheet-Roll Sanding System

The STIKIT sheet-roll sanding system is **among the fastest methods available for hand sanding.** All of the components in the system (see Table 6–5) work together.

1. The continuous sheet roll is shown in Figure 6–9. Take only the length needed for the block or pad used.
2. The dispenser in Figure 6–10 is lightweight and portable. Take it to the job. Abrasives are always handy and save time.
3. Adhesive backed abrasive stays in place without mechanical clips. It prevents slipping and breakage and provides greater control.
4. Hand tools ranging from soft and comfortable to hard and flat are designed for painters and body repairpersons (see Figure 6–11).

To order abrasives or tools, see your local paint jobber.

STIKIT Hand-Sanding Procedure for Plastic Filler

1. Wear a NIOSH-approved dust respirator (TC-21C).
2. Select the proper abrasive system. A two-step system is recommended. Start with 80 grit and finish with 180 or 220 (see Table 6–3).
3. Cut a suitable size abrasive from the roll dispenser and apply it to the sanding block (Figure 6–2a).
4. Do initial sanding with coarser sandpaper until the roughout stage is complete (Figure 6–11).

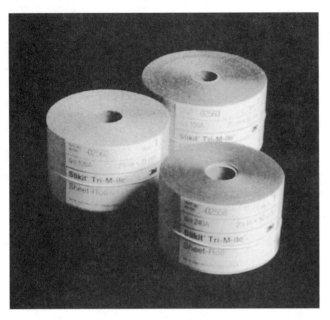

FIGURE 6–9. Sheet-roll sandpaper. (Courtesy of 3M Automotive Aftermarket Division.)

FIGURE 6–11. Hand file board. (Courtesy of 3M Automotive Aftermarket Division.)

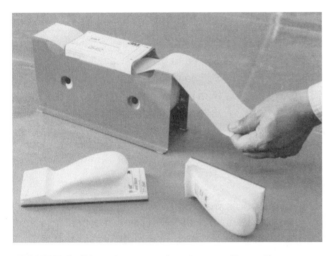

FIGURE 6–10. Dispenser for sheet-roll sandpaper. (Courtesy of 3M Automotive Aftermarket Division.)

CAUTION

When sanding plastic filler adjacent to good paint finish, protect the good paint finish with masking tape until coarse sanding is complete.

---

5. Finish sanding the plastic filler with 180 or 220 until the surface is acceptable for undercoat application.
6. Clean the sanding residue from car and/or floor area as required.
7. If too much plastic is removed or if air pockets are opened, apply a fine glazing coat of plastic and finish sanding as in step 5 above.

POWER SANDERS AND POWER FILES

The big advantage of power sanders and files is that they can do sanding work much faster and much more easily than it can be done by hand. Power tools for surface preparation are available in several forms through paint jobbers. These power tools are made by several paint tool and equipment suppliers. Each tool is designed to do a specific job. Abrasives (sandpaper) for power sanders and power files are available through paint jobbers (see Table 6–5). These abrasives are made available to the refinish trade through several miscellaneous paint material suppliers like the 3M Company. The most common power tools for surface preparation are:

1. Orbital and straight-line air files.
2. Dual action (DA) type sanders.
3. Jitterbug type sanders.
4. Eight-inch disc with a part no. 5600 (3M Company) type of pad on a slow-speed polisher (1500 to 3000 rpm).
5. Variable-speed drill motors with 3-, 4-, or 5-inch sanding pad attachment.

The dual action type sander appears to be the most efficient sander in the trade. Contact your local paint jobber for all details regarding abrasives and tool availability.

SURFACE PREPARATION

The STIKIT Disc Roll Sanding System

Adhesive backed discs are available from 3M Company in 5- and 6-inch diameters and come in roll form (see Table 6–6). They are designed for speed and quality work as follows:

1. Figure 6–12 shows removal of a disc from a disc dispenser. The discs come in a continuous roll as shown.

TABLE 6–6 STIKIT Disc Roll Sandpaper

Part Number	Size	Grit
01420	5″ × NH	320-A
01421	5″ × NH	280-A
01422	5″ × NH	240-A
01423	5″ × NH	220-A
01424	5″ × NH	180-A
01425	5″ × NH	150-A
01426	5″ × NH	120-A
01427	5″ × NH	100-A
01428	5″ × NH	80-A
01435	6″ × NH	320-A
01436	6″ × NH	280-A
01437	6″ × NH	240-A
01438	6″ × NH	220-A
01439	6″ × NH	180-A
01440	6″ × NH	150-A
01441	6″ × NH	120-A
01442	6″ × NH	100-A
01443	6″ × NH	80-A
01501	5″ × NH	80D
01506	6″ × NH	80D

2. Because the discs are adhesive backed, they are ready for attachment, as shown in Figure 6–13.
3. Figure 6–14 shows the STIKIT disc in operation.

STIKIT abrasives are shown in Figure 6–15. They are available through your local paint jobber.

Green Corps Dust-Free (DF) Products

Unique **Green Corps dust-free discs and air files are designed to help eliminate problem-causing dust** and are especially designed to resist loading. Dust-free sanding discs, shown in Figure 6–16, are designed for use with a shop vacuum system. The holes in the disc line up with the holes of the vacuum

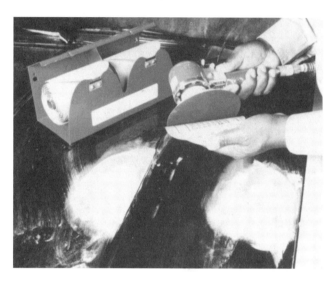

FIGURE 6–13. Disc attachment. (Courtesy of 3M Automotive Aftermarket Division.)

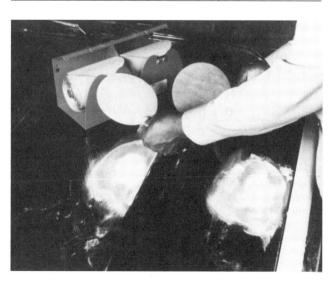

FIGURE 6–12. Disc-roll dispenser (disc selection). (Courtesy of 3M Automotive Aftermarket Division.)

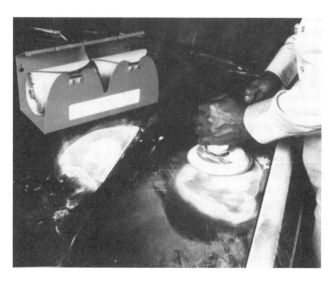

FIGURE 6–14. Disc in operation. (Courtesy of 3M Automotive Aftermarket Division.)

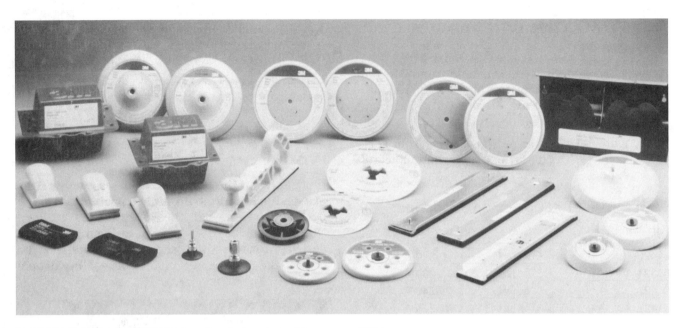

FIGURE 6–15. The STIKIT™ family of abrasives. (Courtesy of 3M Automotive Aftermarket Division.)

FIGURE 6–16. Dust-free STIKIT™ products. (Courtesy of 3M Automotive Aftermarket Division.)

system. In operation, the dust formed by the disc sanding is automatically drawn through the vacuum system.

HOOKIT is 3M's name for "velcro." The HOOKIT line of abrasives features the hook and pile method of abrasive attachment. One side has the male hooks, and the mating side has the pile construction. When the two mating surfaces are pushed together, they form a strong bond. One great advantage of this system is that the mating surfaces stay neat and clean throughout the life of the abrasive.

HOOKIT disc and sheet abrasives (see Figure 6–17):

1. Are available in 6- and 8-inch discs and $2\frac{3}{4}$- × $16\frac{1}{2}$-inch sheets.
2. Have an improved rate of cut.
3. Attach and remove HOOKIT discs easily and quickly.
4. Reattach and reuse the abrasives for their full life.
5. Help eliminate breakage by using clip-on air file sheets.

Disc Sander Operating Technique

1. **Select the proper abrasive.** Always use the two-abrasive system, starting with the coarser grit and finishing with the finer grit.
2. **Be sure to wear safety goggles and a dust respirator.**
3. With a cotton work glove on one hand, check the surface to be sanded by passing the hand lightly over the surface. This step helps the sander to determine how sanding will be done. If sanding plastic filler, see Table 6–4 for grit selection.
4. Holding the sander firmly, turn on the switch and position the sander at a 5° to 10° angle to the work surface.
5. Position the left upper quarter of the disc to the surface when moving to the right in a straight line across the work (see Figure 6–18).

SURFACE PREPARATION

FIGURE 6–17. The HOOKIT™ family of abrasives. (Courtesy of 3M Automotive Aftermarket Division.)

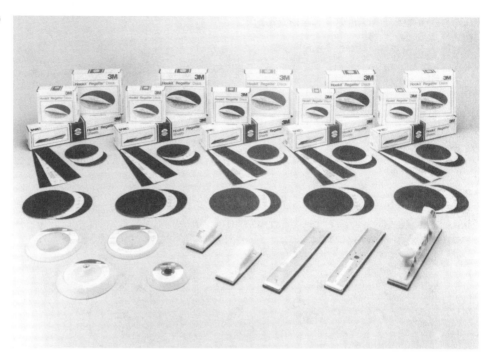

FIGURE 6–18. Disc sander marks: left upper quarter.

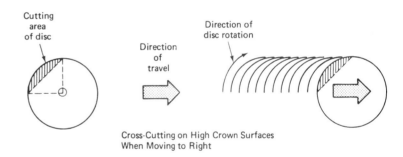

6. Position the right upper quarter of the disc to the surface and move to the left in a straight line across the work (see Figure 6–19).

   Steps 5 and 6 are the proper way to sand high-crown surfaces safely. They also apply to sanding low-crown and flat surfaces. Steps 5 and 6 cause a cross-cutting action (Figure 6–20) on high-crown surfaces without cutting across the high crown. This action produces a maximum leveling of surface. When using power files to level high-crown surfaces, move and sand in the direction of the panel. **Do not cut across the high crown of the panel.** Check and clean the abrasive periodically to ensure that it is cutting properly. If the abrasive becomes clogged with plastic filler prematurely, clean the abrasive with a pointed tool, steel wire brush, or air gun. The plastic filler must be cured sufficiently before it can be sanded satisfactorily.

> **CAUTION**
>
> Do not allow coarse abrasive to cut 90° across a high-crown surface. This action causes very deep sand scratches that are difficult to remove. Also, do not allow coarse abrasive to touch good paint on adjacent surfaces. Protect adjacent paint as required with masking.

7. Position the top center area of the disc (Figure 6–21) to the surface by rotating the bottom away from the surface slightly. This is a buffing action. While maintaining this position, move the sander disc up and down, across the entire work surface. Overlap each pass by 50 percent to 60 percent. This overlap creates an additional and final leveling action. When sanding low-crown or flat surfaces, position the sander as shown in Figure 6–22.

FIGURE 6–19. Disc sander marks: right upper quarter.

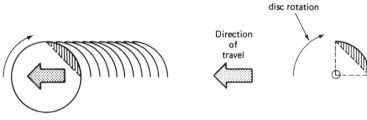

FIGURE 6–20. Disc sander marks: cross-cutting action for high-crown surfaces.

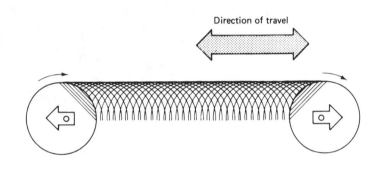

FIGURE 6–21. Disc sander marks: buffing action for high- and low-crown surfaces.

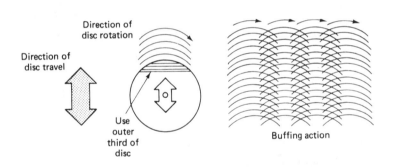

FIGURE 6–22. Disc sander marks: cross-cutting action for low-crown surfaces.

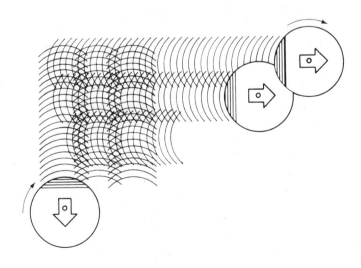

SURFACE PREPARATION

8. Check for high spots by passing a gloved hand over the repair surface. Repeat sanding operations until about three-fourths of the sanding is completed.
9. **On all sanders, change the sandpaper to a finer grit for finishing the plastic filler** (see Table 6–6).
10. Repeat the basic sanding operations over the entire surface, starting with cross cutting and finishing with buffing until the surface is acceptably smooth.
11. Clean the car and floor area as required.

Disc-Sanding Paint from Concave Surfaces

The best way to remove paint from concave surfaces with a disc sander is to use a two-part system consisting of:

1. A **large (9-inch) disc cut to a star shape.**
2. A **small (4-inch) backing pad.**

Use the following procedure to disc-sand paint from concave surfaces:

1. Cut the large disc to a star-shaped pattern, as shown in Figure 6–23.
   a. Mark five equal locations on the back of the disc.
   b. Mark a straight line between locations as shown.
   c. With a suitable tool, cut the disc along the marked lines.
2. Mount the disc and backing pad on the sander as required.
3. Remove paint from the concave surface using the technique shown in Figure 6–25. The disc will flex when the sander is pressed to the surface.

Cautions for Power Sanding

Power sanding is the fastest way of removing paint from a steel surface. However, the method is not too popular for paint removal because of the many cautions that must be exercised and because of the potential problems that power sanding can cause. The Stikit 8-inch disc and disc pad by 3M and other equivalent pads work very well on low-rpm polishers. They are safer to use and present fewer problems on the modern automobile when only paint removal is required.

FIGURE 6–23. Star-shaped disc.

Cautions and Safety Rules

1. Wear safety goggles and a face shield.
2. Wear an approved respirator for dust conditions.
3. Power sanding can be hazardous if not done safely. It should be done only by a person who is qualified to do it properly. When misused or used incorrectly, power sanding can be dangerous.
4. **Never power-sand aluminum or plastic panels with coarse abrasives.** Aluminum and plastic use is increasing. Check the entire panel with a magnet. If the panel does not react to the magnet, **do not power-sand the panel.**
5. **Never power-sand across a high crown of steel or any panel with coarse abrasives.**
6. **Never power-sand across any panel edges or sharp crease lines with coarse abrasives.**
7. **Never power-sand at or over exterior body moldings, into corners, into crevices, or over drip moldings with coarse abrasives.**
8. **Never power-sand deep concave areas with a full pad behind the disc.**
9. **Never power-sand over solder- or plastic-filled body joints or over plastic-filled repair areas with coarse abrasives.**

PAINT REMOVERS

Several different paint removers are available to paint shops through paint jobbers. These chemically prepared removers, known as "**cold strippers,**" are specially designed to remove paint from all automotive substrates and feature various methods of removing the paint. For purposes of this book, paint removers are divided into two very general categories:

1. Table 6–7 lists paint removers for removal of acrylic lacquer finishes only.
2. Table 6–8 lists paint removers that remove all finishes from all automotive substrates. This table includes water rinsable paint removers, which are identified with an asterisk (*).

TABLE 6–7 Paint Remover for Removal of Acrylic Lacquer Only

Product Supplier	Product Name	Product Identity Number
PPG	Acrylic Color Remover	DX-525
DuPont	Paint Remover	5662-S
Klean-Strip	Feather-Edger	FE-387
Klean-Strip	Topcoat Stripper	TC-338

TABLE 6–8 General List of Paint Removers

Product Supplier	Product Name	Product Identity Number
PPG*	Aircraft Paint Remover (for automotive use)	DX-586
Klean-Strip*	Aircraft Paint Remover (qt/gal; for automotive use)	AR-343
	Aircraft Paint Remover (aerosol)	AR-322
Klean-Strip*	Formula A Paint Remover	FA-334
Klean-Strip*	Auto Strip (qt, gal)	OM-235
	Auto Strip (18-oz aerosol)	A-710
Klean-Strip*	Aircraft Paint Remover (for aircraft only)	AR-727
Klean-Strip*	Liquid Sprayable Paint Remover	KS-221
Klean-Strip*	Wet or Dry Auto Stripper	DS-318
Klean-Strip	Dry Peeler blow-off Paint Stripper	FS-459
Klean-Strip	Fiberglass Paint Stripper	AF-354
Klean-Strip	Urethane Parts Paint Stripper	US-366

*Water rinsable types that can be removed by flushing per label directions.

When ordering paint removers from a paint jobber, always specify the type of paint removal you want to do. The paint jobber will assist you in selecting the proper remover.

Paint Removal Methods

Paint removers are divided into the following four categories because the methods of paint removal are quite different:

1. **Conventional method of paint removal:**
   a. Apply by brush or flow it onto the surface per label directions.
   b. Remove paint per label: by scraping or by flushing off with water.

   Product examples in this category are the Klean-Strip Aircraft Remover #AR-343 and the PPG Aircraft Remover #586.

2. **Liquid sprayable method of paint removal:**
   a. Apply Klean-Strip KS-221 (or equivalent) with a liquid-sprayable bottle per label directions. (A trigger spray bottle is included with quart and gallon sizes for easy application.)
   b. Remove paint by scraping and/or flushing off with water per label directions.

   A product example in this category is Klean-Strip Liquid Sprayable #KS-221.

3. **Wet or dry method of paint removal:** Using Klean-Strip DS 318 (or equivalent), apply the material per label directions.
   a. If using the wet method, scrape and remove paint as in the conventional method by scraping, etc., per label directions.
   b. If using the dry method, lift the edges of the loose paint and peel it off per label directions. Paint may be peeled after drying on the same day or the next day.

4. **Dry blow-off method of paint removal:**

   > **NOTE**
   > This method is excellent for factory cars that have experienced peeling problems.

   a. Using Klean-Strip peeler FS-459 (or equivalent), apply the material per label directions.
   b. Simply blow the paint material off the surface with an air hose. Use 30 psi or less air pressure.

SURFACE PREPARATION

General Description

Paint removers are known as cold strippers because they are designed for use in average temperature conditions. Paint removers consist of special strong chemicals that attack and dissolve paint finishes according to design. These chemicals work in conjunction with a special wax as follows:

1. Label directions advise applying paint remover at a very heavy thickness, as heavy as can be applied on a vertical surface.
2. At this thickness, wax floats to the surface to act as a seal or screen. This seal keeps the active chemical ingredients in contact with the paint, and active chemicals enter the paint film.
3. The chemicals attack, soften, and dissolve the paint film. This action takes place within several minutes, as explained in the label directions.
4. When dissolved, the paint film can be scraped off the surface with a squeegee or a suitable scraper.

Removal of stripped paint from a car involves handling a very gummy, slippery, and heavy-bodied material. The material flows like extra-heavy molasses. If controlled as outlined, stripping paint can be a very neat and rewarding experience. However, lack of preparation and poor control of dissolved paint material usually results in a very messy operation. Negligence in allowing dissolved paint and excess paint remover to run into gap spacings, the engine compartment, and the rear compartment gutter causes much extra, uncalled-for work. Also, stripped paint does not clean up too readily from any type of floor. Good planning and proper preparation control the stripped paint and keep it off the floor. Control of stripped paint is the major problem. Good planning is essential to neat, fast work. Haste usually causes waste. It is best to proceed in a planned and well-organized manner. Have all tools and equipment on hand well in advance of their need. Plan stripping paint from each major area individually.

Safety Precautions

1. Use the proper respirator (TC-23C or TC-19C).
2. Use paint removers only with adequate ventilation.
3. Avoid prolonged or repeated breathing of vapor and contact with skin or eyes.
4. Contact with skin may cause irritation. If skin is contacted accidentally by the remover, wash it off immediately with clear water.
5. Contact with flame or hot surfaces may produce toxic vapors.
6. Paint removers contain methylene chloride, propylene dichloride, alcohols, and ammonia.
7. If any paint remover is swallowed, call a physician immediately.
8. Keep the container closed when not in use.
9. Keep out of the reach of children.
10. If the paint remover is splashed in the eyes, immediately flush with clear water for at least 15 minutes. Get medical attention.

Not all paint removers are alike. They appear alike in name only. To know specifically what a paint remover will do, the painter must know by experience, or the painter should ask the paint jobber. Normally, a product will do the job promised by a paint jobber. All paint remover companies endorse their products as stated on the label or in company sales literature.

As a rule, paint removers can stain the finish on anodized aluminum parts and on many plastic parts. Necessary precautions must be taken to prevent contact with these parts. Care must be exercised to prevent paint remover from getting trapped at body side moldings, nameplates, and in door and deck lid gap spacings. This precaution is accomplished through proper masking. Even a diluted paint remover can run down over a fresh paint finish and stain the finish.

Necessary precautions should be taken to keep removed paint off the floor. Paint on the floor makes cleanup difficult. Catch removed paint on protective paper of sufficient size and strength to allow easy control and disposal. For best results when using paint remover, follow the label directions.

Techniques for Using Paint Remover

There are several ways to strip color (and undercoats) off any car. The following technique is generally popular among painters:

1. Mask the floor as required. Double masking is recommended as follows:
   a. Apply a first layer of masking to protect the floor.
   b. Apply a second layer of masking so that it can catch removed paint and is prepared for immediate removal without disturbing the first masking.
2. Have all necessary tools and equipment on hand.

3. Before masking exterior parts of the car, as may be required, apply DART tape (door aperture refinish tape) to all door and/or body openings as required (see Figure 6–25a, b, and c). This step is a big help to the painter when stripping paint at the door or panel openings.
4. Use rubber gloves, safety goggles, and the proper respirator.
5. Fill a smaller container, such as a 1-quart container, from 1 gallon of paint remover. Using a 4-inch (or suitable) paintbrush and a smaller container, apply a generous coating of paint remover on the finish to be stripped. Spread the remover quickly with minimum brushing.

CAUTION

Do not rebrush any area. Simply spread the remover in a very heavy coating as directed and leave it alone.

---

6. Allow the paint remover to work on the paint finish the required amount of time, according to the label directions. Usually, this ranges from 7 to 10 minutes. Do one area at a time.
7. Remove loosened paint with a squeegee or plastic scraper as follows:
   a. From horizontal surfaces:
      (1) The first method involves using a dust pan and squeegee. Collect paint from a 1- to 2-square-foot area at a time, squeegee onto a dust pan, and remove. Wash the dust pan in a thinner bath quickly each time. Thinner must be in a container and sealed with cover when not in use.
      (2) Removed paint is best disposed of by placing paint on several sheets of newspaper, wrapping the newspaper, and placing it in a trash barrel or container that is covered tightly and marked "hazardous materials."
   b. From vertical surfaces: When removing paint from vertical surfaces such as all sides of the car, it is best to take advantage of gravity and let the loosened paint fall straight down, as follows:
      (1) Position necessary masking on the floor below the area being stripped.
      (2) Starting at the top, apply paint remover to the sides of the car slowly by carefully pouring from a paper cup and by spreading uniformly with a large, 4-inch paintbrush. Work by brushing in an upward direction only. Keep brushing to a minimum.
      (3) It may be necessary to apply two coats, one immediately after another, to obtain a generous film.
      (4) Catch the paint remover runoff with a section of cardboard or equivalent, and reapply it to a higher area.
      (5) Remove paint from the car with a squeegee or scraper after waiting the necessary time and by allowing loosened paint to fall straight down.
8. In the case of excess paint film due to a car being repainted one or more times, it is usually necessary to reapply a second coating of paint remover.
9. After the color has been removed, wash the panel surface with a medium-grade thinner. Wash the surface thoroughly and remove the thinner while still wet. Do only a small area, 2-by-2 feet, at a time and overlap as required.
10. Remove the masking. Remove the balance of color hidden by masking as required. On enamel cars, it is best to remove the balance of color by sanding with No. 60- or 80-grit sandpaper.
11. Clean the entire surface from which color was removed with paint finish cleaning solvent to remove any traces of wax.
12. **Place all solvent- and paint-soaked rags in hazardous materials container that is tightly closed when not in use.**

SANDBLASTING AND MEDIABLASTING

Figure 6–24 shows a vacuum sandblaster. Several companies produce vacuum sandblasters and they are becoming increasingly popular because of the work they do while keeping the shop very clean. In addition to sand, other media are used for blasting. Check with your local paint jobber for more information.

Vacuum sandblasters sandblast to remove paint, rust, corrosion, and weld flux completely. It is an economical tool for cleaning any type of surface quickly and thoroughly. It saves money in two ways:

1. It recycles the blast media, so you save on material costs.
2. With this blaster, there is no cleanup.

During operation, the abrasives are separated from the dust and dirt and are recycled for another blast. Depending on the abrasive specifications, you

SURFACE PREPARATION

FIGURE 6–24. Vacuum sandblaster (model F-44). (Courtesy of ALC Sandy Jet Company.)

can recycle your media twenty to thirty times without losing effective blast power.

Model F-44, which is shown in Figure 6–24, operates on the same air supply (80 to 100 psi at 25 CFM) that operates HVLP spray guns. This operation requires a 5-horsepower motor. Follow the manufacturer's guidelines for sandblaster use. For more information on this sandblaster, check with your local paint jobber.

Safety Precautions

Sandblasting has exceptional cutting qualities and the precautions mentioned below must be taken if sandblasting is done near any of the following items:

1. All glass
2. All plastic parts
3. All chrome parts
4. All adjacent paint to be saved
5. Front grille parts
6. All aluminum parts
7. Vinyl top and all soft trim parts

Where practicable, remove parts for protection. Mask and/or protect items as required with suitably strong materials such as heavy cloth-backed tape, plastic, or metal.

Door Aperture Refinish Tape

The purpose of door aperture refinish tape (DART) is to speed cleanup at end of refinish jobs. Its use results in a neater paint job and increased customer satisfaction. The tape is a round, foam plastic construction, $\frac{3}{4}$ inch in diameter (3M part number 06298), with pressure-sensitive adhesive along a straight line for ready installation. The tape is provided on a roll in a suitable box. Unroll the tape one layer at a time and cut it to the length needed. Part Number 06298 consists of 27.3 yards (819 feet) of tape. For application, follow the box label directions or the following procedure:

1. Using a cloth dampened with soapy water and wrung out, wash the surface to be taped. Follow immediately by wiping the surface with a cloth dampened with 3M general-purpose adhesive cleaner (#08984) or the equivalent, and wipe dry with a clean cloth.

2. Pull out the desired length of DART and cut with scissors or a sharp blade. Tape may be butt-joined at any time.

3. Apply the tape to the door jamb, deck lid opening, hood opening, etc., with a $\frac{3}{16}$-inch setback from the outer panel. The ideal alignment has the tape contacting both panels without projecting beyond the flush surface (see Figure 6–25a). Too far in is incorrect (see Figure 6–25b). Too far out is also incorrect (see Figure 6–25c).

4. Run your finger in the reverse direction over the tape for tighter adhesion.

CAUTION

Do not stretch the tape at the corners. Feed sufficient tape into the corners and press in place. Allow a $\frac{3}{16}$-inch setback as shown in Figure 6–25a.

---

5. Close the door, hood, or deck lid and check the alignment of the tape installation before continuing. If tape realignment is necessary, simply lift the tape, reposition it as required, and press it in place.

6. Complete all masking operations on the car.

See Figures 6–26 and 6–27 for proper installation of DART at door opening and at gas tank door opening.

To remove the tape, simply grasp it and pull up and away from painted surfaces. Remove the DART tape when removing the car masking.

To order this tape, check with your local paint jobber.

(a) **CORRECT**

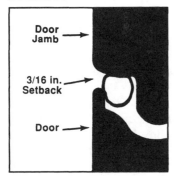

A proper set-back (3/16-inch) is extremely important in obtaining a correct seal and a finish with virtually no paint lines or fuzzy edges.

(b) **INCORRECT**

If 3M™ Door Aperture Refinish Tape is placed too far back, a proper seal will not be attained and will allow overspray to penetrate the area and dust/dirt to escape.

(c) **INCORRECT**

If 3M™ Door Aperture Refinish Tape is allowed to bulge out of the seam, an undesirable paint buildup may occur. This buildup may appear as a fuzzy edge.

FIGURE 6–25. (a) Correct installation and verbal description; (b) incorrect installation and verbal description; (c) incorrect installation and verbal description.

FIGURE 6–26. DART installation at door opening. (Courtesy of 3M Automotive Aftermarket Division.)

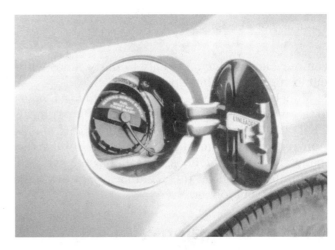

FIGURE 6–27. DART installation at gas tank door opening. (Courtesy of 3M Automotive Aftermarket Division.)

MASKING

The manner in which a job is masked often spells the difference between a top-quality and a poor-quality paint job. Careful masking provides clean paint edges and prevents a ragged appearance at glass, sealing strips, exterior moldings, grille parts, and so on. Careful and precision masking is the mark of a true craftsperson. How to mask a car is largely a matter of understanding fundamentals, using good judgment, and practice.

SURFACE PREPARATION

Before any portion of a car is masked, every part on the car and every surface involved should be washed and cleaned as required. Mask the car under good lighting conditions for a close inspection of the masking job. Masking, like painting, should not be done in temperatures below 50°F.

Masking Materials

Masking materials are designed for fast and efficient application, protection, and removal from surfaces that are not painted. Masking materials consist of the following types of products:

1. Tapes of many widths and varieties.
2. Masking papers of several varieties.
3. Masking covers of cloth and plastic sheets.
4. A new masking liquid applied by spraying.

For the availability of masking materials and the best types to use, check with your local paint jobber or 3M sales representative.

Masking Tape

Masking tape is available in several forms for automotive refinishing. The two most popular are general-purpose masking tape, which is available from several suppliers through paint jobbers, and Fine-Line masking tape supplied by the 3M Company, also available through paint jobbers.

GENERAL-PURPOSE MASKING TAPE

General-purpose masking tape is designed especially for the professional painter. Every feature is perfectly balanced to give maximum masking results. Masking tape is designed to resist breaking, lifting, adhesive transfer, splitting, shrinking, and curling away under normal- and high-temperature conditions. Good-quality masking tape conforms to the surface, hugs curves, is applied quickly and easily, sticks at a touch, and stays put. Masking tape has a good shelf life. The $\frac{1}{4}$- and $\frac{3}{4}$-inch masking tapes are the most popular in refinishing.

FINE-LINE MASKING TAPE

Fine-Line masking tape is a 3M product that is especially suited to custom painting as well as to general automotive refinishing. This tape has a special processed film backing that allows taping over freshly painted surfaces sooner with less chance of imprint damage. The backing for this tape is polypropylene plastic, which provides flexibility for curves and has excellent resistance to solvent penetration. Fine-Line tape provides the finest color separation line possible for several-color striping, for two-toning, and for custom painting. Fine-Line masking tape is excellent for lettering. This tape is not designed for outdoor application. Fine-Line masking tape is available in 60-yard rolls in the following inch sizes: $\frac{1}{16}, \frac{1}{8}, \frac{3}{16}, \frac{1}{4}, \frac{3}{8}, \frac{1}{2}$, and $\frac{3}{4}$. Consult your local paint jobber.

Masking Paper

Masking paper is strong, flexible, and pliable. It is formulated to resist penetration of solvents, lacquers, and enamels. It is especially designed to protect against overspray on every paint job. Masking paper can withstand water from wet sanding. It is available in 1000-foot rolls in several widths. The most popular are 15, 12, and 6 inches. The 15-inch width is popular for windows and windshields. Masking paper is also available in 18-, 24-, 30-, and 36-inch widths.

Draping Film Rolls—Prefolded Sheet Masking

The 3M Company provides several sizes of prefolded drape plastic film (see Table 6–9) rolls plus a hand masking dispenser, which speeds up the task of masking large surfaces more than any other system. Figure 6–28 shows the roll of film and masking tape being applied by the hand masker.

1. While pulling the hand masker with one hand, the other hand aligns and presses down on the tape. This action secures the installation of the masking.
2. The plastic film is then unfolded to cover the protected area of the car.

Draping film rolls—prefolded masking materials have several advantages:

1. The system allows you to mask larger areas in less time. The process requires only one person.
2. The plastic resists penetration of virtually all common paint systems.
3. The plastic clings to the surface.
4. The plastic is translucent and includes "WET PAINT" signs printed on its surface. This feature makes it easy to see the protected surface.
5. The plastic is strong and lightweight.
6. Paint materials stick to the plastic, which prevents paint from flaking during removal.
7. The plastic crunches up tightly for compact disposal.
8. The rolls are available in several popular widths: see Table 6–9.

TABLE 6–9 Draping Film Rolls—Prefolded Sheet Masking

Part Number	Size
SCOTCH™ HAND-MASKER™ DRAPING FILM ROLLS—PREFOLDED[a]	
06844	24 in. × 180 ft.
06848	48 in. × 180 ft.
06852	72 in. × 100 ft.
06853	99 in. × 100 ft.
3M HAND-MASKER™ FILM BLADE	
06802	12-in. cutoff blade
3M HAND-MASKER™ M-3X II DISPENSER	
06788	—
TACK DOWN TAPE DISPENSER	
06789	—

[a]For additional information on the 3M Scotch Hand-Masker draping film rolls—prefolded, check with your local automotive paint jobber.

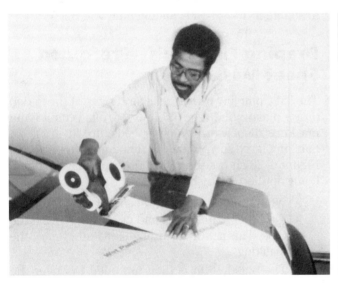

FIGURE 6–28a. Applying drape-film masking. Pulling on hand-masker unrolls drape-film and masked tape at the same time and secures it to the surface as it is pressed down onto previously applied tape.

FIGURE 6–28b. Spreading drape-film masking over car. After securing attached edge of prefolded drape-film plastic (see Figure 6–28a), unfolding the drape-film over the car is quick and easy.

IMPORTANT NOTICE

Automotive Applications:
When moisture or solvent vapors are trapped between the film and masked surface, imprinting may occur. The following measures will reduce the risk of surface imperfections:
- Not recommended for use over uncured coatings.
- Not recommended for use on wet surfaces.
- Removal of film is recommended if moisture/condensation under film occurs or is likely.
- Do not apply printed side of film toward surface to be masked.

Cutting the plastic film requires a Hand-Masker film blade, which is very sharp. Blades intended for cutting masking paper will not cut the plastic properly.

General-Purpose Masking Units

Masking units, sometimes called "**apron tapers**" or "**handy maskers,**" are available through paint jobbers. Masking units provide instant aprons of pretaped paper for masking. These units apply masking tape to the paper as the required amount of masking paper is unrolled. A cutting edge is provided on the masking unit to cut the paper at the desired length. The units are efficient because they cut down masking time.

SURFACE PREPARATION

Some units are designed for mounting on a bench or table, and the more popular units are mounted on casters or wheels and are portable. They can be brought to the car and are a time saver.

Special Masking Tools

Most painters find the following special tools handy when installing and removing masking tape:

1. A 1-inch paintbrush with bristles shortened to $\frac{1}{2}$ inch, which is done easily by cutting the bristles off with scissors. The tool helps the painter to install masking tape on surfaces that cannot be reached with the fingers, such as behind curved moldings and in tight corners.
2. A small scraping stick can be made from a wooden clothespin. Split a clothespin in half to make two sticks. Sand the end of each stick to the desired shape. This tool also helps to position masking tape in tight corners and helps to scrape away dirt and sealers that cannot be removed with normal washing without scratching the paint.
3. A small can opener with a screwdriver-shaped tip and slightly curved at the end can help considerably to remove masking. File the end of the tip to be knife-sharp. The tool is used to help loosen dried, hard-to-remove tape. Simply scrape and lift the edges of the tape to a position where they can be handled. This tool saves the painter's fingernails.

Masking Technique

HOW TO HOLD TAPE

Pick up a roll of masking tape with the fingers inside and the thumb outside the tape roll, and with the adhesive side of the tape away from the painter (when unrolled). This is the conventional way of holding tape.

HOW TO UNROLL TAPE

With the opposite hand, loosen the tape edge, grip the tape, and unroll a several-inch length by pulling the tape while allowing the tape roll to pivot on the fingers. Notice how the tape unrolls naturally while pulling on the free end of the tape.

HOW TO SNAP-TEAR TAPE (OFF JOB)

Snap-tear tape off the roll by pressing on the tape edge of the roll with the thumbnail while pulling the tape quickly in the opposite direction. On the car, the thumbnail works best as a cutting agent. This is the usual way of tearing tape at a precise location when applying masking tape on a car.

HOW TO PRECISION-TEAR TAPE

Prior to applying a pretaped apron of masking on a car, tear several short sections of tape quickly from the roll and stick them to the back side of the hand as follows:

1. Hold the tape roll in the last two fingers and thumb of one hand.
2. Unroll about $2\frac{1}{2}$ inches of tape with the opposite hand.
3. Position the ends of the first two fingers of each hand next to each other on the adhesive side of the tape and place the thumbs on the tape behind the fingers.
4. While squeezing the fingers against the thumb, twist the fingers quickly in opposite directions to snap-tear the tape about 1 inch long. The nickname given to a short piece of tape is "**tack tape.**" Short pieces of tack tape hold large sections of pretaped aprons to a car in spot locations opposite long sections of tape. So the word *tack* means to hold masking in a spot location.

When applying taped aprons to a car, the prepared painter has a series of short sections of tape ready so that when the apron is installed, the tape tacks are ready for use. Properly prepared tack taping is a time saver.

Tips on Masking

1. Apply masking to the part to be protected but avoid contacting the surface to be painted. Contacting or overlapping the surface to be painted is a poor practice that leads to edge lifting of finish when the tape is removed. Positioning and securing tape properly the first time is a matter of painter judgment and practice. To be sure that the tape is applied properly, **look at each detail** as the work progresses. **A safe distance to maintain between the tape edge and the paint surface is $\frac{1}{32}$ inch.**
2. Masking tape is designed with a certain amount of stretch. Make use of this characteristic by pulling on the tape edge as tape is applied at the corners and on curved surfaces.

3. **When overlapping tape over tape,** the edges at the painted surface should be **flush** for a straight, uniform line. When overlapping **paper over paper,** start installation at the bottom to create a shingling effect. This technique prevents the possible seepage of paint materials toward the protected parts.
4. Tape can be removed as soon as the paint is no longer sticky or tacky. This condition is determined by rubbing the back of a finger on a lower rear corner of a paint job.

> **CAUTION**
> Never allow masking tape to contact fresh painted surfaces when removing the tape on the same day a car is painted. Masking tape will take fresh paint right off the surface.

5. Remove the tape by pulling in the opposite direction and on the same side of any curved edges.
6. On freshly painted and on lacquer surfaces when contact between the paper and the surface cannot be avoided, use two or three thicknesses of masking paper. Also, remove this masking as soon as painting is completed.
7. Care should be exercised never to oversoak a single layer of masking paper if the surface underneath will be affected by solvent bleed-through.
8. Use the proper width of tape for each masking job. Using too wide a tape where a narrower tape will do is a waste of money.
9. Always unroll a length of tape that can be positioned and secured most easily. Observe the installation carefully. If any tape installation is faulty, stop and make the correction immediately.

> **CAUTION**
> Never use newspaper to mask lacquer-painted surfaces when adjacent surfaces are being painted with a lacquer repair system. Newspaper ink is dissolved by lacquer solvents and fumes, and the ink can cause serious staining problems. Use of newspaper should be limited to nonpainted surfaces such as headlamps, bumpers, and wheels. These surfaces can be cleaned readily with a solvent after painting is finished.

> **CAUTION**
> Always disconnect the battery when working on door jambs.

10. For best results, remove the tape:
    a. At temperatures above 60°F.
    b. At a 90° angle to the surface.
    c. With a slow to medium pace without tearing.

Molding Masking

To mask a molding with a curved edge (see Figure 6–29):

1. Using $\frac{3}{4}$-inch masking tape, position the tape along the edge of the molding so that the tape protrudes beyond the molding edge sufficiently to cover the molding when the tape is slicked down.
2. Unroll the tape with one hand while positioning the tape with the other hand.
3. Using a short-bristled brush or other suitable tool, lay the tape down flush to the molding at edges unreachable with the fingers.
4. When masking corners, it is sometimes easier to mask with a fresh end of tape. Position the tape to the molding curved edge while pulling on the tape with one hand and conforming the tape to the molding with the other hand. Make the tape stretch to fit the corner.
5. Follow this procedure to tape the complete perimeter of the molding and apply filler tape on the molding as required.

To mask a molding with a flush edge (see Figure 6–30):

1. Using the proper width of tape, apply tape to the molding around the complete perimeter of the molding. Position the tape to the side of the molding so that the tape is $\frac{1}{32}$ inch from the paint surface.
2. Lay the tape, applied in step 1, flush to the molding and apply filler tape as required. *Note:* Use of a small special tool may be required to lay the tape flush to the molding or nameplate. A short-bristled paintbrush or small stick is very helpful.

Perimeter Masking

Perimeter masking (see Figure 6–31) is used around the outside edge of a large area that is to be

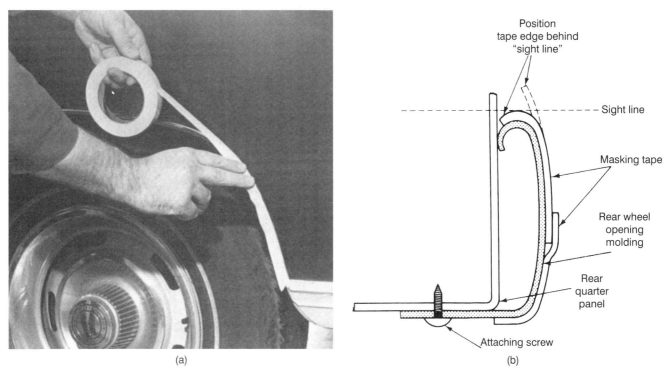

FIGURE 6–29. (a) How to mask molding with curved edge. (Courtesy of 3M Automotive Aftermarket Division.) (b) Masking molding with curved edge (cross-section).

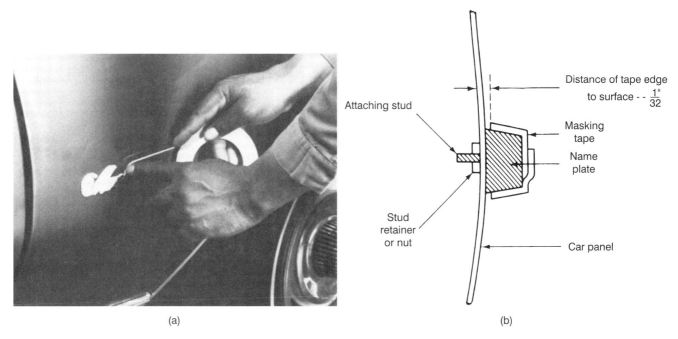

FIGURE 6–30. (a) How to mask molding with flush edge. (Courtesy of 3M Automotive Aftermarket Division.) (b) Masking molding with flush edge (cross section).

FIGURE 6–31. Windshield perimeter and aerial masking. (Courtesy of 3M Automotive Aftermarket Division.)

masked. Once this masking is completed, the masking operation is simplified. Perimeter masking is used on the following areas and parts: windshields, back windows, vinyl tops, two-tone colors, woodgrain transfers, sun roofs, side windows, wide decals, front-end and rear-end flexible plastic parts, wheels, door jambs, front grille work, door handles, and accent colors.

Filler Masking

Filler masking (Figure 6–31) is done in conjunction with perimeter masking. As the term indicates, filler masking completes the masking within a given perimeter. All filler masking, if composed of sections, must be sealed by taping all joints completely. There should be no leakage of paint spray fumes and paint materials through filler masking.

One can see why apron tapers are so important to a paint shop. While detail masking is always important and requires the most time, filler masking covers the greatest amount of square foot area quickly.

FILLER MASKING WITHOUT AN APRON TAPER

Automatic masking equipment is not always available to the painter. In these situations, painters do filler masking using the following procedure:

1. Lay a large section of paper on a clean, flat surface such as a hood or bench.
2. Apply tape, starting at one edge of the paper, lengthwise so that half of the tape goes on the paper and the other half goes on the hood or bench.
3. Prepare about 12 or 18 inches in length at a time and slick the tape down on the paper and surface. Tear the tape at the end of the paper and raise the taped apron by lifting the end of the secured paper and tape.
4. Prepare three or four tape tacks and place them on the back of the hand.
5. Install taped filler masking on the car and tack tape in place.

The preceding operations apply to masking large sections of paper. The same operations apply to preparing a small apron, or use the following procedure:

1. Tear the paper to a suitable size. Lay it on a flat surface.
2. Apply tape at one straight edge of the apron. Tear the tape at the paper edge.
3. Apply masking to the car as required.

Masking Parts of the Car

WINDSHIELD (AND/OR BACK WINDOW)

Follow the steps below:

1. Apply masking tape to the perimeter of the windshield moldings as required. If glass is retained

by a rubber channel, clean the channel as required with paint finish cleaning solvent. Then apply perimeter masking to the rubber channel.
2. Apply filler masking as required. Apply masking across the bottom first, then mask the sides and finish across the top.
3. Apply masking to all exposed joints to make a unitized, sealed mask.
4. Make a special "window" in the masking on the driver's side if the car is to be moved to the spray area after masking.

SIDE WINDOWS

Side windows require precision masking of components around the perimeter of the opening. Closed-style (sedan) windows are masked individually. On most hard-top styles (without pillar), windows can be filler-masked as a unit. Hard-top styles require special attention to perimeter masking around the window openings because of the window and sealing strip design.

1. Apply masking to the perimeter of the side windows as required:
   a. On sedan styles (Figure 6–32a), use care to mask the glass run channels properly.
   b. On hard-top styles (Figure 6–32b), apply perimeter masking as required to sealing strips, weatherstrips, and retainers around the opening. Complete detail masking on the body and door lock pillars before closing the door and before filler masking.
2. Apply filler masking to the side windows as required. Seal all overlapping joints with tape.
3. If the car must be driven or moved from the surface preparation area to the spray painting area, special masking attention must be given to the driver's door:
   a. Mask the driver's door (including the door window and sealing strips) to allow the door to be opened and closed without disturbing the masking.
   b. Make a special opening in the windshield masking so that it can be closed after the car is moved.

DOOR HANDLES

A similar masking procedure can be used on protruding (push button) and flush-type (pull-out) door handles. Also, the perimeter and filler masking can be done in reverse.

1. Use $\frac{1}{4}$-inch tape around the base perimeter of the handle at the paint surface. Maintain the proper distance, $\frac{1}{32}$ inch. Keep the tape off rubber or plastic gaskets at the base of the handle.
2. Use $\frac{3}{4}$-inch tape. Apply filler tape lengthwise to the handle. Apply $\frac{3}{4}$-inch tape flush around the push button and check its operation on the left door only. Double-mask the pull-out handle and check its operation on the left door only. "Double" means tape *under* and *over* the pull-out section on the left door only. This technique allows operation of the handle after masking.

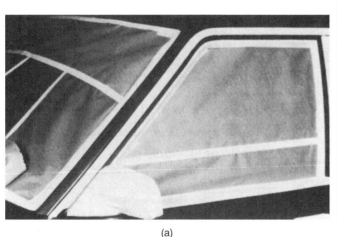
(a)

(b)

FIGURE 6–32. (a) Masking side windows (sedan styles); (b) masking side windows (hard-top styles). (Courtesy of 3M Automotive Aftermarket Division.)

LOCK CYLINDERS

Follow the steps below:

1. Use $\frac{1}{4}$-inch tape around the base perimeter of the cylinder.
2. Apply filler tape as required.

AERIALS

The base is masked with $\frac{1}{4}$- or $\frac{3}{4}$-inch perimeter masking tape. Make a long sleeve (envelope) of the proper length by rolling up a long section of paper, sealing it with tape, and slipping it over the aerial. Seal the envelope to the base masking with tape.

WHEELS

1. Using $\frac{3}{8}$- or $\frac{1}{2}$-inch tape, apply masking to the tire completely around the perimeter of the wheel. Keep the tape within $\frac{1}{16}$ or $\frac{1}{8}$ inch of the wheel behind the edge of the wheel rim.
2. Using sections of pretaped aprons (6 inches wide), complete filler masking of the tire for wheel painting.

FRONT ENDS, GRILLES, AND BUMPERS

On most cars, front ends, grilles, and bumpers can be masked with a combination of perimeter and filler masking as shown in Figure 6–33. Gather up the loose edges of the masking along the bottom of the bumper, tape them together, and secure them at the bottom of the bumper.

Masking the Car for Painting Exterior Flexible Plastic Parts

When used, flexible plastic parts are located between the front and/or rear bumper and the car. These parts are usually named **filler panels** because they fill the space between the bumpers and the car. They are made of flexible plastic and are designed to flex when the bumpers are pushed in, such as during incidental contact upon parking the car. These parts are painted with a special paint system that must flex with the panels when the parts are bent or flexed in normal or cold weather without cracking. Masking the car to paint these parts is done as follows:

1. Mask the front and/or rear bumpers as follows:
   a. Mask around the perimeter of the bumpers next to the filler panels only: across the edges of the bumper nearest the car only.
   b. Then filler-mask the balance of the bumper.

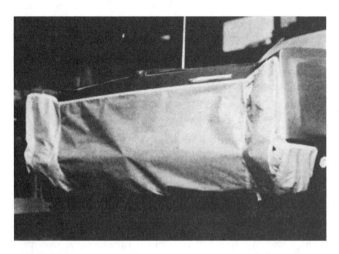

FIGURE 6–33. Masking front ends, grilles, and bumpers. (Courtesy of 3M Automotive Aftermarket Division.)

2. Mask off the car front and/or rear end. First, perimeter-mask the adjacent parts; then apply filler masking to the necessary height (12 or more inches).

Crease-Line or Reverse Masking

Various methods of crease-line masking have been used in the refinish trade since crease-lines became popular in the 1960s. Crease-line masking consists of using natural panel crease-lines, where present, for masking purposes to minimize paint repair areas. Crease-lines must run all the way across a panel for this type of masking to be most effective. Use of crease-lines for sectional panel repairs reduces the volume of refinish paint material used, and it cuts down paint labor because a smaller area is painted.

> **IMPORTANT**
>
> Change the position of masking tape at crease-lines just slightly outward after each coating application to aid in feathering the material at blend areas. For this reason, do the masking at crease-lines last, just before each new coating application. When masking at crease-lines is done properly, perfectly feathered edges are achieved.

An alternate method of masking along panel crease-lines is the reverse masking procedure described in Figure 6–34.

1. Using a 6-inch or wider premasked apron of masking paper, position the centerline of

SURFACE PREPARATION

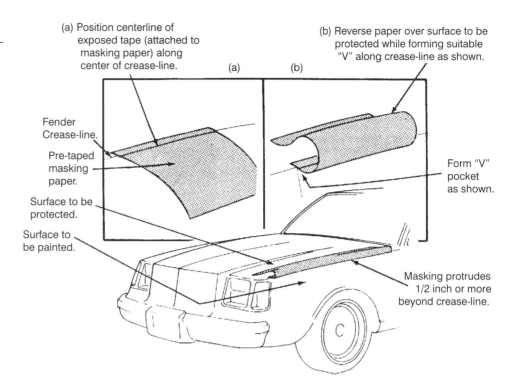

FIGURE 6–34. Reverse masking along panel crease-line. (Courtesy of 3M Automotive Aftermarket Division.)

exposed tape attached to the masking paper along the center of the crease-line (Figure 6–34a) and slick down the tape for the length of the crease-line as required on the surface to be protected. Use one or more sections of pre-masked paper as needed.

2. Reverse the position of the masking paper over the crease-line by raising the paper onto the panel to be protected and secure the paper to the panel as shown in Figure 6–34b.

Masking Door Jambs

The term *door jambs* means all the facing panels around the doors, such as the door hinge pillar, lock pillar, bottom, and upper frame. Door jambs include all facing panels on the body around the doors, such as the body hinge pillar, lock pillar, rocker panel (bottom), and side roof rail (upper). When door jambs are to be painted, it is necessary to mask off all adjacent weatherstrips, door lock and striker components, body number plates, and all trim components.

Door jambs are masked best by first detail-perimeter-masking adjacent parts and then completing the job with a 6-inch or wider pretaped apron.

Door jambs are best spray-painted with acrylic lacquer because of the paint's quick-dry characteristics, even on enamel paint jobs. On complete paint jobs, door jambs are always painted before the exterior paint finish is applied.

DETERMINE THE SURFACE CONDITION

Before starting work on a car, always examine the surface carefully. The best paint in the world will not provide the durability and appearance expected if it is applied over an improperly prepared surface. Analyze the original paint finish to determine the nature of a particular problem, if any, and to determine the type and extent of surface preparation required. Chapter 17 covers description and remedies for common paint problems. By calling on his or her personal experience and good judgment, the painter should determine if the surface to be refinished is in good condition, poor condition, or a combination of the two.

Surface in Good Condition

A surface in good condition is one that consists primarily of surface problems such as off-color, excessive orange peel, and localized rust conditions due to chipped edges, stone bruises, and so on. A surface in good condition can be repaired by the spot-repair method or by correction of the color coat only, because the paint, where not damaged is basically sound.

Surface in Poor Condition

A surface in poor condition is one that has problems that affect the entire thickness of the paint film, color, and undercoats over large areas of the panels in several locations of the car. This poor condition is particularly evident when such characteristics as excessively deep scratching, rust under paint, and lifting and peeling are encountered. Normally, a finish in poor condition must be removed down to the bare metal for satisfactory refinishing.

Aids for Thorough Paint Analysis

Any one or combination of the following aids will assist the painter in analyzing a paint finish to determine if the finish is in good or poor condition.

1. Adequate light is most important to a good inspection.
2. Wash the car for a better visual inspection.
3. Solvent-clean the finish to remove any deposits not removed by soap and water.
4. Use a magnifying glass to observe suspected "crazing" or paint cracking conditions more closely.
5. Compound the finish to determine the gloss restoration and depth penetration of the particular condition.
6. Sand the finish as required to determine the depth penetration of the particular condition.
7. Use a fine-pointed tool to examine blisters and peeling paint more closely.

The life of a paint finish and the appearance of that finish depend directly on the condition of the surface over which the paint is applied. A thoroughly clean and level surface for paint to grip is essential. All dirt, oil, grease, wax, silicone, moisture, and rust must be removed. Any of these contaminants can affect the durability and appearance of the final finish.

Surface preparation, including the application of undercoats, involves all steps necessary to get good adhesion. Experienced painters know that refinish color coats are not designed with special surface-filling qualities. A finished paint job is no smoother than the surface over which color coats are applied.

REVIEW QUESTIONS

1. Explain the purpose and procedure for using paint finish cleaning solvent.
2. What is the procedure for removing silicones?
3. Metal conditioners should be used in two parts. What is the purpose of each part, and what is the correct use of each? (Include reduction of conditioner.)
    a. On steel surfaces.
    b. On galvanized surfaces.
    c. On aluminum surfaces.
4. What is a quick, simple way to determine if the base, which is painted, is steel?
5. What is the purpose and correct use of a tack rag?
6. What is the purpose and correct use of a squeegee?
7. What is the purpose of a sanding block?
8. What is the purpose of sandpaper? (Give four uses.)
9. What does "open coat" mean? Why is it used?
10. What does "closed coat" mean? Why is it used?
11. What is the safety rule for eye protection when using a disc sander?
12. Why is most sanding done in two stages: a coarse followed by a fine sanding?
13. Explain the block sanding technique.
14. Explain the hand sanding technique (without a sanding block).
15. Explain the finger sanding technique.
16. Explain the circle sanding technique.
17. Explain the cross-cutting technique.
18. Explain wet-sanding technique by hand (with and without a sanding block).
19. Explain the scuff-sanding technique.
20. Explain the featheredging technique.
21. Explain the power sander operating technique.
22. Explain how to protect a floor when using paint remover.
23. Explain how to keep paint remover from gap spacings.
24. What is the purpose of masking?
25. Name or describe two types of masking tape.
26. List several qualities of a good masking tape.
27. List several qualities of a good masking paper.
28. What two widths of masking tape are most popular in the refinishing trade?
29. What special tools assist the painter in applying and/or removing masking tape?
30. Explain how to hold and unroll masking tape.
31. Explain the technique for snap-tearing masking tape.
32. What is a safe distance to maintain between the tape edge and the paint surface?
33. How soon after painting can masking tape be removed?
34. What is perimeter masking?
35. What is filler masking?
36. How is filler masking done most quickly when there is no apron taper or masking machine on hand?
37. Explain how to mask a complete windshield or back window.
38. Explain how to mask the side windows of a car.
39. Explain how to mask the complete door outside handle.
40. Explain how to mask the outside vertical aerial.
41. What is crease-line masking? How is it done?

Automotive Refinishing Solvents

CHAPTER 7

PURPOSE OF SOLVENTS

The purpose of a solvent is to **dissolve** the various ingredients of paint like resins, pigments, etc., allowing them to go into solution when paint is made. Thus, solvents serve many purposes in the industry, in the:

1. Manufacture of paint.
2. Storage and transportation of paint before it is packaged.
3. Reduction and application of paint.
4. Cleanup of paint application equipment.

TYPES OF SOLVENTS

The federal government classifies all solvents into three general classes (see Figure 7–1):

1. **Exempt solvents** are a specific list of compounds that **have been proven not to cause pollution.** When these solvents evaporate into the air, they are nonphotochemically reactive with anything in the sun. They are considered good solvents because they do not affect clean air. However, exempt solvents are not true solvents for paint because alone they have a poor ability to dissolve anything. They may be used as dilutants to dilute the viscosity of paint materials for application.
2. **Nonexempt solvents** are known as bad volatile organic compounds (VOCs) because they **cause pollution.** Nonexempt solvents are the backbone of the paint industry because they are true solvents and they make possible the best quality paint resins and paint finishes. However, when nonexempt solvents evaporate

FIGURE 7–1. Role of solvents in paint and how to determine VOC content. (Courtesy of California EPA and California Air Resources Board.)

into the air, they are photochemically reactive with various compounds in the sun and produce ozone, a major component of smog. Nonexempt solvents must be reduced in refinish usage to reduce pollution.

3. While not a true solvent, water is used in the refinish trade in two ways:
   a. De-ionized (condensed) water is used to reduce waterborne (waterbased) finishes prior to application.

b. Standard tap water is used to help clean spray equipment after application of water-borne finishes. It is recommended that the spray system be flush-cleaned with an appropriate cleaning solvent following the use of tap water for cleanup.

AUTOMOTIVE REFINISHING RULES LIMIT VOC CONCENTRATION

Figure 7–2 shows how automotive refinishing rules control a paint's VOC content by limiting the VOC concentration of mixed paint. Note that going from pint to gallon and from gallon to five gallons does not change the concentration as the volume of mixed paint increases. **The VOC concentration does increase when adding VOC solvents** (most thinners and reducers). A permit condition may limit the total pounds of VOC that a shop can use daily. This limit is calculated by multiplying the volume of mixed paint used (minus any water and exempt solvents) times the VOC concentration (pounds per gallon or grams per liter minus the water and exempt solvents). **Look for the VOC content on the paint can label or the manufacturer's paint specification sheet.** Only some material safety data sheets (MSDSs) have VOC information.

> **NOTE**
>
> A paint's VOC content = VOC concentration = VOC limit (or below).

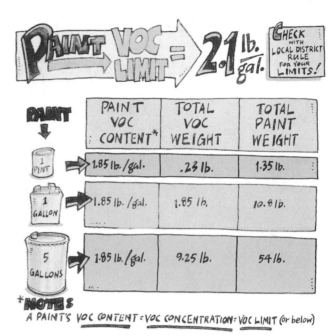

FIGURE 7–2. How automotive refinishing rules limit concentration. (Courtesy of California EPA and California Air Resources Board.)

TYPES OF NONEXEMPT SOLVENTS

Color Blenders

Not true solvents or true additives, color blenders are specially designed products for certain lines of paint. They act like solvents because they provide increased flowout of a paint finish. Color blenders are used in the repair trade to achieve excellent blending of topcoats and clearcoats and also single-stage finishes. The purpose of color blenders is to achieve the utmost flowout of a particular paint finish. They are designed to work within specific lines of refinish products as recommended by your paint jobber. Follow paint company and paint jobber guidelines to avoid refinish problems.

Experience has proved that no one solvent will meet every refinishing requirement. To do quality refinishing on a year-round basis, a paint shop should be supplied with four different types of solvent for each paint system used.

Slow-Drying Solvent

A slow-drying solvent is designed for the application of acrylic colors in average and higher shop temperatures, 65°F and up. This solvent produces a high gloss because it possesses excellent flowout characteristics. In metallics, it produces standard color shades. The flash-time evaporation rate for slow lacquer thinner ranges from $3\frac{1}{2}$ to 5 minutes when applied wet (with an eye dropper) on a red oxide primer-surfaced panel at about 75°F with no exhaust system operating. (This procedure demonstrates how dry time is determined.) For slow enamel reducer, the flash time is slightly longer. Slow-drying solvents are used in the following ways:

1. For applying acrylic colors in average shop temperatures.
2. For mist-coating blend areas of spot repairs when a slow evaporation rate is needed.
3. For mist-coating in its pure form over an entire spot or panel repair (within 10 to 20 minutes after thorough flash-off) to achieve additional

gloss and/or slight darkening of single-stage metallic colors.
4. For reduction in a modified conventional system, which involves a blend with retarder and the use of less solvent by volume to produce darker color shades in metallics.
5. For mist coating, by blending with retarder and adding 5 to 20 percent clear acrylic, to produce darker shades in a given metallic color. Use after thorough flash-off, but within 20 to 30 minutes after color application.
6. Slow-drying solvent can be speeded up by blending it with a faster solvent. A 50-50 blend of slow and medium solvents makes an excellent mist-coat solvent for blending in average shop temperatures (65° to 85°F).

Medium-Drying Solvent

A medium-drying solvent is designed for primer–surfacer application in average shop temperature conditions, 65°F and up. This thinner produces good flowout of the primer–surfacer and dries more quickly than the slow, high-gloss solvent. The average flash-time evaporation rate for medium thinner is about 2 minutes when applied wet (with an eye dropper) on a red oxide primer-surfaced panel at about 72°F with no exhaust system operating. The flash time for medium reducers is slightly longer.

The evaporation rate of medium thinner slows down as the temperature drops. To obtain equal wetting action and flowout of primer–surfacer in 60° to 65°F temperature, a medium-fast solvent can be developed. A medium-fast solvent is a blend of about 50 percent fast solvent and 50 percent medium solvent. Medium-fast solvent evaporates in about 1 minute.

Uses for medium-drying solvent include:
1. Application of primer–surfacer (65° to 85°F)
2. Solid colors (60° to 70°F)
3. Mist coating (60° to 70°F)
4. Speeding up slow solvent
5. Slowing down fast solvent

Medium-fast solvent is used for:
1. Primer–surfacer (60° to 65°F)
2. Solid colors (60° to 65°F)
3. Mist coating (60° to 65°F)

Application of primer–surfacer with too fast a solvent for the temperature results in a very porous undercoat, which could result in poor holdout, a rough surface, excessive shrinkage, and less gloss.

Fast-Drying Solvent

A fast-drying solvent is necessary for the repair of original acrylic lacquers, which can craze if topcoated with slower solvents in average shop temperatures (65° to 85°F). **Fast thinners** evaporate so quickly in average temperatures that they do not penetrate aged acrylic lacquers and thus are classified as nonpenetrating. The flash-time evaporation rate for fast thinners ranges from 15 to 20 seconds when applied wet with an eye dropper on a red oxide primer-surfaced panel at about 72°F with no exhaust system operating. The flash time for fast reducers is slightly longer.

Uses of fast-drying solvent include the following:
1. To prevent crazing in average temperature (65° to 85°F), use in a conventional manner and first apply two or three color coats somewhat on the dry side. Then topcoat as required.
2. For spot or panel repairs, if metallic lacquer is involved and the color shade is too light, use initial color coats as sealers by applying them dry and mist-coating lightly. Allow to dry and scuff-sand if necessary. Then use slower thinner as required and color-coat to match.
3. Use for primer–surfacer application in temperatures below 60°F.
4. Many paint shops feel they save money by using fast-drying thinner slowed down with retarder for color application. Although this system of painting has its advantages, it often leads to comebacks, particularly in metallic color matching.
5. The use of fast-drying thinner for primer–surfacer application in temperatures over 70°F results in a very porous, spongy undercoat and thus to problems such as poor holdout, rough surface, and excessive shrinkage.
6. Fast-drying thinner may be used to speed the evaporation rate of the slower thinners when the need arises.

Retarder

Retarder is not a true thinner. It is a pure form of a single ingredient. Cellosolve acetate is an example. It is added to reduced material or to another solvent when a slower drying time is desired. This solvent can be used effectively in both acrylic lacquer and acrylic enamel paint systems. Retarder prevents blushing because it has an inherent affinity for water. As condensation of water takes place on the surface of freshly applied acrylic paint with retarder, the retarder absorbs the water and keeps it from mixing with the paint ingredients, which would cause a kickout or a blushing condition. The flash-time evaporation rate of retarder is about 30 minutes when applied on a red oxide undercoated panel at about 72°F with no exhaust system operating.

Uses of retarder include the following:

1. Prevents blushing (see Chapter 17).
2. Makes possible color application in hot, humid weather.
3. Provides better flowout by lengthening the evaporation time.
4. Enriches the solvent action when added to a slow solvent.
5. Enriches a slow solvent for mist-coating in spot repairs, to eliminate overspray, to provide more flowout, and to create more gloss.

Temperature can affect the final smoothness of the finish in terms of flowout, gloss, orange peel, and other properties that depend on drying of the finish. Temperature and humidity problems are overcome most easily by the proper selection of solvents. The ideal spraying temperature range is between 68° and 75°F. Paint suppliers usually recommend this temperature range for spray painting. Repair of acrylic finishes should not be done in temperatures below 60°F.

When paint supplier recommendations are made in label directions, they are made for "average" temperature and humidity conditions: 68° to 75°F temperature and 50 percent to 80 percent humidity. There is no set formula or method for determining the "ideal solvent mixture" that works perfectly on all colors in all temperatures and all humidity conditions. Each painter must use his or her good judgment and knowledge of solvent behavior and selection in all temperature and humidity conditions. The painter must be able to select and to adjust solvents to meet the needs of the job being painted. A painter must be able to assemble the right combination of solvents to do high-quality work in all types of weather conditions. A painter must be able to use:

1. The fastest evaporation solvents on cold days.
2. The slowest evaporation solvents, including retarder, in extra hot and humid weather.
3. A combination of solvents between the above two extremes.

The proper solvent is a key tool in successful refinishing. The painter must determine and use the solvent combination that is proper for any spray painting application. Sometimes a painter does not realize the full importance of temperature and humidity on the refinish process. Much of the time a painter is busy thinking about finishing the jobs he or she is working on; ordering color and supplies for the next job; and while he or she is doing these, a color-matching problem crops up. The higher the temperature, the faster the solvents evaporate between the spray gun and the work surface. Solvents evaporating too fast has a marked effect on the painting results, particularly when you are color-matching metallic colors. This predicament is the same as dry spraying or not getting enough solvent on the part being painted for proper flowout.

The painter can control spraying techniques, which are important to the final results of a paint job. The essential items a painter can control when painting are:

1. Thinner selection and reduction
2. Proper spray gun adjustments
3. Distance of gun from surface
4. Speed of spraying
5. Overlap
6. Air pressure at gun

Three things a painter cannot control are temperature, humidity, and ventilation. The rate of evaporation on a hot summer day (90° to 110°F) can be approximately 50 percent faster than it is on an average day of 72°F. But what the painter can do is select the proper solvent that does a high-quality job under the circumstances. The chart of variables in Table 7–1 was assembled as a result of many years of experience matching metallic-type colors. Table 7–1 charts three shop variables (temperature, humidity, and ventilation) and the effect these variables have on metallic colors. Although the painter cannot control these variables, he or she must learn how to adapt to the variables and still do required paint work. As a variable changes, corresponding changes also take place in the final shade of metallic colors, as shown in Table 7–1. Ventilation is included as an uncontrollable variable because generally the painter is not responsible for providing shop ventilation. This is a responsibility of the paint shop. Not all paint shops have 100 percent foolproof ventilation.

PAINT VISCOSITY CUP

Over- or underreduction of paint results in numerous problems, such as mismatches with metallics, runs or sags, excessive orange peel, and generally poor paint repairability. The best way to check the viscosity of a reduced paint material is with a viscosity cup.

AUTOMOTIVE REFINISHING SOLVENTS

TABLE 7-1 Variables Affecting Color of Metallics

Variable	To Make Color Lighter	To Make Color Darker
Shop conditions:		
a. Temperature	Increase	Decrease
b. Humidity	Decrease	Increase
c. Ventilation	Increase	Decrease
Spray equipment and adjustments:		
a. Fluid tip	Use smaller size	Use larger size
b. Air cap	Use air cap with greater number of openings	Use air cap with lesser number of openings
c. Fluid adjustment valve	Reduce volume of material flow	Increase volume of material flow
d. Spreader adjustment valve	Increase fan width	Decrease fan width
e. Air pressure (at gun)	Increase air pressure	Decrease air pressure
Thinner usage:		
a. Type of thinner	Use faster evaporating thinner	Use slower evaporating thinner
b. Reduction of color	Increase volume of thinner	Decrease volume of thinner
c. Use of retarder	(Do not use retarder)	Add proportional amount of retarder to thinner
Spraying techniques:		
a. Gun distance	Increase distance	Decrease distance
b. Gun speed	Increase speed	Decrease speed
c. Flash time between coats	Allow more flash time	Allow less flash time
d. Mist coat	(Will not lighten color)	The wetter the mist coat, the darker the color

Temperature influences viscosity directly. Body shop thermometers are available to the trade through paint jobbers. Figure 7–3 is an example of a thermometer that paint suppliers have available. Primarily designed for body shop use, the −40° to 120°F thermometer features a solvent scale for automotive paint reduction, as well as an exact percentage humidity scale.

If cold paint is brought into an average temperature room, it will be thicker and more viscous. Simply adding solvent to make it sprayable is not always the best option. It is best to allow paint to reach workable or average room temperature. However, if paint must be applied in cool temperatures, below 70°F, the viscosity should still read within proper time limits. Viscosity is adjusted as required by adding solvent or unreduced paint to the reduction.

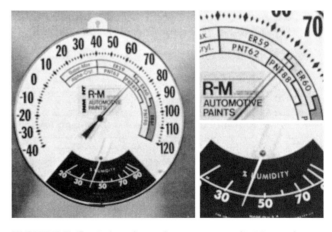

FIGURE 7–3. Paint shop thermometer (R-M type). (Courtesy of BASF Corporation.)

Directions for Using a Viscosity Cup

1. Dip the cup into reduced paint until it is full. Have a stopwatch ready.

2. Remove the cup and **as the bottom of the cup clears the surface of the paint, start timing the flow of paint** from the small hole in the bottom of the cup (Figure 7–4). (Use of a stopwatch is preferred for this step.)

3. Stop the timer when a continuous stream of paint breaks. Check the time in seconds with your local paint jobber for the type of material and make adjustments as required.

PERCENTAGE OF REDUCTION AND MIXING RATIO

No matter how experienced a painter is, he or she should **always read the label directions before using a paint material.** Products go through changes on a continual basis. Reading the labels before they become illegible is one way to keep up with the product.

When label directions make a statement such as "Reduce 125 percent" or "Reduce 33 percent," the **percentage figure always applies to the solvent system.** Table 7–2 is a reduction ratio guide. It explains the mixing ratio for each common reduction percentage figure from $12\frac{1}{2}$ percent to 300 percent. Also included is an expansion of the mixing ratio that simplifies each reduction into parts by volume of each material used. Painters can usually tell how much mixed paint they need by the size of the paint repair.

How to Reduce Paint in a Paint Cup

Graduated or marked paint mixing paddles are always available through paint companies and local paint jobbers. These marked paint paddles have rows

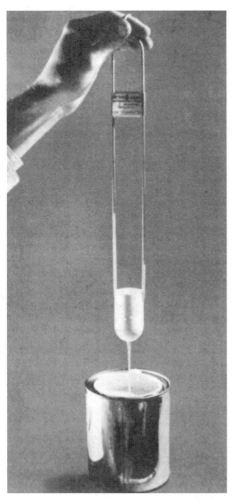

FIGURE 7–4. Zahn no. 2 paint viscosity cup. (Courtesy of General Electric Company.)

TABLE 7–2 Reduction Ratio Guide

Percentage of Reduction	Mixing Ratio	Paint Material	Solvent Material
$12\frac{1}{2}$	8:1 =	8 parts	1 part
25	4:1 =	4 parts	1 part
33	3:1 =	3 parts	1 part
50	2:1 =	2 parts	1 part
75	4:3 =	4 parts	3 parts
100	1:1 =	1 part	1 part
125	4:5 =	4 parts	5 parts
150	2:3 =	2 parts	3 parts
175	4:7 =	4 parts	7 parts
200	1:2 =	1 part	2 parts
225	4:9 =	4 parts	9 parts
250	4:10 =	4 parts	10 parts
275	4:11 =	4 parts	11 parts
300	1:3 =	1 part	3 parts

FIGURE 7–5. Graduated Glasurit paint mixing paddle supplied by R-M paint jobbers (R-M and Glasurit are subsidiaries of BASF). (Courtesy of BASF Corporation.)

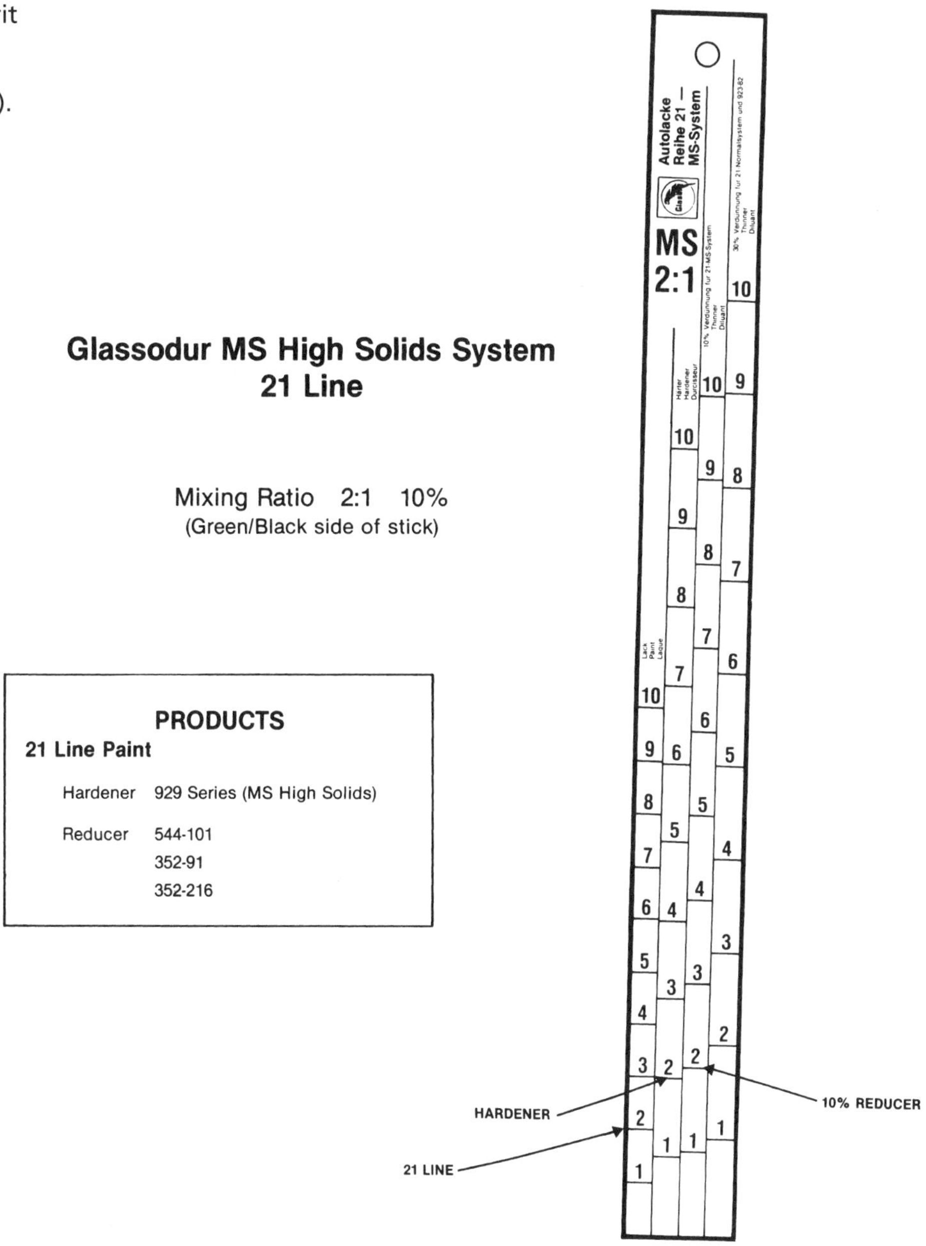
Glassodur MS High Solids System
21 Line
Mixing Ratio 2:1 10%
(Green/Black side of stick)
PRODUCTS
21 Line Paint
Hardener 929 Series (MS High Solids)
Reducer 544-101
352-91
352-216
Autolacke
Reihe 21 —
MS-System
MS
2:1
HARDENER
21 LINE
10% REDUCER

of figures and graduated markings to help the painter measure precisely the exact amount of:

1. Color (first)
2. Hardener (second)
3. Solvent (third)

For directions on how to use a particular paint paddle, obtain the instructions from the paint jobber who supplied the paddle.

Figure 7–5 shows a **graduated (proportionally marked)** paint mixing paddle available through R-M paint jobbers, which are subsidiaries of the BASF Corporation. The paddle can be used to help reduce any refinish paint material. Familiarity with the stick can be achieved by checking with the paint jobber who supplied it.

Figure 7–5 indicates how a typical Glasurit 21-Line paint material is reduced when the mixing ratio is 2:1 and 10 percent (2 parts color, 1 part hardener, and 10 percent reducer).

1. The 2 stands for color. With the paint stick held vertically in the paint cup, add thoroughly agitated 21-line color to the first 2 line.
2. While holding the stick still, add hardener to the 2 in the second column, as shown.
3. Also while holding the stick still, add the 10 percent reducer to the 2 in the third column, as shown.

4. Mix the reduction thoroughly and strain into the spray gun cup.

RECORDKEEPING

When recordkeeping is required, it consists of two categories:

1. Records for **coating and solvent use.**
2. Records for **equipment cleaning solvent use.**

Keep daily records of:

1. The **amount and type of each coating** (including catalyst and added solvent) used. Coatings include, but are not limited to, precoat, primer, primer–surfacer, primer sealer, and topcoat.
2. The **amount and type of each surface preparation** (including oil and grease removal) solvent and cleanup solvent used.
3. The **volatile organic compound (VOC) content of each VOC-containing material used.**
4. The **vapor pressure of each cleanup solvent.**
5. The operating **temperature of the oven.**
6. Any other data required by your permit conditions.

Once recorded, **keep the daily records for at least two years** before disposing of them. For example, keep records for July 10, 1997, until at least July 10, 1999. Records may be kept on your district's standard form, by computer program, or on your own form as long as the required information is recorded. See Figure 7–6 for an example.

DAILY COATING LOG

Date: _____

Facility: _____   Booth #: _____

Manufacturer Name	Product I.D. Number	List Catalyst & Reducer Used	Mix Volume* P	C	R	Coating Type	VOC Content As Applied	Quantity Used	Spray Method ☆	Substrate & Part	Baked· Temp	Operator Initials	Comments:
							Total Coating Quantity				Total Clean-up Solvent Usage		

* P = Paint   C = Catalyst   R = Reducer   (Use Actual Volumes)
· If Air Dried write A/D
☆ A = Air Spray   AL = Airless Spray   AA = Air-assisted Airless Spray
   B = Hand Application   D = Dip Coat   E = Electrostatic Spray   F = Flow Coat   H = High Volume Low Pressure

FIGURE 7–6. Track your paint use with recordkeeping. (Courtesy of California Air Resources Board.) *Note:* Many districts require daily recordkeeping of your shop's paint and solvent use. Record your coating mixtures *exactly.* Include the manufacturer's name and paint identification number, the amount of each component used, and the amount of other solvents added. Accurate records prove your compliance with rule limits. Follow recordkeeping instructions—be consistent and be accurate!

REVIEW QUESTIONS

1. What is the purpose of solvents?
2. What is an exempt solvent?
3. What is a nonexempt solvent?
4. How does air pollution affect society and the environment?
5. What does VOC stand for?
6. Explain how the federal government regulates VOCs resulting from automotive refinishing.
7. What four major categories of ingredients are used in the makeup of a gallon of paint as covered in this chapter?
8. How is a paint's VOC content determined?
9. How do automotive refinishing rules limit VOC concentration?
10. In terms of evaporation rate, what four types of solvents are needed in the refinish trade?
11. What is viscosity?
12. How is viscosity affected by solvents?
13. What is the purpose of a viscosity cup?
14. How is a viscosity cup used?
15. Explain how solvents can be blended for mist-coating or color-blending purposes.
16. Explain how retarder prevents blushing.
17. Why does the federal government require recordkeeping in compliant areas?
18. How long must a paint shop keep records before they can be destroyed?
19. Can recordkeeping be kept by a computer program?
20. You are mixing paint in a paint cup. Of the three ingredients mixed, solvent, color, and hardener, which ingredient goes into the cup first? Second? Third?
21. When label directions state, "Reduce 33 percent," to which refinish ingredient does the percentage apply: the paint or the solvent?
22. How does a painter determine the VOC content of a paint material?
23. What is meant by nonphotochemically reactive?
24. What is meant by photochemically reactive?

CHAPTER 8

Undercoat Materials and Application

GENERAL TYPES OF UNDERCOATS

In recent years, undercoats as well as topcoats have undergone significant changes in chemistry because of a desire to improve durability and performance while at the same time meeting the requirements of clean air regulations. The newly developed and popular undercoat systems consist of five basic types:

1. Epoxy and etching primers.
2. Tintable undercoats.
3. Two-component (2-K) undercoats.
4. Low VOC urethane and epoxy primers.
5. Waterborne undercoats.

The safest choice of undercoat selection is the one recommended by the paint system you are using because it may be necessary that it comply with warranty requirements. **When using any undercoat system, be sure to use the respirator system** (TC-23C or TC-19C) recommended in the label directions.

HISTORY

Early undercoat systems were primarily of single-component, solvent-based construction. These systems are still in use and are basically composed of acrylic lacquer and/or enamel. Because of the EPA national rule, **the use of acrylic-lacquer products will be abolished as soon as existing supplies run out.** No new acrylic-lacquer products will be available in cans from paint companies. Lacquer products will continue to be available only as specialty coatings. For specialty coating use, see Specialty Coatings in the Glossary of Terms.

Single-component undercoats are reduced with solvent and are applied with conventional or HVLP spray equipment. A general-purpose cartridge-type respirator (TC-23C) may be used by the painter during preparation and application of single-component undercoats, as has been done in the past.

In recent years, two two-component undercoat systems have been developed. These two systems have become increasingly popular because they are tougher and more durable, and they dry reasonably quickly.

1. **In system 1, the hardener is an isocyanate catalyst.** To prepare and apply this material requires that the painter use an air-supplied respirator (TC-19C).
2. **In system 2, the hardener is a nonisocyanate catalyst.** To prepare and apply this material, the painter has the option of using a general-purpose (cartridge-type) respirator (TC-23C) or an air-supplied respirator (TC-19C).

Excellent waterbase, ready-to-use undercoats are available for priming, filling, and sealing with excellent durability properties. For the application of waterbase products, painters are urged to use the proper respiratory protection as called for in label directions. While waterbase and Low VOC two-component undercoat systems are designed specifically for highly regulated areas, they also can be used anywhere in the country.

PURPOSE OF UNDERCOATS

Automobiles are painted for two primary reasons: **appearance and protection.** Appearance is one of the main factors that helps to sell cars, and all car

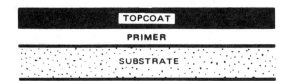

FIGURE 8–1. Primer is required to make color stick to metal. (Courtesy of DuPont Company.)

owners like to have their cars in good repair. The smallest pinhole or any deep scratch in paint over steel results in a rust spot, so the steel needs protection.

In general, paint finishes are made up of two parts: color and undercoats. The color coat is formulated to produce lasting beauty, but it requires undercoats to provide the required durability through good adhesion and protection to the metal. Undercoats are designed to protect and to adhere to the metal. Also, they provide the proper surface to which color can adhere (Figure 8–1). Color alone will not adhere properly to bare metal.

To provide its intended usefulness under all weather conditions, a paint finish must be:

1. **Hard and durable:** To have excellent gloss and color retention and to withstand the elements over a long period of time.
2. **Water resistant:** To withstand continued exposure to water, rain, snow, dew, and high-humidity conditions without breakdown and to prevent moisture from reaching the metal surface.
3. **Adhesive:** To cling permanently with a strong bond to the surface to which it is applied.
4. **Elastic:** To expand and contract to an equal degree with the surface to which it is applied when exposed to hot and cold temperatures without breakdown.
5. **Chip-resistant:** To resist damage by flying sand, gravel, and other forms of contact without chipping under normal driving and use conditions.
6. **Repairable:** To be repaired efficiently and economically in a highly competitive market.

UNDERCOAT CATEGORIES

Undercoats are divided into the following general categories:

1. Primers (one- and two-component types)
2. Primer–surfacers (one- and two-component types)
3. Primer–sealers
4. Barrier coat (one-component systems)

FIGURE 8–2. Straight primers are drawn into the valleys and irregularities of the surface. (Courtesy of PPG Industries.)

5. Polyester glazing putties
6. Adhesion promoters
7. Converters
8. Stone chip protectors

Straight Primers

The purpose of straight primers is to produce adhesion for the top coats and the best protection to the steel available. Straight primers offer the best protection to steel because they are mostly a thin liquid that penetrates into the valleys and cavities of sand scratches and surface irregularities most completely. When applied wet, as it should be, primer is drawn into the valleys of sand scratches and surface irregularities by capillary attraction. **Capillary attraction** is the attraction of the liquid primer by the molecular forces of the metal as the fluid is drawn into 100 percent firm contact with all the metal surface. This superior wetting action is achieved when primer is applied in thin, wet coats (Figure 8–2). Straight primers are very tough and flexible.

Epoxy Waterborne Primer (Glasurit 76-22)

A precoat corrosion protection primer is a special coating applied to bare metal primarily to deactivate the metal surface for corrosion resistance to a subsequent waterbase primer or primer–surfacer.

CAUTION

Before applying a waterbase primer–surfacer over bare metal, it is necessary first to apply a precoat coating. See the paint manufacturer's Low VOC wall chart to determine the proper precoat coating. Also, do not use water or paint finish cleaner on dried waterbase primer–surfacer before topcoating. Use R-M 909 Final Wipe or an equivalent.

Glasurit's primer is 76-22 Glassohyd 2K Precoat Primer. (Competitor companies have products equivalent to this material.) The primer is a two-component, epoxy resin, catalyzed primer that can be used over bare metal, body putty, plastic filler, and/or original equipment manufacturer (OEM) finishes. Primer must be applied on bare metal before applying waterbase topcoats. The mixing ratio is 2 parts 76-22 waterborne primer to 1 part 76-4 hardener.

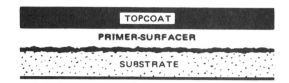

FIGURE 8–3. Primer–surfacer is a primer that also fills small flaws. (Courtesy of DuPont Company.)

> **IMPORTANT**
>
> **These components must be mixed by weight, not by volume.** Use 3 percent to 5 percent 90 VE water if desired. Apply one light coat at 0.7 to 0.9 mils and allow it to dry per label instructions.

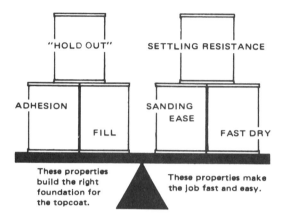

FIGURE 8–4. Good primer–surfacer is a balance of many properties. (Courtesy of DuPont Company.)

Primer–Surfacers

The purpose of primer–surfacer is to do a twofold job for paint systems as follows:

1. To prime the bare metal.
2. To fill small surface irregularities quickly so that, when sanded, the surface is smooth and ready for color application (Figure 8–3).

Desirable characteristics of a good primer–surfacer are (Figure 8–4):

1. **As a primer:** It should have excellent adhesion to all automotive finishes and good anticorrosion protection for steel.
2. **As a surfacer:** It should have good filling and rapid buildup of film thickness.
3. **Sanding:** It should be easy to sand.
4. **Drying:** It should dry in 30 to 60 minutes.
5. **Holdout:** The holdout should be good and not porous.
6. **Settling:** It should not settle in the gun between jobs. Moderate settling should be stirred back into solution quickly.

FILM THICKNESS FOR PRIMER–SURFACERS

For best results, the thickness of primer–surfacer, after sanding, should be about $1\frac{1}{2}$ to 2 mils, to be on the safe side. Excessive thickness of primer–surfacer, beyond 2 mils or more, under refinish acrylic top coats can lead to premature crazing and/or cracking conditions (see Chapter 17).

(Glasurit) 76-92 1K Waterborne Primer Filler

Glassohyd 1K Primer/Filler 76-92 is a high-build waterborne primer filler and is designed to provide excellent adhesion to most substrates while yielding maximum topcoat holdout. (Competitor companies have products equivalent to this material.) Formulated as a low-solvent coating, 76-92 Primer/Filler is a ready-to-spray, one-component product that performs as well as most two-component primer–surfacers. It also functions as a barrier coat over acrylic enamel or old paint finishes and prevents staining from body fillers.

Apply over all types of OEM or cured, air-dried paint finishes. When priming over bare steel; galvanized, zinc-plated, or aluminum metals; body filler; or fiberglass, first apply a light coat of 76-22 Glassohyd 2K Precoat/Primer.

Glassohyd 1K Primer/Filler 76-92 comes ready to use and requires no mixing or reduction. Do *not* reduce with 90-VE water.

> **NOTE**
>
> Do not shake. Stir product gently before using.

Apply two to three coats of 76-92 1K Primer/Filler and allow 10 minutes flash time between coats. Air dry for 1.5 hours at 68°F and 50 percent humidity (ranging to 3 hours at 86°F and 80 percent humidity.

> **IMPORTANT**
>
> Keep 76-92 1K Primer/Filler from freezing.

Primer–Sealers

A sealer is used in the automotive refinishing trade for one of two purposes:

1. To provide adhesion between the paint material to be applied and the repair surface. Sanding of the surface may be required before application.
2. To act as a barrier that prevents or retards the mass penetration of refinish solvents from the color and undercoats being repaired.

Sealers also provide the following desirable benefits:

1. Improve adhesion of the repair color to very hard undercoats and enamel surfaces (see Figure 8–5).
2. Retard sand-scratch swelling (see Figure 8–6).
3. Improve gloss.
4. Improve color holdout.
5. Prevent bleeding (when designed for this purpose).
6. Can be used on small, clean, bare metal surfaces.

Two general types of sealers are available:

1. **One type of sealer is designed for acrylic-lacquer paint systems.** This sealer, when so stated on the label, is not to be topcoated with enamel finishes.
2. **A second type of sealer is designed for air-dry enamel paint systems.** This sealer, when so stated on the label, is not to be topcoated with acrylic-lacquer finishes.

Because of differences in refinish paint systems and the severe hardness of most factory paint finishes, particularly the basecoat/clearcoat finishes, special sealers are needed for the proper repair of all the finishes on the road. No single sealer can serve all purposes. Painters must be thoroughly familiar with the popular paint systems and the sealers required for their repair.

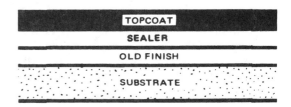

FIGURE 8–5. A sealer bonds the top coat and the old finish. (Courtesy of DuPont Company.)

FIGURE 8–6. Sealers reduce sandscratch swelling problems. (Courtesy of DuPont Company.)

Adhesion Promoter

Adhesion promoter is a specially designed, neutral colored, ready-to-spray coating that provides a chemical or mechanical bond to previously applied coatings for the purpose of increasing the adhesion characteristics of subsequent coatings. No topcoat, primer, primer–sealer, or primer–surfacer can be classified as an adhesion promoter.

Converter

A converter is an acid-type liquid used in specific primers to promote chemical adhesion and the etching abilities of the primer. It also reduces the viscosity of a primer reduction to aid the spraying ability of the primer. A converter is *not* a true solvent, hardener, catalyst, or activator.

Putty

The purpose of putty is to fill pits, deep scratches, file marks, and other isolated irregularities on the surface of the metal that cannot be filled satisfactorily with primer–surfacer (see Figure 8–7 and 8–8). Putty is designed for application to spot locations rather

UNDERCOAT MATERIALS AND APPLICATION

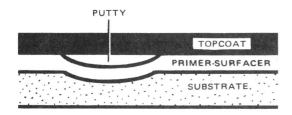

FIGURE 8–7. Glazing putty fills small flaws. (Courtesy of DuPont Company.)

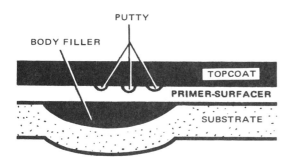

FIGURE 8–8. Body filler is used to fill large flaws. (Courtesy of DuPont Company.)

than large areas. Also, putty should always be applied over primer–surfacer.

2K Lightweight Glazing Putty

Early glazing putties were made of lacquer-base materials. Even when properly applied, these glazing putties caused the painter many problems, chief of which were sandscratch swelling and low spots due to material shrinkage. Glazing putty manufacturers have developed an answer to these product concerns: a two-part product known as 2K lightweight glazing putty. It consists of a polyester base, which requires a catalyst or hardener. When applied according to label directions, the putty spreads easily, dries quickly, sands easily, and is trouble-free with minimal shrinkage. 2K glazing putties are available through your local paint jobber or supplier.

UNDERCOAT APPLICATION

All surface preparation operations up to this point must be completed, including:

1. All washing and cleaning
2. All sanding and/or paint removal
3. All masking
4. All metal conditioning as recommended by the paint company

Although metal conditioning was covered in Chapter 6, it should be done just before primer application. It is a surface maintenance operation that should never be neglected. Some primers in the trade do not require metal conditioning. Follow label directions.

Conventional Primer Application

1. Prepare the primer according to label directions. In cold weather, check it with a viscosity cup.
2. Apply a thin, wet coat of primer to bare metal surfaces.
   a. Apply with spray equipment to large surfaces.
   b. Apply with a brush to very small or tiny spot surfaces. Most quality paint shops have a top-quality primer (and brush) ready for use at all times every day for small spot area applications.
3. Allow the primer to dry before applying primer–surfacer. See label directions.

CAUTION

Most good primers are enamel based. **Never sandwich enamel primer between coats of lacquer** because lifting will result.

Conventional Primer–Surfacer Application

Follow the label directions to reduce and prepare primer–surfacer. Label directions usually stress the following important operations:

1. Stir the material thoroughly at package viscosity to obtain a homogeneous mixture before reduction. A material on the shelf for several weeks or months requires several minutes of mixing by hand or on a paint shaker.
2. Select the proper solvent for the weather conditions. For average temperature conditions (65°F and up), use medium-evaporation solvent. Use a premarked container or paddle for fast and accurate reduction of primer–surfacer according to label directions. In cold weather, below 65°F, use fast-evaporating solvent. Spraying in too cold weather (below 55°F) is not recommended.
3. Reduce the material, stir it thoroughly in a mixing container, and strain it through a coarse strainer into a paint cup.

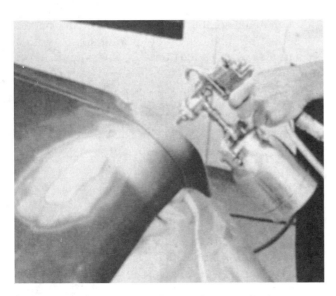

FIGURE 8–9. Primer-surfacer is best applied to medium-wet coats. (Courtesy of PPG Industries.)

4. Usually, label directions stress that material should be applied with the shop and metal temperature at least 65°F. Use a low air pressure.

See Figure 8–9 and follow the application guidelines listed below:

1. a. Check the spray gun adjustment with a flooding test pattern and adjust the spray gun as required.
   b. Check application of the primer–surfacer on a spray stand or other suitable object before applying the primer on a car.
   c. Spray a thin, wet, first coat as the primer coat. Allow to flash.
   d. Spray up to three or more thin to medium wet coats for additional film buildup. Allow each coat to flash off before applying the succeeding coat. Never speed up flash-off by blowing air on wet film. Normally, for spot repair, apply each medium coat of primer–surfacer with low psi (at the gun). For panel repair, use a suitable air pressure between 3 and 4 psi (at the gun). **The film thickness of undercoats should total between 1 and 2 mils after sanding.** The film thickness may be 3.2 mils before sanding.

2. Allow to air-dry. A normal application of primer–surfacer (before sanding) will dry in 30 minutes or less. The thickness of the film, the solvent used, and the temperature of the shop will influence the drying time.

3. Do not apply extra-heavy coats of primer–surfacer in one pass or in many passes of the spray gun to speed up the operation. Primer–surfacer applied in this way requires more time to dry. The film will appear hard at the surface but will not be dry next to the metal. This condition can lead to difficulty in sanding, poor holdout, pinholes, mud cracking, and crazing.

4. Water-sand the surface with No. 400 sandpaper for best results. Use a sanding block when sanding large flat surfaces and a sponge pad on concave surfaces. Water sanding is preferred over dry sanding as a general rule. However, there are times when it is quicker and more convenient to dry-sand. Guard against cut-throughs at all edges, high ridges, and crease-lines. Do not press normally hard at these locations. Just wipe across them carefully or stay away from them completely. Repair cut-throughs when dry with additional primer–surfacer or primer–sealer. Check the progress of your work with a squeegee. Clean as required when finished.

Conventional Guide-Coat Filling and Sanding

The purpose of a guide-coat system is to fill and sand a featheredged repair to ensure a level surface and avoid a low spot from showing through the final color. This system is particularly helpful when filling one or more featheredged spots during the repair of enamel finishes.

Undercutting (removing too much primer–surfacer when sanding) happens between two hard surfaces of original enamel when sanding is done and the softer surface (new primer–surfacer) is removed accidentally and unknowingly (see Figure 8–10). Guide-coat sanding helps the painter to **know when to stop sanding or when to become extra cautious.** All surface preparation operations before this must be completed.

Use the following procedure:

1. Using light (or dark)-colored primer–surfacer:
   a. First apply two medium-wet coats to the area within a sanded featheredge. Allow flash time between coats.
   b. Next apply two coats slightly beyond the first two coats. Allow each coat to flash before applying the next coat.

2. Use dark (or light)-colored primer–surfacer (of contrasting color) and reduce to label directions.
   a. Apply one wet coat of primer–surfacer within the sanded repair area.

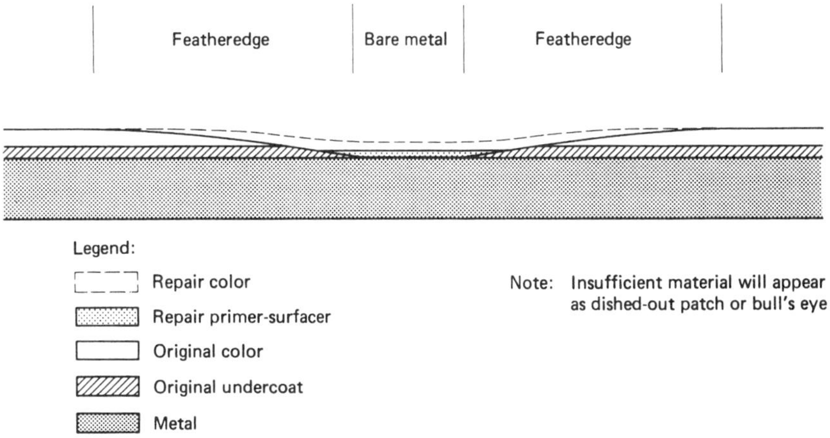

FIGURE 8–10. Causes of low spots in refinishing. (Courtesy of General Motors Corporation.)

b. Apply the next wet coat to extend slightly beyond the previous coat. Allow each coat to flash before applying the next coat.
3. Allow the primer–surfacer to dry.
4. Water-sand the primer–surfacer with No. 400 sandpaper and a suitable sanding block as follows:
   a. If sanding on adjacent acrylic-lacquer, keep off the lacquer as much as possible.
   b. Start sanding at the perimeter and use plenty of water. Continue sanding toward the center as the last-applied primer–surfacer is removed.
   c. Finish sanding across the center of the repair spot using uniform, light sanding strokes. Stop and check the progress of sanding frequently with a squeegee.

CAUTION

Do not sand the center of the repair too much. As soon as the orange peel is removed across the center of the spot, stop sanding.

---

5. Clean and allow to dry before the next operation.

Improper Uses of Conventional Primer–Surfacer

Condition. The excessive application of primer–surfacer over an entire car that has a good surface. This kind of application can cause excessive film thickness, which can lead to checking and cracking.

Remedy. Apply primer–surfacer only to damaged areas that are being repaired. A single coat of highly reduced primer–surfacer may be used as a sealer over the original acrylic-lacquer color to minimize sandscratch swelling if the final film thickness does not go over 8 mils.

Condition. Application of primer–surfacer over poorly prepared or dirty surface, which results in poor adhesion, poor drying, and other problems.

Remedy. Remove all foreign matter, visible and hard to see, from the surface before the application of primer–surfacer by using paint finish cleaning solvent and metal conditioner with approved sanding practices.

Condition. Insufficient reduction of primer–surfacer when the painter attempts to obtain greater filling over heavy file marks. In those cases, the painter uses insufficient solvent in reduction and higher air pressure to get thicker material through the gun. This condition causes excessive orange peel, a rough surface, and sandscratches to show up after the color is applied. Application of primer–surfacer in this manner usually traps air and vapor at the base (in the valley) of sandscratches, as shown in Figures 8–11a and b. When color is applied, the solvent penetrates and resoftens the primer–surfacer, which then sinks into the sandscratch valleys (Figure 8–11c). This condition becomes most visible when paint films shrink upon drying.

When correctly reduced and applied over coarse file marks, the file marks become evident as the primer–surfacer dries (Figure 8–12).

Remedy. Reduce the primer–surfacer according to the label directions. Apply the required amount of material in thin, wet coats with sufficient flash time between coats. **The best remedy is to remove all coarse file marks before painting is done.**

Condition. The use of an improperly formulated solvent in the primer–surfacer. In many cases, the

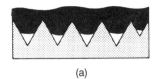 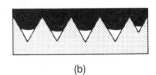 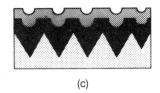

(a)   (b)   (c)

FIGURE 8–11. A common cause of sandscratches (cross-section through painted file marks): (a) insufficiently reduced primer–surfacer bridges sandscratches; (b) after sanding, bridges and voids remain; (c) solvent in color coats rewets the primer–surfacer, which sinks into sandscratch valleys. Scratches are now visible. (Courtesy of PPG Industries.)

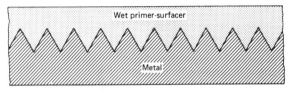
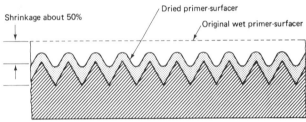

FIGURE 8–12. Coarse file marks require additional filling. (Courtesy of General Motors Corporation.)

painter uses leftovers from other paint work and/or paint systems. This practice can cause any number of problems, depending upon the contamination of solvents used.

Remedy. Follow the label directions and use the correct brand of solvent for the weather conditions.

Conventional Sealer Application

1. In the area to be color coated, clean the finish thoroughly with paint finish cleaning solvent.
   a. When topcoating over cured OEM enamels, use #400, #600, or ultra fine sandpaper plus cleaning solvent to wet-sand the complete OEM color to be sealed.
   b. When topcoating over acrylic-lacquer, compound and reclean the entire area to be sealed with paint cleaning solvent.
2. Using the correct sealer for the repair situation, apply one or two coats of sealer at the correct air pressure, per label directions, slightly beyond where color will stop.
3. Allow the sealer to air dry 30 minutes or longer, per label directions, before applying color.

Polyester Putty Application

Perform all necessary surface preparation operations through the application of primer–surfacer and allow to dry. Putty is designed to be applied over primer–surfacer.

1. Prepare a suitable amount of polyester putty on a flat working surface, and apply hardener to the putty per the label directions.
2. Mix the putty to a homogeneous mixture with a flat-bladed tool.
3. With a squeegee or plastic applicator, apply a tight, uniform, and thin first coat of putty to the surface. Add additional putty with moderate pressure, and smooth the putty application to the desired surface.
4. Allow the putty to cure per the label directions.
5. Using a proper respirator, dry-sand the repair area to a smooth surface with a coarser sandpaper, like 220, followed by a finer sandpaper, like 320.

If necessary, mix and apply a very thin glaze coat of putty, followed by curing, and fine-sand with 320 sandpaper to complete the repair.

REVIEW QUESTIONS

1. What are the two primary reasons why cars are painted?
2. List six qualities built into a paint finish.
3. What does repairability mean?
4. What is a precoat coating?
5. How does a painter determine the VOC content of an as applied paint product?
6. What types of products are listed on a paint company's Low VOC wall chart?
7. What is the purpose of a primer?
8. What is the purpose of a primer–surfacer?
9. List several positive characteristics of a primer–surfacer.

UNDERCOAT MATERIALS AND APPLICATION

10. When should the metal-conditioning operation be done in the repair sequence?
11. How thick should a film of conventional primer–surfacer be after application and sanding?
12. What is the recommended method for sanding conventional primer–surfacer?
13. What is meant by guide-coat filling and sanding? How is this repair technique done?
14. What is the purpose of glazing putty?
15. What is the purpose of a sealer?
16. What happens to a paint shop when a clean air inspector finds that it is using paint products in excess of VOC limits?
17. What type of primer must be applied on bare metal before waterbase primer–surfacer can be applied?
18. What is the purpose of an adhesion promoter?
19. What is the capillary attraction of an undercoat?
20. How does a paint shop determine the VOC content of a particular paint system in a compliant area? Explain how this determination is done.

CHAPTER 9

Automotive Topcoats

PURPOSE OF COLOR

The purpose of color in a paint finish is to beautify the surface painted and, with the help of undercoats, to protect the metal against corrosion. The appearance of a car is important to every car owner. One of the most important features that makes a car appealing to the owner is the beauty and gloss of the paint finish. A paint job in good repair makes a car look its best. The appearance of every factory-produced car, van, and truck is vital to the sale of that vehicle. People know that a good paint finish is essential to the lasting beauty and durability of a car.

WHAT IS COLOR?

Pigments have the ability to reflect and to absorb light rays. When a person sees blue, the pigment absorbs all the light rays with the exception of blue, which it reflects. When a person sees red, the pigment absorbs all light rays except red, which it reflects. When a person sees white, all the colors of the spectrum are reflected. When a person sees black, all the colors of the spectrum are absorbed. The eye is like a tiny radio receiver and picks up the waves of light as they are reflected.

> **NOTE**
> Complete descriptions of color appear in Chapter 14.

A Typical Gallon of Paint

A typical gallon of automotive paint consists of the following ingredients, as shown in Figure 9–1:

1. **Solids**
2. **Nonexempt VOC solvents**
3. **Exempt VOC solvents (dilutents)**
4. **Water**

The solids portion of the paint (what remains on the car after the paint dries) consists of the following components, each described in detail.

FIGURE 9–1. Components of a gallon of paint.

Resin or Binder

Resin is the principal film-forming part of a paint system into which all the other film-forming ingredients are added. The resin system of a paint is the primary ingredient that determines the identity or type of paint. The resin system provides the following qualities in a paint:

1. Hardness
2. Durability
3. Gloss
4. Adhesion
5. Water resistance
6. Chip resistance

Added features of urethane enamels are:

7. Super high gloss
8. Chemical stain resistance
9. Greater chip and scratch resistance

Pigments

Pigments perform two important functions for a paint film:

1. They provide the color to the paint.
2. They make a paint film more durable to the extent to which they screen out the ultraviolet light of the sun.

Metallic Particles and Mica Flakes

Metallic particles are most often constructed of aluminum flakes and are used in metallic colors to achieve more glamorous color effects with depth. Mica flakes are used to produce the most glamorous colors available.

Ultraviolet Screener

The purpose of a screener, as the name implies, is to stop ultraviolet light at the point of contact with the paint film. Without a screener of this type, a paint film suffers from fading, chalking, crazing, and general paint film breakdown much more quickly.

Extenders

Extenders, as the word implies, give flexibility to a paint finish.

Other Additives

Other additives are used in the manufacture of automotive colors, depending on the color. Metallic colors sometimes use antimottling agents. These agents make possible a more trouble-free paint material by preventing the mottling condition when the paint is applied. Sometimes special agents are used to prevent gloss. They are called **flattening agents** and are used when repainting interior parts of cars in certain locations.

Solvents

Solvents may be considered the backbone in the refinish industry because they **serve so many purposes,** for example, in the:

1. **Manufacture of paint.**
2. **Transportation and storage of paint** (when paint is made and packaged).
3. **Reduction and application of paint.**
4. **Cleanup of paint application equipment.**

In other words, **without solvents, automotive refinishing would be impossible.** Chapter 7 covers solvents and their use in greater detail.

In paint manufacture and application, solvents are used in three forms:

1. **Exempt solvents** are made of compounds that are proven to be non-causing agents of ozone or pollution. They are the **nonharmful solvents.**
2. **Nonexempt solvents** are made of compounds that are known to be or are suspected of being **toxic or potentially toxic,** or causing other environmental problems. They are known as the VOCs, but they are required by the industry to make paint. The federal government seeks to regulate VOC solvents.
3. **Water** is used more and more by paint manufacturers in paint formulations to aid the solvent systems. Several Low VOC products are waterbased and are proving to be successful in the refinish trade. Pure or condensed water is used to aid reduction in refinishing and is available under a specific part number. Ordinary tap water may be used during cleanup after applying waterborne paint finishes. However, the use of ordinary tap water should be followed by a recommended solvent cleanup.

TYPES OF AUTOMOTIVE TOPCOATS

All automotive topcoats are divided into two general classes: solid colors and metallic colors.

Solid-Color Topcoats

Solid colors are made with a high volume of opaque pigments. Opaque pigments block the rays of the sun and absorb light in accordance with their type of color. Features of solid colors include the following:

1. Opaque pigments do not let light pass through them.
2. Ultraviolet light is blocked from the paint film.

Metallic-Color Topcoats

The first metallic colors were designed with a high volume of aluminum particles, special pigments that provide transparency, and a clear resin. **Aluminum metallics** are still very much in use.

The most recent metallic effect was developed with the use of **mica particles,** special pigments that provide transparency, and a clear resin. Features of metallic colors include the following:

1. More sales appeal and more depth.
2. Allow light to penetrate beyond the surface.
3. Light penetration, reflecting off or through the flakes and passing around pigment particles of varying density, produces the final color shade and effect.
4. Change in color when viewed under different types of light.
5. Change in color when viewed from different angles.

Aluminum flakes do not allow light to pass through them. They act like tiny mirrors to reflect light, and this feature provides glamour and brilliance to single-stage topcoats (see Figure 9–2).

Light is affected by mica in two ways (see Figure 9–3):

1. Light reflects off a mica flake at specific coating thicknesses to produce a pearlescent color effect in red, green, etc.
2. Transmitted light passes through the mica flake and through each specific coating thickness on the mica to produce additional pearlescent effects of many colors like red, green, blue, etc.

Mica possesses the following characteristics:

1. Micas produce very bright colors with transparent depth.
2. Micas produce high metallic, pearlescent tones when viewed under direct bright light.
3. Under weak light, micas appear as nonmetallic colors.

FIGURE 9–2. Description of a single-stage metallic color.

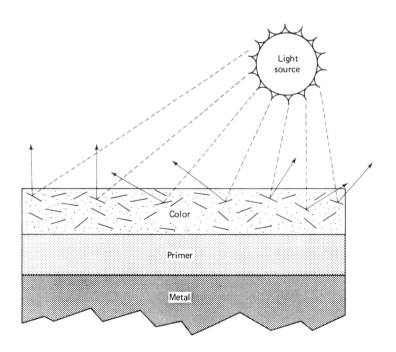

FIGURE 9–3. Cross-section of a typical mica flake with titanium dioxide (TiO$_2$) coating, which illustrates how a pearlescent effect is achieved. The mica flake shows how light reflects off its many surfaces and how light is transmitted through a flake to cause additional pearlescent effects. Chemical coatings are applied to mica flakes in precise thicknesses, and this process controls the quality of the designed color. Mica flakes are available in many colors and are used in specific color formulations. (Courtesy of BASF Corporation.)

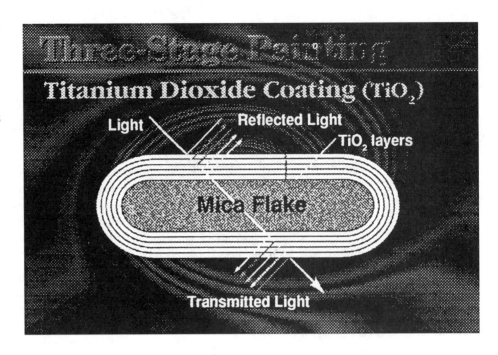

4. Mica flakes are very heavy and settle quickly. An application gun with an agitator cup is recommended for their application.
5. Color match is greatly influenced by the number of coats applied.
6. While micas can be used with any paint system, they are used mostly in the multicoat systems such as basecoat/clearcoat, tricoat, and quadcoat.

Transparency is a characteristic of metallic finishes that allows rays of light to penetrate beyond the surface so that the metallic particles can be seen distinctly. **Translucence** is the property of allowing rays of light to penetrate into the paint film without permitting the aluminum or mica flakes to be seen distinctly. Translucence is a semitransparency. **Refraction** of light is the breakup of a light beam into its component parts. In sunlight, light rays include all the colors in a rainbow: red, orange, yellow, green, blue, indigo, violet, etc.

TOPCOAT CLASSIFICATIONS

All topcoats are classified in the following categories:

1. Single-stage solid-color topcoats.
2. Single-stage metallic-color topcoats.
3. Single-stage, metallic, multicolored topcoats.
4. Multistage topcoats: basecoat/clearcoat finishes.
5. Multistage, multicolored topcoats: tricoat finishes and quadcoat finishes.
6. Specialty coatings.

Single-Stage Solid-Color Topcoats

The term *single-stage finish* means that the colorcoat is of a single-film construction, although it may be applied in several coats. This film averages about two mils in thickness. Solid colors were the most popular until the development of metallic colors.

Single-Stage Metallic-Color Topcoats

Single-stage metallic color means that (Figure 9–2) the colorcoat is of a single-film construction, although it may be applied in several coats. This film averages about 2 mils in thickness. Single-stage metallics became popular when special pigments, clear resins, and metallic flakes were first developed. They remained the most popular until the development of multicolor and multistage finishes.

Single-Stage, Metallic, Multicolored Topcoats

Single-stage metallic colors become **multicolored when flakes are added** to them. The colors may be applied in several coats and are about 2 mils thick. The mica flakes add beauty and brilliance to the finish.

Multistage Topcoats

The term *multistage* applies to basecoat/clearcoat finishes (Figure 9–4) because two stages are involved to produce one final colorcoat. In the first stage, the basecoat appears with a low gloss. The second stage is the clearcoat application, which produces the high gloss and enhances the beauty of the color. The thickness of the complete colorcoat varies with the minimum and maximum film thicknesses shown in Figure 9–5.

Multistage, Multicolored Topcoats

Multistage, multicolored topcoat systems are basecoat/clearcoat systems in which the basecoat portion is a multicolored topcoat. Examples are the tricoat and quadcoat finishes. **In tricoat finishes:**

1. A basecoat color is applied over undercoats.
2. A midcoat color with special mica pigments is applied over the basecoat.
3. A final clearcoat is applied over the midcoat.

In quadcoat finishes (at the factory):

1. A basecoat color is applied over undercoats.
2. A clearcoat is applied over the basecoat and dried.
3. A midcoat color is applied over the clearcoat.
4. A final clearcoat is applied over the midcoat.

In field repair, quadcoat finishes are repaired with a tricoat repair procedure as follows:

1. The basecoat color is applied.
2. The midcoat color application is determined by the let-down panel procedure.
3. A final clearcoat is applied over the midcoat.

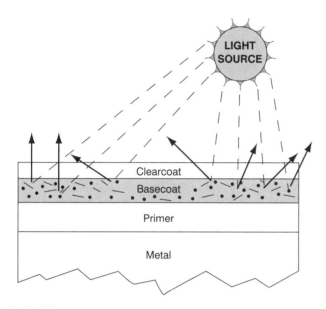

FIGURE 9–4. Description of basecoat/clearcoat finishes.

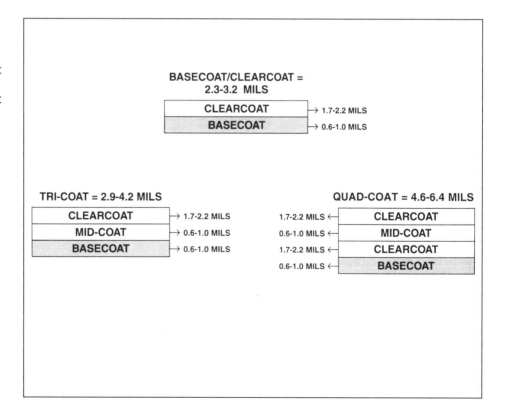

FIGURE 9–5. Film thickness comparison of basecoat/clearcoat and multicoat color finishes. Most paint technicians are familiar with basecoat/clearcoat paint technology. The multicoat color is sandwiched between the basecoat and clearcoat and is commonly referred to as the "midcoat" or "sandwich coat." (Reprinted with permission of General Motors Corporation.)

Specialty Coatings

Specialty coatings are a special class of coatings consisting of a long list of products that serve special needs on a car. Specialty coatings include:

1. Chip resistant coatings
2. Flexible plastic parts coatings
3. Low-gloss finishes
4. Vinyl-top finishes
5. Uniform finish blenders, etc.
6. Acrylic-lacquer coatings

AMERICAN CAR FACTORY PAINT USE

Until the mid-1980s, most car factories used single-stage finishes of solid or metallic colors on most production cars. In the mid-1980s, most car factories started using basecoat/clearcoat finishes more and more. By the late 1990s, most car factories adopted a multistage system of one form or another. Millions of factory painted cars on U.S. roads have been painted with one of the following paint systems:

1. Single-stage acrylic enamel (solid and metallic colors).
2. Single-stage acrylic lacquer (solid and metallic colors).
3. Single-stage acrylic urethane enamel (solid and metallic colors).
4. Multistage acrylic enamel (basecoat/clearcoat systems).
5. Multistage polyester urethane enamel (basecoat/clearcoat systems).

Factory Thermosetting Enamel Topcoats

Although each car factory identifies its enamel under different company names, the basic enamel system as used by the American car factories is thermosetting acrylic enamel. The following describes how a factory thermosetting enamel system works:

1. Car bodies (or complete car shells) are painted as required with enamel. On basecoat/clearcoat finishes, the color is applied and, after a short interval, the clearcoating is applied.
2. The bodies then proceed through a factory high-bake oven system.
3. At a certain temperature, some solvents are driven out of the paint.
4. At a higher temperature, surface tension is created on the paint surface, similar to that on a glass of water, producing a high gloss.
5. At this temperature, the paint film sets up or hardens and cures very quickly and the gloss remains. This paint cannot be redisolved.

Field Paint Repair Materials

Field paint repair materials can be divided into two categories. Low VOC paint systems are designed for the most highly regulated areas, which have the most severe pollution in the country. The special low VOC paint systems are required in California. Paint systems for the rest of the country are regulated by the National Rule VOC Limits, which are shown in Table 1–1, Chapter 1. In this rule, paint suppliers must meet VOC limits before packaging the paint. The system aids painters in selecting the correct paint.

All automotive colors are described by their general construction, as shown in Figures 9–2 through 9–5. To simplify a discussion of these same general classes of color systems, the following field paint repair color systems are available:

1. **Air-dried alkyd and acrylic enamel colors** (single-stage). Acrylic enamel colors may be catalyzed.
2. **Acrylic urethane enamel colors** (single-stage) require a hardener. This is the most popular of the single-stage enamel colors.
3. **Urethane enamel colors.** Some types require a hardener and some types do not.
4. **Basecoat color finishes** are very popular and are used almost exclusively by OEM. Basecolors are available in the following different paint technologies:
   a. Acrylic urethane type
   b. Acrylic enamel type
   c. Polyester acrylic type
5. **Clearcoat finishes** are very popular. They are used principally as a finish topcoat over basecoat colors. Available clearcoat types are the following:
   a. Acrylic urethane enamel
   b. Acrylic enamel
   c. Polyester acrylic enamel
   d. Nonisocyanate
6. **Waterbase paint systems** are least hazardous to the environment, but they are not widely used because of slow drying characteristics. They are used principally in the most highly regulated areas.

Lacquer paint systems for single-stage and clearcoats will disappear almost completely because lacquer systems cannot meet federal regulations. They may be used, however, as specialty coatings.

All leading automotive paint companies have several paint technologies for painters to meet the National Rule and the toughest quality standards imposed by the federal government. Each paint company distributes several types of conventional solventborne paint systems as well as an excellent waterbase paint system. All paint systems are designed to meet federal regulations and various warranties.

LOW VOC PAINT SYSTEMS

All major paint companies have low VOC paint systems available for all areas of the country. No two paint companies' systems are alike. Each company has special lines of undercoats and colorcoats that are guaranteed when they are applied by a qualified or certified painter in accordance with company guidelines.

Space in this book does not allow full coverage of all low VOC paint systems available in the trade. Only some systems are covered. To become thoroughly familiar with any specific waterbase or other low VOC paint system, it is strongly recommended that you enroll in and attend the training center of a major paint company of your choice. Contact your local jobber for enrollment information about a paint company training center that covers the low VOC paint system technology in which you are interested.

DESCRIPTION OF SINGLE-STAGE METALLIC COLORS

Standard Shade of Single-Stage Metallic Colors

Figure 9–6 is a picture of the uniform dispersion of aluminum flake and pigments. The small dots in the picture represent pigment particles. The lines that look like little logs are a cross-section of metallic particles. The concept of metallics is easier to explain in this manner.

Each metallic particle is like a tiny mirror. Note that the metallic particles are positioned in different directions. They are about equally spaced. On any given line, the metallic particles rotate. They point in all directions of a 360° circle and at the same time are distributed very uniformly throughout this film. This description is a standard color shade of metallics.

Light Shade of Single-Stage Metallic Colors

Figure 9–7 shows the flat orientation of metallics. They lie in a plane almost parallel to the surface. Metallic flakes oriented as shown reflect the greatest amount of light, like tiny mirrors, and thereby produce the lightest shades of that color. Another factor that helps to produce light shades of color is that the metallic particles, because they are flat, cover a high percentage of the pigments.

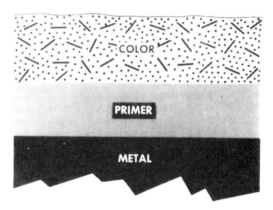

FIGURE 9–6. Standard color shade of metallic color. The uniform dispersion and density of aluminum flake and pigment particles shown result in a standard color shade. (Courtesy of General Motors Corporation.)

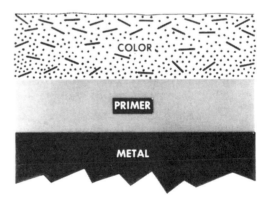

FIGURE 9–7. Light color shade of metallic color. An accumulation of aluminum flake dispersed nearly horizontally at the top of the paint film and obscuring most pigment particles beneath results in a light color shade. (Courtesy of General Motors Corporation.)

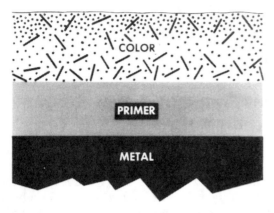

FIGURE 9–8. Dark color shade of metallic color. Dense flotation of pigment particles near the surface combined with aluminum flake dispersed nearly perpendicularly results in a dark color shade. (Courtesy of General Motors Corporation.)

Dark Shade of Single-Stage Metallic Colors

Figure 9–8 shows the perpendicular orientation of metallics. In a wet-coat application of single-stage metallic colors, the metallic orientation may be described as shown. Notice that a high percentage of particles lie in a plane almost perpendicular to the surface. This condition causes a maximum of light penetration into the film and less metallic reflection. This condition exposes a higher-than-normal percentage of pigments at the surface and thus produces more richness or darkness of color shade.

PAINT COMPATIBILITY

Paint compatibility is the ability of paint coating to work well with other coatings when applied on them or under them during refinishing. Automotive paint coatings and solvents are designed by a paint supplier for a specific purpose. Some products are designed to be used in several different lines of paint made by the same paint company. This statement does not mean that the products can be used universally with other company products. The modern automobile requires several different paint systems to service the paint on the complete car.

In the refinish trade the painter cannot always tell just by looking at a car what type of finish it has. Sometimes painters encounter cars that have been repainted since production. A painter must identify the type of paint on a car before attempting repairs.

Nonbasecoat or single-stage colors and basecoat paint repair systems are not interchangeable. Single-stage refinish colors should not be used to repair basecoat/clearcoat finishes, and basecoat colors should not be used to repair single-stage colors.

A painter must know three pieces of information to identify a proper paint repair procedure for any car:

1. Model year, type of car, and series.
2. Paint code and/or WA number (if GM).
3. Type of paint on car.

To find the body number plate and color identity, see Chapter 2. If in doubt about the paint type and code, check with the paint jobber or perform the test to determine the type of finish outlined below.

NOTE

As a reminder to the painter, basecoat colors are marked boldly on the container as follows: **THIS IS A BASECOAT COLOR.**

TEST TO DETERMINE THE TYPE OF FINISH

Generally, the painter can determine the type of finish on a car with the following test. Using acrylic-lacquer removing solvent (or a good grade of lacquer thinner), soak a section of clean cloth with the solvent and rub a spot in an area to be repaired.

1. If the color washes off readily onto the cloth, the finish is acrylic-lacquer.
2. If the finish does not wash off, even with vigorous rubbing, the finish is enamel.
3. To determine if the finish is a single-stage enamel or an OEM basecoat/clearcoat finish, perform a sanding test as follows:
    a. Use No. 220 or No. 320 sandpaper and dry-sand over a small spot in the area to be repaired.
    b. Check the sanding residue particles:
        (1) If the sanding residue is colored like the surface, the finish is a single-stage enamel color.
        (2) If the sanding residue is whitish, powdery, and clear, the finish is a basecoat/clearcoat finish.

PAINT THICKNESS LIMITS

All car factories and the refinish trade are concerned with the durability and appearance of all cars. Factories have certain standards and minimum and maximum thickness limits to guide the refinish trade. It is the job of the paint suppliers and refinish trade to learn about these guidelines. For this reason, paint thickness gauges have become a required tool for the painter.

Minimum Paint Thickness Limits

Since the basecoat/clearcoat finishes became so popular in the mid-1980s, certain types of repairs can be made on these finishes (and on monocoat, single-stage finishes) without refinishing. These repairs are made possible by removing some problem conditions from the paint with special **finesse repair systems** that remove some of the paint film while maintaining paint film integrity and acceptable film thickness. To perform these repairs properly within factory specifications, the paint shop must be guided with an approved paint thickness gauge. (Finesse repair procedures are described in Chapter 13.)

Car company paint engineers generally agree that paint finishes can serve their purpose when paint removal operations are limited to the following:

1. **On basecoat/clearcoat finishes,** .5 mil may be removed.
2. **On monocoat (single-stage) finishes,** .3 mil may be removed.

If repairs involve removal of more paint than recommended above, additional color or clear must be added to maintain paint film integrity and acceptable film thickness.

Allowable Paint Thickness Limits

Generally, a factory paint finish can be re-colorcoated once with a complete colorcoat finish and without difficulty if paint supplier guaranteeable paint finish materials and procedures are followed. Figure 9–5 compares typical paint film thicknesses of factory basecoat/clearcoat and multicoat paint finishes. Minimum thickness range figures are on the left, and maximum thickness range figures are on the right.

1. As shown in Figure 9–5, a complete refinish of a basecoat/clearcoat system adds 3.2 mils of topcoats, for a total of 6.4 mils. This total is within maximum thickness limits.
2. A complete refinish of a tricoat finish adds 4.2 mils of color and clearcoats, for a total of 8.4 mils. This total is also within maximum thickness limits.
3. A complete refinish of a monocoat (single-stage) finish would involve the following. Factory monocoats range in thickness from 4.1 to 5.1 mils, with about half being undercoats and half being colorcoats. The average factory film thickness is about 4.6 mils. Adding 2.3 mils of refinish color to 4.6 mils of original finish brings the total film thickness to 6.9 mils. This total is well within maximum thickness limits.

Maximum Paint Thickness Limit

When excessive thicknesses (exceeding 15 to 20 mils) of paint are applied and when the finish is exposed for long periods in subfreezing temperatures and then in warm temperatures, alternately, checking and/or cracking problems may develop. When cars come to a paint shop for refinishing, the shop should check the paint thickness on every panel before starting. Paint thicknesses at 15 mils or higher should be reduced by sanding or removed entirely by chemical stripping (paint remover). This fact should be entered on the repair order so the car owner is informed before he or she gives final approval.

> **NOTE**
>
> **The maximum paint thickness limit does not apply to the gravel chip areas of a car.** These areas range from about 12 to 24 inches in width along all lower areas of a car. A special antichipping coating measuring 15 to 20 mils thick is applied at the factory in these areas before factory colorcoating.

PAINT THICKNESS GAUGES

The purpose of a paint thickness gauge is to measure the thickness of the paint finish on a car. Two types of paint thickness gauges are available in the refinish trade:

1. The **electronic type** can be used on steel and aluminum substrates.
2. The **magnetic type** can be used only on steel substrates.

FIGURE 9–9. ETG-1 electronic paint thickness gauge. (Courtesy of Pro Motorcar Products, Inc.)

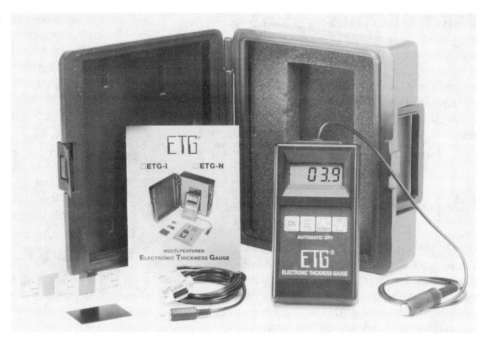

Electronic Paint Thickness Gauges

The best and most accurate paint thickness gauges are the electronic types because they produce the same accurate results every time they are used. If paint shops expect to do well in the refinish trade, they must have a good paint thickness gauge.

Figure 9–9 shows the ETG-1 electronic paint thickness gauge; Figure 9–10 shows the Elcometer 345 electronic paint thickness gauge. These gauges measure paint thickness on steel and aluminum substrates. They range from the most basic, which read only paint thicknesses, to the more sophisticated types, which have memory systems, can store paint readings, and can produce printouts like a personal computer. Several companies sell electronic gauges. For the availability of electronic paint thickness gauges, see your local paint jobber.

The ETG electronic thickness gauge in Figure 9–9 is a microprocessor electronic thickness gauge. The unit is powered by a battery. ETG's circuitry measures to .00001 of an inch (which is ten millionths of an inch) and automatically rounds off to .0001 (one tenth of a mil) for a stable digital display. To operate the electronic paint thickness gauge:

1. Press the on key and then release.
2. Hold the remote probe by the black rubber boot and lightly press the probe against the paint surface to be measured. Use enough pressure to prevent the probe from rocking.

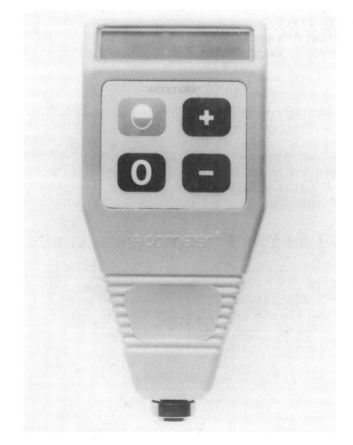

FIGURE 9–10. Elcometer 345 electronic paint thickness gauge. (Courtesy of Elcometer Instruments Ltd.)

The tool automatically measures and displays the film thickness in digital numbers to one-tenth of a mil.

3. The unit will turn off automatically after 90 seconds when not being used.

Please note the following when operating an electronic paint thickness gauge:

1. "   " is displayed in the measuring (run) and calibration (CAL) modes when the readings are not within the 0 to 40 mil (0 to 999 microns) range.
2. If the last digit alternates, the thickness is halfway between those values.
3. The decimal point is displayed when the gauge is set to read in mils.
4. Mil is an English unit commonly used in the paint industry. It equals one thousandth (.001) of an inch. The metric unit used is the micron, which is one thousandth of a millimeter (one millionth of a meter).

Magnetic Paint Thickness Gauges

Magnetic paint gauges are the least expensive gauges. They can measure paint thickness to within .1 of a mil. A mil is one one-thousandth of an inch, which is the industry standard. Magnetic gauges can be used only on steel substrates. No gauges are available in the trade for measuring paint thickness on plastic substrates. Several magnetic paint thickness gauges are available in the refinish trade. Two popular types are:

1. The **pencil gauge,** so-called because of its shape and size (see Figure 9–11).
2. The **banana gauge,** so-called because of its shape (see Figure 9–12).

The Pro Gauge II shown in Figure 9–11 is a popular, low-cost magnetic thickness gauge.

1. The gauge has a permanent magnetic center shaft on which two scales are mounted:
   a. The top scale is used when measuring paint thickness on horizontal or flat surfaces like roofs, deck lids, or hood panels.
   b. The side scale is used when measuring paint thickness on vertical surfaces or sides of a car, like doors and sides of fenders, quarter panels, etc.
2. The gauge housing is made of dent-resistant thermoplastic.
3. A durable, stainless steel spring attaches the magnetic shaft to the housing.

For the availability of a pencil paint thickness gauge, consult your local paint paint jobber.

To operate the Pro Gauge II paint gauge:

1. Hold the gauge very loosely by the cap. Allow the magnet to self-center to ensure accurate, repeatable readings. If the gauge is held by its body, it will not be perfectly square to the panel and thus will give erratic, higher readings.
2. Take three readings. After taking the first reading to learn the approximate thickness, take two more readings but pull the gauge very slowly as it nears the first reading.
3. The top scale is used on horizontal surfaces such as roofs, hoods, and deck lids.
4. The side scale is used on vertical surfaces such as doors and the sides of fenders.
5. Long lines indicate the thickness value printed next to the scale. On mil gauges, these lines are even numbers such as 2, 4, 6, etc.
6. Short lines indicate odd numbers such as 1, 3, 5, etc.

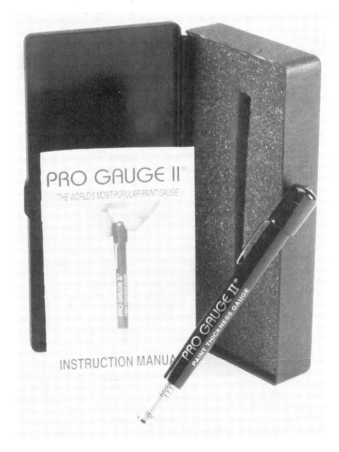

FIGURE 9–11. Pro Gauge II "pencil" magnetic paint thickness gauge. (Courtesy of Pro Motorcar Products, Inc.)

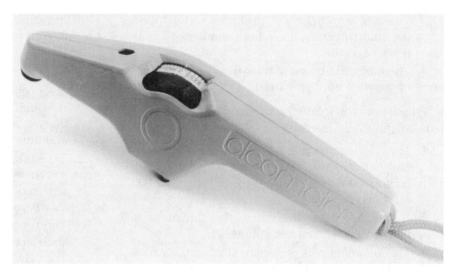

FIGURE 9–12. The 211 "banana" magnetic paint thickness gauge. (Courtesy of Elcometer Instruments Ltd.)

With its dent-resistant housing and corrosion-resistant components, the Pro Gage is nearly maintenance-free. Store the gauge in its case to keep it clean and safe. Avoid extreme heat. Do not leave the Pro Gauge II on car seats or on instrument panels.

The Pro Gauge II will last indefinitely under normal use. It has no wearing parts, and neither its high-energy magnet nor its specially designed stainless steel spring will lose calibration if the cautions listed below are observed:

1. **Inspection:** Inspect the magnet closely because it attracts small rust particles and metal filings.
2. **Cleaning:** Particles can be wiped off with a clean damp cloth or blown off with low-pressure air. Use only water for cleaning. Wipe gently to prevent damage to the scale.

CAUTION

When letting someone else use the gauge, instruct them to observe the following warnings:

1. **Do not pull off the cap.**
2. **Do not pull the scale out past the stop warning.**
3. **Do not touch the magnet to any other magnet.**

Doing any of the above will void the warranty.

REVIEW QUESTIONS

1. What is color?
2. What is the principal film-forming part of a paint system?
3. What is the purpose of solvents in paint manufacturing?
4. What are two primary functions of pigments in a color?
5. What is the purpose of an extender?
6. Of what material are most metallic particles made?
7. Describe solid colors. List several of their characteristics.
8. Describe metallic colors. List several of their characteristics.
9. What is transparency?
10. What is translucence?
11. What does the word *thermosetting* mean?
12. With what basic color systems are American-made cars painted at the factory?
    a. Chrysler
    b. Ford
    c. GM
13. How do acrylic enamels dry?
14. Explain the orientation of metallic particles and pigments in a standard color shade.
15. Explain the orientation of metallic particles and pigments in a light color shade.
16. Explain the orientation of metallic particles and pigments in a dark color shade.
17. What is the ideal temperature range for spraying automotive finishes?
18. What is the lowest temperature below which spray painting is not recommended?
19. What is meant by paint compatibility in automotive refinishing?
20. Explain how to determine if original paint being tested is acrylic lacquer, acrylic enamel, or basecoat/clearcoat finish.
21. What is a paint thickness gauge? How does it work?
22. What is a single-stage color?
23. Describe a two-stage color.
24. What is a three-stage color and what are its components?
25. What are the two types of paint thickness gauges?

CHAPTER 10

Complete Vehicle Refinishing

INTRODUCTION

Before preparing colors for application, certain basic preparation operations should be done first so that the spray painting operation can proceed smoothly from start to finish without interruption. First, all tools and equipment for mixing, handling, and straining paint and refilling the paint cup should be ready, as follows:

1. Have a large covered container with a sufficient supply of reduced paint ready for use.
2. The car must be inspected and approved as ready for painting.
3. The spray booth must be clean and ready for use.
4. The spray gun and all equipment must be ready.
   a. The air regulator should be checked to see that it works properly.
   b. Check the regulator drain to ensure that no water is in the unit.
   c. Have a place to hang the spray gun while refilling the paint cup.

TOOLS AND EQUIPMENT NEEDED IN THE PAINT MIXING ROOM

1. Mix paint in a room or area that meets OSHA standards.
2. Clean the workbench. An unpainted metal surface is easy to keep clean and is easy to maintain.
3. The room or area should be properly lit.
4. The room should be continuously ventilated.
5. All metal benches and shelves should be grounded.
6. Special tools should hang at the front of the workbench:
   a. Pliers (for opening cans, etc.).
   b. A hanger for hanging the spray gun while refilling the paint cup with primer, etc.
7. A supply of clean shop towels and a roll of paper towels.
8. A small covered glass or cup for soaking and cleaning the air cap.
9. A 1-inch paintbrush for spray equipment cleanup.
10. Another 1-inch paintbrush cut down to $\frac{5}{8}$ inch in length. This brush is needed for additional cleanup.
11. A woven plastic scouring pad for cleanup.
12. A supply of paint strainers.
13. A paint strainer stand.
14. A paint shaker should be outside the paint mixing room. (This step is a great time saver. Every paint shop should have a paint shaker.)
15. A supply of paint paddles. These paddles should be about $1\frac{1}{2}$ inches wide by $\frac{1}{8}$ inch thick and 12 inches long. Aluminum or brass paddles are best. They should be flat-bottomed.
16. Masking tape.
17. A screwdriver and a hammer.

Every paint shop should be well stocked with medium-grade thinner, which is used as the primary cleaning solvent, whether painting in lacquer or enamel. Acrylic-enamel reducers that should be kept on hand for immediate use by the gallon are:

1. Slow-grade color reducer
2. Medium-grade color reducer
3. Fast-evaporating reducer
4. Acrylic enamel retarder

153

The best-grade primer (and catalyst) and the best-grade primer–surfacer should also be on hand ready for use.

All the items listed should be stored in a paint mix room for immediate use. When these items are not on hand, they must be procured from a store, which takes time. Time cuts down on jobs, and time is money in a paint shop.

CHECKLIST FOR PREPARING A CAR BEFORE COLOR COATING

Check, Repair, and Complete All Masking

Before giving the final go-ahead for masking, inspect all masking with a careful eye to quality. Start at a given point on the car, make a complete circle around the car, and end in the spot where inspection started. Look over every inch of masking. Make sure by visual inspection and by feeling specific joints that they are down tightly, flush, and smooth. Be sure that all masking on the side windows, windshield, and back window is sealed, with no open joints.

Check and Prepare the Spray Booth As Required

There are two basic ways to spray paint. Spraying enamel finishes requires much cleaner sanitation conditions than does spraying lacquers.

1. **Lacquer application.** Spraying acrylic lacquers does not require as clean a spray booth as for spraying enamels. However, the spray booth should be kept as clean as possible because less booth preparation is required when an enamel job is to be done. Floors should be kept clean by sweeping and watering down periodically.
2. **Enamel application.** For enamel application, the walls, ceiling, and floor should be cleaned by hosing them down and blowing them dry. All booth intake filters should be blown clean and vacuumed (if not replaced). All openings in filter doors, walls, ceiling, and all booth joints should be checked with a positive air flow and sealed as required. All door seals lose their sealing qualities as they age and should be repaired or replaced as required. Air exhaust filters should be replaced according to OSHA standards. **No flying dirt particles whatsoever should enter the booth** when an enamel job is being sprayed.

The previously mentioned maintenance operations do not have to be done with every paint job but should be done as part of a regular maintenance program two or three times each year. However, the spray booth should be hosed down before every enamel paint job.

PREPARATION OF COLOR

Automotive colors are prepared in two basic ways for refinishing: hand mixing and machine mixing.

Hand Mixing

The most thorough way to mix the paint by hand is as follows:

1. Open the can carefully. Use the proper can opener.
2. Make a pouring spout on the can with masking tape. Use several sections of $\frac{3}{4}$-inch tape if wider tape is not available.
3. Pour about half of the paint into a second clean container.
4. Mix each half of the container thoroughly using a wide, flat-bottomed paddle.
5. Combine both halves in the original container and mix them thoroughly. Remove the tape spout when finished, and the can is clean.

This method results in a faster and more thorough way of mixing paint with less mess.

Machine Mixing (Use of a Paint Shaker)

Not many painters complete the following three steps, but they should:

1. Open the can of paint and, with a flat-bottomed paddle, loosen all pigments at the bottom, including pigments trapped at all corners. Use the edge of the paddle to scrape pigments loose at the lower corner edges.
2. Using a suitable clean paintbrush, return paint from the paddle to the container. Install the cover on the can by positioning the cover, covering the can with a suitable cloth, and tapping the cover down tightly with light hammer blows.
3. Put the can of paint on a paint shaker (see Figure 10–1 for an example) and secure firmly in place. Check the shaker operation for 1 or 2 seconds, and stop the machine. Recheck the

COMPLETE VEHICLE REFINISHING

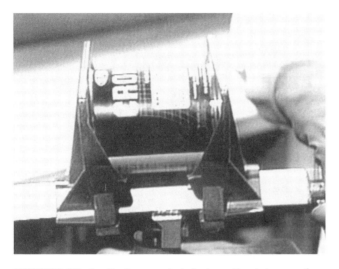

FIGURE 10–1. Cyclone paint shaker. Containers of paint are held horizontally in retaining jaws that secure containers at each end while shaking. (Courtesy of DuPont Company.)

tightness of the clamps, can, etc., and when all is in order, turn the shaker on for the required amount of time. About 5 to 10 minutes of shaking does a thorough job.

Loosening the hard, settled paint in the can before shaking cuts down mixing time considerably. To speed up mixing time even more, some painters throw a small amount of solvent-washed tacks, small screws, or bolts into the can before putting it on the shaker. The objects are caught later in a paint strainer. The painter washes and rinses them in thinner and saves them for the next use.

COLOR REDUCTION (CONVENTIONAL METHOD)

Reducing color by the conventional method means reducing color according to label directions. Label directions change every so often because paint technology changes. If preparing 1 quart or less of reduced paint material, prepare the reduction in an extra container, mix it thoroughly, and then strain the reduced material into the regular paint cup.

USE OF HARDENER (CATALYST) IN ACRYLIC AND URETHANE ENAMEL

Chapter 9 explains the growth and increasing popularity of catalyzed acrylic enamels. All paint companies are uniform on the amount of catalyst to use per gallon of acrylic and urethane enamel, which is 1 pint, before reduction. A gallon has 8 pints, so the ratio is 8 to 1. Follow label directions.

Most painters catalyze only the required amount of enamel to be used. Then they reduce and use the amount prepared. Painters reduce more paint only as required. Once acrylic enamel is catalyzed, it must be used within the required time indicated in the label directions. Otherwise it spoils.

Procedure for Reduction

Figure 10–2 shows a graduated (proportionally marked) paint mixing paddle available through R-M paint jobbers, who are subsidiaries of the BASF Corporation. Equivalent paint paddles are available at all paint jobbers. Familiarity with the paddle can be achieved by checking with the supplying paint jobber. Figure 10–2 indicates how a typical Glasurit 21-line paint material is reduced when the mixing ratio is 2:1 and 10 percent (translated this means 2 parts color, 1 part hardener, and 10 percent reducer).

1. The first 2 stands for color. **Always add color first.** With the paint stick held vertically in the paint cup, add thoroughly agitated 21-line color to the first 2 line.
2. While holding the stick still, add hardener to the 2 in the second column, as shown. **Hardeners are added second.**
3. Also while holding the stick still, add the 10 percent reducer to the 2 in the third column. (See Figure 10–3.) **Reducers are added last.**
4. Mix the reduction thoroughly.
5. **Check the viscosity** of the reduced paint material (see Figure 10–4).
6. **Strain the reduced paint material** into the paint cup (see Figure 10–5).

> **NOTE**
> Follow this procedure to prepare 1 cup of reduced material. When painting a complete car, the required amount of material is prepared all at once in a suitable container. Then, as each quart (or two) is refilled, the reduced paint is strained at that time. Each reduction is made in accordance with the size of the job, that is, size of the car, type of paint, and amount.

FIGURE 10–2. Graduated Glasurit paint mixing paddle supplied by R-M paint jobbers (R-M and Glasurit are subsidiaries of BASF.) (Courtesy of BASF Corporation.)

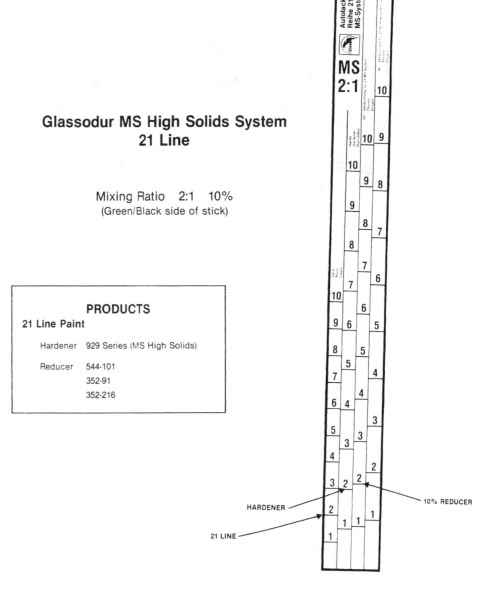

**Glassodur MS High Solids System
21 Line**

Mixing Ratio   2:1   10%
(Green/Black side of stick)

```
                PRODUCTS
21 Line Paint
    Hardener   929 Series (MS High Solids)
    Reducer    544-101
               352-91
               352-216
```

CHECK THE SPRAY OUTFIT BEFORE PAINTING

1. Attach the cup to the spray gun and tighten the attachment securely.
2. Check the cup lid vent hole with a pointed tool to make sure that it is fully open.
3. Check the spray gun by holding it horizontally as when spraying a roof panel to determine if it will leak at the gasket. If the cup is the old type and even remotely liable to leak, install a "diaper" around the cup lid. Simply wrap a several-inch-wide band of cloth or absorbent paper towel around the cup lid and secure with masking tape.

> **NOTE**
> Various modern spray guns have special gaskets and leak-proof lids that make possible spraying on horizontal surfaces without dripping. This precaution is not required on vertical surfaces.

4. **Adjust the spray gun for a full-open spray fan.** Check a flooding test pattern on a vertical wall. (This check could be on a section of the wall or spray stand suitably masked off for this purpose.) See "Spray Gun Adjustments" in Chapter 4.

COMPLETE VEHICLE REFINISHING

FIGURE 10–3. Pouring the reducer from a square container.

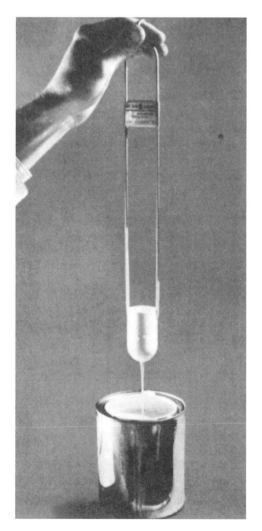

FIGURE 10–4. Straining reduced paint into paint cup. (Courtesy of PPG Industries.)

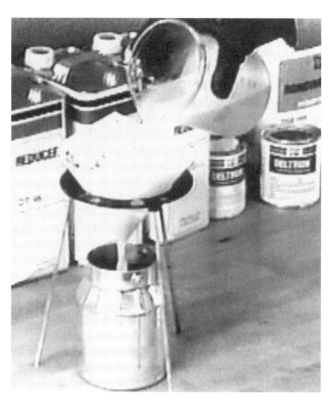

FIGURE 10–5. Checking viscosity.

5. Check and adjust the air pressure as follows:
   a. For acrylic lacquers, use 30 to 40 psi (at the gun). Allow for pressure drop (see Table 5–4), which could be plus or minus about 5 psi. The complete range is 25 to 45 psi.
   b. For enamels, use 50 psi for most acrylic enamels unless otherwise recommended in the label directions. The complete range is 50 to 70 psi for other types of enamels.

The best air pressure for any paint system is one that provides the best atomization at the lowest air pressure. The proper spray gun adjustment is one that provides a level and uniform flooded test pattern.

Before starting to apply color, be sure the following are in the booth or are easily accessible outside the booth:

1. Reduced color and paint paddle.
2. Paint strainers and strainer stand.
3. Cleaning solvent (medium thinner).
4. Cleaning equipment (pan, brush, and wiping rags).

HOSE CONTROL WHEN SPRAYING EXTRA-WIDE SURFACES

An item that a painter must be able to control at all times and that is not covered in Chapter 4 is the hose. He or she should be able to handle the hose and control it so that the hose does not interfere with the spray painting operation. Simply holding the hose with the opposite hand is not sufficient control, particularly when spraying the roof, the hood, or any large area. The best way to control the hose is to run it under one armpit, over the neck, and allow the hose to extend to the spray gun being held by the other hand. In this fashion, the hose is prevented from scraping on the edge of the roof or car panel when making an extended reach during paint application.

PAINT APPLICATION SYSTEMS

Figures 10–6 and 10–7 provide the painter with road maps about how to paint a complete car from start to finish. Figure 10–6 shows a popular procedure to follow when painting a car in a **cross-draft booth.** In a cross-draft booth, the air comes through the door filters or side filters and exits the booth through filters on the opposite side. Figure 10–7 shows a suggested procedure when painting a complete car in a **down-draft booth.** In this booth, the air comes into the booth through ceiling filters and exits the booth through floor filters before exiting the building.

PAINT APPLICATION PROCEDURE

The final operation before spray painting is to use an air blow gun and a tack rag to blow off and to tack-wipe the entire car (Figures 10–8 and 10–9).

1. With the spray booth exhaust on, start blowing off each surface, starting at the rear of the car and working forward.
2. Blow off, then tack-wipe each panel, overlapping each wiping area with the next one. Use new sections of the tack rag as the operation progresses.

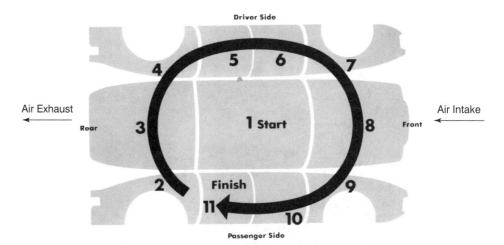

FIGURE 10–6. Spraying in a cross-draft spray booth. (Courtesy of DuPont Company.)

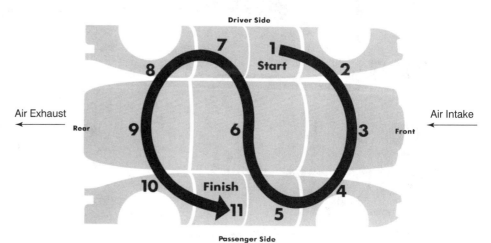

FIGURE 10–7. Spraying in a down-draft spray booth. (Courtesy of DuPont Company.)

COMPLETE VEHICLE REFINISHING

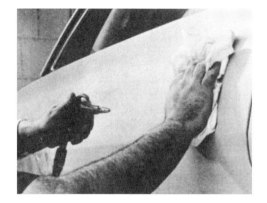

FIGURE 10–8. Dusting surface with a blow gun.

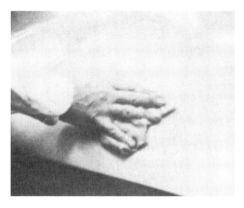

FIGURE 10–9. Tack wiping.

NOTE

A suitable bench or stable boxes should be on hand in the spray booth for painters who need them to reach all areas of passenger car roofs. Also, special ladders or scaffolding is required to paint van roofs and other high vehicles.

Roof Panel

With the paint spray cup full and a respirator in place:

1. Apply a banding coat at the windshield and back window.
2. Apply the first coat, starting at the near edge of the roof, by making a pass from left to right *and* from right to left with a medium wet application. Always start edges in this fashion.
3. Then, because of the limited angle of a *full* spray gun cup, overlap 60 percent to 70 percent and apply each succeeding pass working toward the center of the roof. As each pass is made, observe the paint for the proper gloss to show. Observation is done by looking at the applied paint from an angle with proper lighting in the background. This observation is an indication of medium wetness of application, which is desired.

Follow these fundamentals as closely as possible:

1. Maintain a uniform distance; do not waver the arm inconsistently. **Stay parallel to the surface.** Keep 6 to 8 inches from the surface.
2. Maintain a uniform speed for wetness of application.
3. Use adequate lighting. **Do not spray in the dark.**
4. **Keep the overlap uniform,** stroke to stroke.
5. **Keep as perpendicular as possible.**

When one side of the roof is finished, go immediately to the opposite side and, with benches in place, **continue applying paint, working from the center to the near side.** This system provides a wet-on-wet application that minimizes overspray. Apply banding as required at the front and rear. When a roof is painted in ideal fashion, it is uniformly wet in all areas. This uniform wetness can be seen from the reflected gloss. Refill the cup for spraying the next area.

Entire Hood and Tops of Front Fenders

With no blowing of air, tack-wipe the hood area to be painted (see Figure 10–10):

1. Apply a banding stroke along the base of the windshield. No banding stroke is needed along the front of the hood unless it is short and reachable.
2. Apply the first medium-wet coat starting at the near edge over the fender by making a pass from left to right and from right to left with a medium-wet application. Always start edges in this fashion.
3. Then, because of the limited angle of a full spray gun, overlap as required and apply each succeeding pass working toward the center of the hood. Observe each pass for the proper gloss to show.
4. Working from the opposite side, start the spray application along the center of the hood and continue the application toward the near side of the hood.
5. Then proceed to the front of the car and, starting at one side, complete the application of a

FIGURE 10–10. Spray painting a hood panel. (Courtesy of DuPont Company.)

single coat to the front of the car. Use about a 4-inch overlap when applying color to new areas. Apply color to valance panels if present below the bumper. The application of sound fundamentals is what produces the highest quality results.

Entire Rear Deck Area

Work in a similar fashion to painting the roof and hood. Tack-wipe but do not air-blow the rear deck area. Check and refill the spray gun cup as required.

1. Apply a banding stroke along the base of the back window to the center.
2. Standing at the side of the rear fender, apply the first medium-wet coat, starting at the near edge. Make a medium-fast pass from left to right and from right to left as the first pass.
3. Continue applying single coats with proper overlap to the center of the rear deck.
4. Switch sides and, continuing from the center, apply single coats as directed toward the near side of the car. Observe the reflection of the applied paint to keep it wet at all times. **Move only as fast as the reflection shows wetness of application.**
5. Proceed to the rear of the car and, starting at one side, complete the application of a single coat to the rear of the car. Use about a 4-inch overlap when blending color to previously applied color. Apply color to the valance panels below the bumper, if they are present.

Sides of the Car

Tack-wipe but do not air-blow the sides to be painted. Check and refill the spray gun cup as required.

1. Apply banding strokes vertically down the sides at a comfortable distance toward the front of the car.
2. Starting at the top or bottom, apply a double pass, from left to right and from right to left, to start an edge. Continue to apply a single coat (at a 50 percent overlap) to the reachable area until the area is fully coated.
3. Move to the next forward area and make another vertical banding pass at the area that can be reached comfortably. Using a 4-inch overlap (see Figure 10–11), fill this area with a single-coat paint application, making sure that the paint is applied wet and that you observe what is applied.

CAUTION

When in doubt about whether to apply another coat of paint to a questionable area, it is best not to apply it. The reason is that too much paint will result in a run. A missed area can always be color-coated later when it is determined that more paint is needed. **It is up to the painter to know specifically and exactly where color was last applied.** When necessary, provide proper, safe lighting.

FIGURE 10-11. How the sides of a car are spray painted.

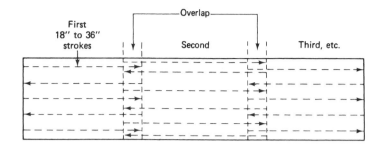

4. Painting wheel-opening flanges:
   a. Reduce the fan size as required. Use 50 percent or less trigger pull.
   b. Position the gun at the center of the opening and at right angles to the surface.
   c. Apply color lightly to the flange while pivoting the gun at the center of the opening. Avoid flooding, which is accomplished with "feel." Color can always be applied. When applied too heavily, it is too late.
   d. First do one-half of the wheel opening; then position the gun and do the opposite half of the wheel opening, overlapping a few inches at the center.
5. Painting rocker panels (if convenient, raise the side of the car several inches with a car jack, or set the wheels on 4-inch wooden blocks):
   a. Use a medium-size fan and controlled trigger pull as required. The painter must improvise and deviate from standard practice.
   b. While leading in one direction with the wrist, allow the hand and gun to trail to obtain the best possible application angle. Apply color to the rocker panel carefully with this motion. Strive for a medium-wet application. Avoid flooding. Observe what is being applied. This operation must be done with "feel" and care, and good lighting.
6. Painting windshield pillars and sedan door headers:
   a. Use a small fan and controlled trigger pull as required.
   b. Apply color lightly with several passes until the proper wetness of color is seen.

DIFFERENCES BETWEEN ACRYLIC LACQUER AND ENAMEL PAINT SYSTEMS

The color application system covered in detail earlier applies to both paint systems, acrylic lacquers and enamels. The same basic application procedure and spraying techniques apply to both paint systems, but there is a big difference between both paint systems. The differences are explained in Table 10–1.

Even though acrylic enamels and acrylic lacquers are totally different paint systems, the refinishing principles of (1) surface preparation (cleaning, sanding, primer, and primer–surfacer application) and (2) color preparation and application of first color coat are highly similar. Basic refinishing principles that apply to both paint systems have been covered just once. The principles that are different are so labeled and explained. The basic differences between acrylic lacquers and acrylic enamels (which also includes alkyd enamels) are shown in Table 10–1. The differences in lacquer and enamel refinishing operations are explained next.

APPLYING SINGLE-STAGE ACRYLIC AND URETHANE ENAMEL PAINT SYSTEMS

1. Apply a second coat of enamel when the first coat has dried sufficiently.
   a. The painter can rub the back of a forefinger (between the first two knuckles) on the corner of a panel section painted with the first coat and the paint does not rub off to determine dryness.
   b. Drying usually takes place by the time the painter completes application of the first coat when painting a complete car.
   c. Apply a second coat extra wet, which is much heavier than the first coat. The heavier second coat will not run, as the first coat helps to absorb much of the solvent.
   d. Three typical ways a painter can apply a wetter application of paint are as follows:
      (1) Slow down each stroke the required amount. Make slow, deliberate passes, but do not make them too slowly.
      (2) Use about a three-quarter overlap pass instead of a 50 percent overlap.

TABLE 10–1 Differences Between Application of Acrylic Lacquers and Acrylic Enamels

	Acrylic Lacquers	Acrylic Enamels
Reduction (see label directions)	From 100% to 200% (125%[a])	Sometimes no reduction Normally 25%; 33%[b]; 50%
Air pressure (at gun)	Normal: 40 psi May be 20 to 30 psi May be 45 psi	Normal: 60 psi[b] May be 50 psi May be 35 psi (mist)[b]
First coat	Medium-wet coat (0.3 to 0.4 mil)	Medium-wet coat ($\frac{3}{4}$ to 1 mil)[b]
Second coat	Medium-wet coat (0.3 to 0.4 mil)	Extra-wet coat ($1\frac{1}{4}$ mils or more)[b]
Total coats needed	6 to 8 coats needed; 4 to 6 coats to go	2 enamel color coats[b] completed
To produce final gloss	Lightly sand; compound and polish	Gloss comes up automatically
Recoating (e.g., repairing next day)	Can be recoated or repaired anytime	Catalyzed coating can be recoated anytime[b] Uncatalyzed coating requires *recoat sealer* in sensitive period[b]
Dirt control	Not too much of a problem	Greatest problem[b]

[a]Most popular for acrylic lacquer system.
[b]Applicable to acrylic enamel system.

(3) Cut down slightly on the distance from the surface. The painter must become experienced at using any of the above systems before attempting to combine any two of them. In enamel painting, lack of experience can lead to a run very quickly.

> **IMPORTANT**
>
> If an uneven gloss or overspray problem is encountered, it can be remedied by careful application of mist coating. The painter must remember that excessive mist coating of enamels can affect the gloss and durability of enamels in the long run. Excessive mist coating tends to wash the protective resin off the pigments.

2. If mist coating is required, proceed as follows:
   a. Use 95 percent solvent and 5 percent color, or use straight 100 percent reducer of the same type as that used in the reduction.
   b. Use reduced air pressure of about 20 to 30 psi at the gun.
   c. Adjust spray gun in midrange adjustment and check spray pattern. (See "Spray Gun Adjustments" in Chapter 4.)
   d. Apply uniformly wet mist-coat material to affected surfaces to remedy the problem. Avoid application of too much solvent. Allow a little flash time for the condition to clear as solvent levels surface and evaporate.
3. Allow the finish to become dust-free before removing the paint job (30 to 60 minutes) from the spray booth. Never place the car in the hot sun to dry, whether the finish is acrylic lacquer or enamel. Ideal drying conditions include plenty of fresh, warm air. Avoid polishing new paint jobs with sealer type polishes for at least 30 days.

APPLYING THE SINGLE-STAGE ACRYLIC LACQUER PAINT SYSTEM

1. Apply a second medium-wet coat of acrylic lacquer to the car. Follow the same sequence outlined for applying the first coat.
2. When a complete car is being painted, color coats are applied continually, as complete flash-off takes place between coats.
3. Apply the third through eighth coats (or the number of coats per the label directions) until complete hiding is achieved (see Table 10–1). Proper **hiding** means that when a color is applied to a thickness of 2 mils (when dry), the color will effectively and equally cover and hide a light and a dark shade of undercoat.

> **NOTE**
>
> Painting an acrylic lacquer car takes many more coats than painting an enamel car. This process requires many more refills of the paint cup, each of which must be strained. One strainer can be reused several times on the same color on the same day. A diaper on large flat horizontal surfaces is an option that, if used, must be redone with each refill. That fact is why having all reduced color and sufficient strainers in the paint booth saves time.

4. Allow the car to dry sufficiently before moving it to another location.
5. Remove masking as follows:
   a. On catalyzed enamel jobs, after 2 hours.
   b. On uncatalyzed enamel jobs, after 4 hours.
   c. On acrylic lacquer jobs, after 1 hour.
6. Clean the car as required: windows, front and rear ends, and bumpers and wheels. Also clean the interior as required.
7. Allow the finish to air-dry several days, preferably 1 week, before compounding. Or force-dry the finish with banks of infrared heat lamps at 180°F for 30 minutes or 160°F for 1 hour.
8. Compound and polish as required.

REPAIRING MOTTLING DURING SINGLE-STAGE LACQUER AND ENAMEL PAINT APPLICATION

A condition that may occur when wet spraying any metallic color is called **mottling.** Mottling appears as dark-shaded or off-color areas or streaks within a paint finish. Mottling is described more fully and discussed in Chapter 17.

If a mottling condition appears during the final color application, it can be remedied with the following fog-coating technique:

1. Move the spray gun back 18 to 24 inches from the affected surface.
2. Use the same reduction as in applying color on a car.
3. Use a three-quarter to full pull on the trigger, and apply a fog coat to the entire affected area as follows:
   a. Maintain the gun at 18 to 24 inches from the surface.
   b. Keep the trigger pulled and hold it open continually over the affected area while swirling the spray gun in a continuously moving circular motion. After several seconds of fog coating, observe the appearance of the mottling condition, which gradually disappears with an application of fog coating.
   c. As the mottling or streaking condition disappears, stop the application and move to another affected area.
4. Allow the repair application to air-dry in the spray booth with the air exhaust on for 30 to 45 minutes.
5. Then apply a final mist coat with a high-volume solvent for final gloss on enamels or lacquers.

USE OF SILICONE ADDITIVE

The purpose of a silicone additive is to speed the completion of a refinishing operation after all surface preparation operations have been completed, and when fisheyes appear after the first color coat is applied (see Table 10–2). The purpose of a silicone additive is not as a substitute for the usual surface preparation operations. Silicone additives should not be used freely in every color application. Use the product only when necessary.

Surfaces to be refinished may be contaminated with silicone, which may come from one of the following in a shop: car polishes, rubber lubricants, oil, grease, and shop towels. Silicone is very slippery. When used in car polishes, it results in less dirt and road film sticking to the finish.

When a surface is contaminated with silicones and is painted, fisheyes can result as follows:

1. The fresh paint solvent dissolves a particle of silicone that is on the paint finish.
2. The silicone, being quite light, floats to the surface on horizontal surfaces. On vertical surfaces, the circulation of solvent as it evaporates carries silicones with it.
3. Silicones, being very slippery, cannot hold the surface tension of the new color coat as it is applied wet. The lack of surface tension at the silicones opens up and causes fisheyes, which are always round in shape and appear as small craters.

Use

Before applying color coats to a surface, perform all the required surface preparation, including washing with paint finish cleaning solvent; sanding

TABLE 10–2 Additives for Prevention of Fisheyes

Description of Product	DuPont	PPG	Rinshed-Mason	Sherwin-Williams	Martin-Senour
For all acrylic lacquer and enamel colors	FEE	DX-66	809	V3-K-265	77B
For all urethane and urethane-catalyzed finishes	259-S	DX-77	809	V3K-780	87

with No. 500, No. 600, or finer sandpaper; and recleaning with solvent. This procedure is the best way to remove silicones. Silicones are sometimes very difficult to remove.

Active particles of silicone may still remain on the surface, which would not affect adhesion of newly applied color but could cause fisheyes. Prevention of fisheyes under these circumstances has led paint companies to provide silicone additives to the refinishing trade. When a silicone additive is used in repair color, the new color becomes 100 percent saturated with silicones. This saturation causes a uniform tension on the surface of the new color coat, thus preventing fisheyes.

Application

For best results, follow the label directions. Using too much silicone additive leads to other refinishing problems. Also, use proper additive for the specified paint system (see Table 10–2).

Fisheye Prevention (Without a Silicone Additive)

If a painter encounters a fisheye problem when spraying the first coat, the problem can be remedied without a silicone additive by proceeding immediately using the following technique:

1. Adjust the spray gun for dry-spray application (see "Spray Gun Adjustments" in Chapter 4).
2. Apply the equivalent of one full dry-spray coat of color on areas where fisheyes occurred and on other surfaces to be painted. The fisheyes practically disappear because of only one coat of paint. Allow the dry-spray application to dry for 10 to 15 minutes.
3. a. If applying acrylic lacquer color, proceed with medium to light color coats before applying the final two coats of color extra wet.
   b. If applying acrylic enamel color, apply two *very light* coats (which together would make up the first coat). Allow full flash-off between coats. Then apply a final extra-wet color coat.

In effect, the dry-spray color coats act as a sealer and prevent the covered silicones from reaching the surface. An appropriate sealer applied on the semi-dry side would serve the same purpose.

Insect or Dirt Lands on Wet Paint

If a flying insect or a large piece of dirt lands on wet paint during a painting operation, pick it off within 30 to 90 seconds as follows:

1. Apply a short section of masking tape in reverse on a pencil or a section of wire. Sometimes a painter will reverse-roll a short section of tape by itself, with the adhesive side out.
2. Quickly, but carefully, press the tape lightly on the insect or dirt and pick it off. Expertly done, the damage to the finish is negligible.
3. If tape does not work and the finish is setting up, use suitable tweezers to pick off the dirt.

Accidental Hose or Clothing Contact on Fresh Paint

If an accidental smudge occurs on a fresh paint finish during paint operations, most conditions can be remedied as follows:

1. Apply a dry spray (see "Spray Gun Adjustments" in Chapter 4) to the affected area for several seconds to attempt flattening and coating the spot. When very dry, a coating can be applied continually with a circling motion.
2. With midrange spot-repair spray gun adjustments, apply two color coats and allow to flash 5 to 10 minutes.
3. Apply a mist-coat solvent to the repair and surrounding areas to blend in the gloss.

REVIEW QUESTIONS

1. List several important items that should be done before spray painting starts so that the operation can proceed smoothly from start to finish without interruption.

2. List tools, equipment, and containers that should be in a paint mixing room (or equivalent area) to speed the preparation of paint materials before painting starts.
3. What solvent is used most in a paint shop for cleanup purposes?
4. List the basic solvents that should be in a paint mixing room (or equivalent area) to do paint work on a year-round basis:
   a. Acrylic lacquer system solvents.
   b. Acrylic enamel system solvents.
5. What primers and primer–surfacers should be on hand for immediate use?
6. On a complete refinish, what is the procedure to check out masking before painting starts?
7. What is the procedure for checking out a paint spray booth before spraying an enamel job?
8. Explain the procedure for mixing a can of paint by hand.
9. Explain the procedure for mixing a can of paint on a paint shaker.
10. Explain how to check a spray gun for adjustment and for air pressure setting to apply:
    a. Acrylic lacquer.
    b. Acrylic enamel.
11. What should the air pressure setting (at the gun) be for applying a 95 percent to 100 percent mist-coat solvent?
12. Explain how to control the hose during spray application when painting a roof, hood, and rear deck of a car requiring an extended reach.
13. What is the purpose of a silicone additive?
14. What is the procedure for using a silicone additive?
15. Explain how silicones cause fisheyes to occur.
16. Explain how to prevent a fisheye condition by the dry-spray method in the event that no silicone additive is on hand.
17. Explain the method for spray painting a roof panel (starting, etc.).
18. Explain the method for spray painting the hood and tops of the front fenders (starting, etc.).
19. Explain the method for spray painting the rear deck of a car (starting, etc.).
20. Explain the method for spray painting a complete car when the air exhaust system in the spray booth is:
    a. A cross-draft system.
    b. A down-draft system.
21. How many single coats are generally required when applying a complete recolor coating with:
    a. Acrylic lacquer system.
    b. Acrylic enamel system.
22. In mils, or thousandths of an inch, how thick must the color coat be to give complete hiding when applying standard acrylic lacquers?
23. Explain the remedy for correcting mottling that appears during final heavy color application.
24. Explain how and when to pick off a large dirt nib or insect that lands on wet paint during a painting operation.
25. Explain how to repair quickly an accidental hose or clothing smudge on fresh paint during painting operations.

CHAPTER 11

Panel and Sectional Panel Refinishing

INTRODUCTION

The repair of all automotive finishes can be divided into five general categories:

1. Complete vehicle refinishing is covered in Chapter 10.
2. Panel and sectional panel repair are covered in this chapter. (11)
3. Compounding and polishing repairs are covered in Chapter 13.
4. Spot and spot/partial panel repairs are covered in Chapter 12.
5. New service panel finishing is covered in this chapter. (11)

The purpose of this chapter is to review the basic fundamentals of panel and partial panel repair. To do all types of panel repainting properly, the painter must be familiar with the following:

1. How to identify the paint finish on any car before starting repairs. **The single-stage and basecoat/clearcoat finishes are entirely different and are not to be intermixed.** (See Chapters 2 and 9.)
2. The proper refinish paint systems and refinishing methods available.
3. Car factory recommendations for repainting panels on new and used cars.
4. Paint supplier recommendations for repainting panels on new and used cars.

Painters can become familiar with these paint requirements and procedures by continual communication with:

1. Car factory and paint company paint training schools
2. Insurance companies (when doing collision repairs and reviewing estimates with insurance companies before doing the work)
3. Paint jobbers (who are continually updated on new products and factory requirements)
4. Car dealers (dealers who do not have their own paint shop work closely with independent paint shops)

DESCRIPTION OF COMPLETE PANEL REPAIR

When 50 percent or more of a given panel needs repainting, the complete panel is generally repainted while it is on the car (see Figure 11–1). When a service replacement panel is involved, initial refinish operations are done on the panel before installation (see Figure 11–2). The painter thus has access to all surfaces and can duplicate factory painting. After installation, the outer surfaces of the panel are painted and, as explained in this chapter, the color coats are often blended into adjacent surfaces.

NEED FOR COLOR ACCURACY

When a complete car is painted, color accuracy is not a problem because the same color is used on all panels of the car. However, when only a single panel on a car is repainted, the need for color accuracy becomes much more important and color-matching problems are most likely to be encountered.

Panel-to-panel mismatches show up most readily because side-by-side comparison displays them most vividly. Any differences between two colors are noted easily when each color is cut abruptly on either side of a line.

FIGURE 11–1. Painter showing proper form while spray painting a hood panel on a past model car. Note the safety gear: air-supplied respirator, Tyvek suit, gloves, safety shoes, and air-line support belt. (Courtesy of DuPont Company.)

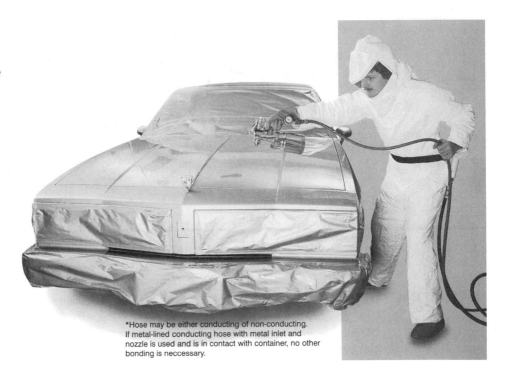

*Hose may be either conducting of non-conducting. If metal-lined conducting hose with metal inlet and nozzle is used and is in contact with container, no other bonding is neccessary.

FIGURE 11–2. Painting a deck lid panel (off car). Note the special spray stand holding the panel. (Courtesy of BASF Corporation.)

Before completing a panel repair, raise the masking on the adjacent panel and compare the repair color with the original color under proper lighting conditions (see Figure 11–3).

1. If the color match is satisfactory, remask the adjacent surfaces and complete the panel repair.

2. If the color is a mismatch, determine the problem, adjust the repair color to make it blendable, and then blend the adjusted final color into the adjacent surfaces or panels.

The following information on blending is provided through the courtesy of PPG Finishes. While the information highlights the need for color and

PANEL AND SECTIONAL PANEL REFINISHING

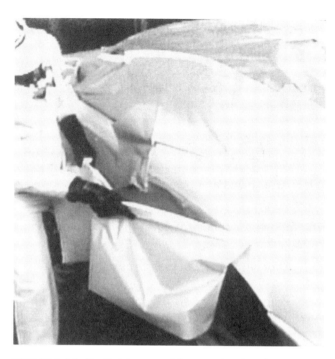

FIGURE 11-3. Raising masking to check color match of panel repair. (Courtesy of PPG Industries.)

clearcoat blending, several essential PPG blending products are mentioned and needed. Refinish painters know that all paint companies produce equivalent products to meet the needs of the refinish trade. These products must meet minimum factory standards of performance. It is impractical to list all the products of all paint companies in one book. If you work with a paint company other than PPG, simply go to your paint jobber and explain specifically what you need and she or he will be glad to fulfill your needs.

Color Blending

We now describe methods of spraying either single-stage or basecoat color onto a vehicle so that the edge of the sprayed color blends into the surrounding area rather than stopping at a hard breakline such as a door gap or molding. In most cases, color is blended to create the illusion of a perfect match, so that the panel or car looks as though it had never been painted. The purpose is not to deceive but to make the repair so invisible that even someone who knows it is there can't see it.

Because of the many color variances, or **color drift,** on original equipment manufacturer (OEM) cars from the same factory, matching a color's true appearance in a refinish repair with a "butt" match would be due almost exclusively to luck.

PPG has excellent factory and intermix color matches historically, but because we want to match so many aspects of a color, blending is a necessity. Consider the following list of appearance factors when spraying a color:

- **Color** (hue, value, and chroma).
- **Texture** (amount of orange peel).
- **Gloss** (or distinctness of image [DOI]).
- **Metallic** distribution (on the face, flash, and flop angles).

Because color itself is not the only item that must match to produce an invisible repair, blending to some degree of almost every color is expected.

Blend Preparation

Proper preparation in blending is always important. After making the required repairs to a vehicle and priming and sanding as necessary, you are ready to prep for a blend.

Before spraying any color on the vehicle, the area surrounding the spot to be painted should be **thoroughly cleaned and sanded with 1200- to 1500-grit wet** or equivalent. Sand just beyond the area where color will be applied for a single-stage blend. For basecoat, prep to the ends of all the panels involved, or at least to the area just beyond where the clearcoat will be applied (in the case of a clearcoat blend.)

> **NOTE**
> A few other acceptable methods of preparing a blend area can be done with special abrasive cleaners, Scotch Brite Pads, and very fine grit DA sandpaper. **Care should always be taken to clean the area thoroughly in every case.**

Color Matching Blending

- Blending a color creates an illusion that minimizes a color, texture, or metallic flop mismatch.
- **Always be prepared to blend.** It should be standard procedure to blend all repairs. **Blending is the most effective color-matching tool in the painter's arsenal.**
- Prepare adjacent panels for blending at the start of a repair rather than doing so in the middle of the repair.

Color Blending Technique

After cleaning, abrading, and cleaning the surface again, check the color against a sprayout. For basecoats and clearcoats, the **sprayout must be clearcoated.** DBU 500 will act as an acceptable clearcoating on a sprayout card for comparison purposes. If the color is close enough to blend (and in almost every case it should be), start spraying. The closer the color is in the first place, the easier it is to blend. Pick an appropriate variant color for the vehicle.

Unless you are spraying lacquer, an **adhesion promoter is not necessary.** Just apply the basecoat or single-stage urethane directly over the primer or sealer and out onto the surrounding area as you blend.

Color Blenders

Other tools available to help in basecoat color blending are DBU 500 for DBU colors and DBC 500 for DBC colors. These items are unpigmented basecoat resins. After reducing and adding color blender to the basecoat, the resin is sprayed on the edge of the color blend area to help blend the refinish color with the OEM color. It is especially useful on high hiding colors because it reduces the hiding power of the color with which it is mixed and allows for a more translucent progression in the blend.

Universal Blender Solvent

Straight solvents have always been used for melting in overspray on color and clearcoat. Now there is a solvent specially designed for blending. DX 830 Universal Blender is a custom blend of solvents that can be used in all types of shops and conditions and with a variety of products.

Clearcoat Blending

When there is no clear edge or breakline to stop clearcoating, the clear edge can be blended.

NOTE

Blend clearcoat only when you cannot clear an entire panel, as in the sail panel shown in Figure 11–4. Clearing an entire panel is the preferred recommendation. Also, some car companies specify that blending the clear on certain models is a violation of the warranty. This specification is usually the exception to the rule, but confirm before proceeding.

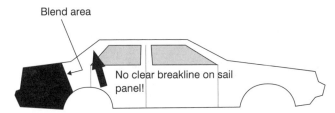

FIGURE 11–4. In the lowest level of basecoat/clearcoat warranty repair, a small repair on a quarter panel may require clearcoating on most of the quarter panel and blending the edges, as shown.

The following procedure lists the steps for blending clearcoat:

1. Choose as small an area as possible to minimize the blend area.
2. Prepare the area for the blend. A light sanding with 1200- to 1500-grit paper is recommended.
3. Spray the basecoat to hiding and step out the edges where necessary.
4. Apply the first coat of clear. Stop the edge in the designated blend area. Dust the edge with DX 830 Blending Solvent.
5. Apply the second coat of clear. Extend the edge just beyond the first clear blend edge, about 1 to 3 inches. Extending the blend edges too far may lead to accidental removal of some of the clear when the edge is polished because the film build may be too thin.
6. Dust the edge with DX 830 Blending Solvent (see Figure 11–5). You may lower your air pressure somewhat but too low an air pressure will not atomize the blending solvent properly. If you lower the pressure appreciably to work in a tight area, turn the fluid nozzle in to cut back on fluid delivery.
7. Apply a third coat of clear, if necessary, as already outlined.
8. When it is dry, polish the edge lightly by hand or machine. Use a foam or similar pad and a finessing type polish so you do not buff the clear edge and leave a line. (The only way to fix the line is to apply more clear!)

OEM Warranty and Clearcoats

All car companies today offer fairly extensive paint system warranties. The warranties are considerably more comprehensive than they were in the past, which can be partially attributed to the following factors:

1. Car consumers now demand a higher level of quality in their vehicles' appearance. The aver-

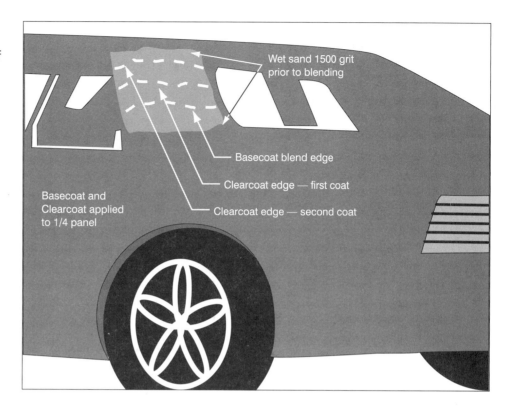

FIGURE 11–5. Blending clearcoat of quarter panel into roof panel. (Courtesy of PPG Industries.)

age price for a family style, midsize sedan is over $20,000, so the car should look good for a long time.

2. Most cars today are painted with basecoat/clearcoat systems at the factory. These finishes are more durable and have higher gloss than any paint system used in the past.
3. The OEM clearcoat on the vehicle's painted exterior panels provides protection against UV rays, rock chips, and the environment in general.
4. The OEM car companies know how important the clearcoat is to the long-term life and beauty of the vehicles they sell and warranty. Some car companies have changed their warranty recommendations on refinishing specifically on the application of clearcoats. Essentially, what they are recommending is something like the following:

> When performing warranty repairs on an OEM basecoat/clearcoat finish and if no clean breakline or body line exists for stopping the application of clearcoat, extend the application of clearcoat to the nearest panel edge.

Remember that these types of repairs are specified only by certain OEM car companies and are **strictly for warranty repair of basecoat/clearcoat OEM paint systems.** If your shop does warranty repairs for OEM dealerships, you should check the technical bulletins regarding warranty repair to be sure you are following the proper procedures (see Figure 11–6).

Panel Repair Procedure

The panel repair procedure outlined here is for basecoat/clearcoat finishes. Because panel-to-panel color matches are the most difficult to achieve, most painters in the trade have adopted the technique of achieving an acceptable color match by blending repair colors into adjacent panels. The procedure requires using a paint color judged by the painter as blendable. A blendable color is one that, while not 100 percent perfect, is a close match and can be blended into adjacent panels successfully.

SURFACE PREPARATION

1. Wash and clean the affected panel and adjacent panels with detergent and water, followed by solvent cleaner. Figure 11–7 shows a panel that has been vandalized with a deep scratch extending to the metal substrate. Note the scratch just below the hand of the painter, who is in the process of cleaning the panel with a solvent cleaner.
2. Mask adjacent panels and featheredge the damaged area with a suitable sandpaper and finish

FIGURE 11–6. In the highest level of basecoat/clearcoat warranty repair (a small repair on a quarter panel where there is no clean breakline or body line for stopping the application of clearcoating), the application of clearcoat has to extend to the nearest panel edge. This extension could mean clearcoating both quarter panels and the entire roof panel, as shown.

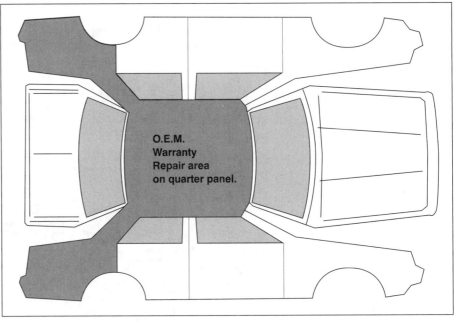

Apply last coat of clear to **quarter, roof, and opposite quarter.**

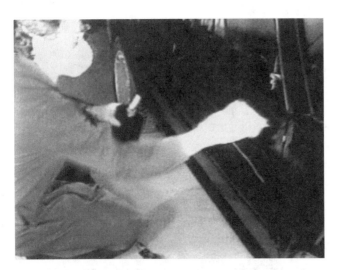

FIGURE 11–7. Technician cleaning panel with solvent cleaner. (Courtesy of Engelhard Corporation.)

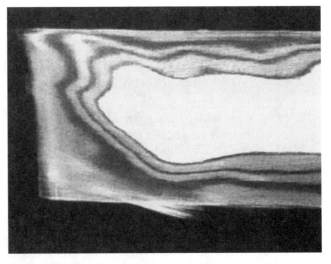

FIGURE 11–8. Close-up view of damaged area completely featheredged. (Courtesy of Engelhard Corporation.)

with a finer grit. Chapter 6 gives full details about featheredging. Figure 11–8 shows a close-up of the featheredged area on the repair panel.
3. Degloss all adjacent panels by sanding with 1200- to 2000-grit sandpaper. Solvent wash all repair surfaces and wipe them dry.
4. Mask the repair area for the undercoat application.
5. Because white metal has been exposed, use two-part chromate etch primer. Restrict the primer to the bare metal, per label directions. Figure 11–9 shows the application of two-part primer–surfacer to the repair area. Allow the primer–surfacer to dry and sand smooth.
6. Apply an appropriately colored sealer to all surfaces to be colorcoated. Extending the sealer a short distance into the adjacent panels (see Figure 11–9). Allow the sealer to dry per label directions.

BASECOAT COLOR APPLICATION

1. Apply the first coat of basecoat color over the repair area and follow with an application over

PANEL AND SECTIONAL PANEL REFINISHING

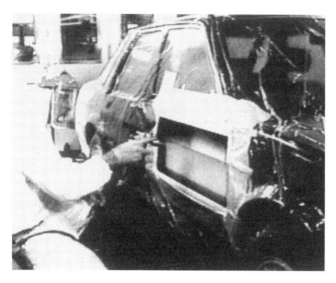

FIGURE 11-9. Application of etch primer followed by application of primer–surfacer after repair area is masked off. (Courtesy of Engelhard Corporation.)

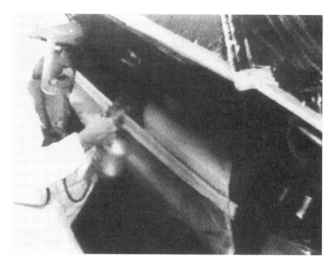

FIGURE 11-10. Application of basecoat color to repair area and blending color into adjacent panels. (Courtesy of Engelhard Corporation.)

the complete affected panel according to the paint manufacturer's specifications. Allow the application to flash.

2. Apply the second coat of color over the complete affected panel and extend the application into the adjacent panels for several inches. Allow the application to flash. Featheroff all application of color at the blend area edges.
3. If necessary, apply a third coat of colorcoat over the entire repair area and extend the application beyond the previous coat. No further blending is required at this time. The clearcoats will remedy this situation (see Figure 11-10).

CLEARCOAT APPLICATION

Before applying clearcoats, allow the basecoat color to dry according to the paint manufacturer's recommendations.

1. Mix the clearcoat with the required components, per label directions.
2. Strain the clearcoat.
3. Use the correct spray equipment and safety gear recommended.
4. Apply the first coat of clear to the repair panel and extend the application beyond the blend area, where it is feathered off (see Figure 11-11).
5. Apply the second coat of clearcoat to the entire affected panel and to the entire front and rear adjacent panels.

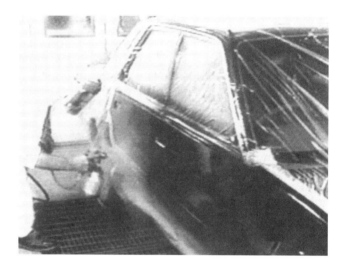

FIGURE 11-11. Application of clearcoat to repair panel and to complete adjacent panels. (Courtesy of Engelhard Corporation.)

6. Apply a third coat of clearcoat to all repair areas if polishing is required.

Figure 11-11 shows the clearcoat application on the front door and on the adjacent panels, which are the rear door and front fender. This procedure shows how the front door on a four-door sedan is painted with basecoat/clearcoat finish. Figure 11-12 shows the completed job.

FIGURE 11–12. The finished job. (Courtesy of Engelhard Corporation.)

TRICOAT PAINT REPAIRS

Tricoat paint finishes are an advanced form of basecoat/clearcoat finishes. The finishes are described in Chapter 9. The tricoat paint repair information below is based on the latest data available at the time of publication approval. Car factories reserve the right to make product or publication changes at any time without notice.

The successful repair of pearl luster effects depends on the painter's spray technique, the painter's experience with such colors, and the color itself. While some painters may be able to repair certain pearl luster effects successfully, other painters may experience a great amount of difficulty with them.

Some pearl luster effect colors are more difficult to repair than others. All colors are rated for repairability, ranging from easy to repair to difficult to repair. However, all production colors are designed to be repairable.

Mica or pearl flake, a component part of pearl colors, weighs more than comparable aluminum flake. The use of agitator cups is highly recommended to keep the flake in suspension for a consistent color match.

When selecting and using tricoat paint materials:

1. Use only top quality materials.
2. Follow the manufacturer's label instructions.
3. Do not mix different brands of materials.

The success of a tricoat paint repair depends on:

1. An acceptable color match.
2. A good blend with the original color.
3. Proper buildup of the sandwich coat.
4. A similar gloss and uninterrupted smoothness of the finish.

TABLE 11–1 After-Market Tricoat Paint Repair Materials

Manufacturer	Material Name
1. BASF (R-M Products)	Diamont
2. DuPont	Lucite or Cronar
3. PPG	Deltron

The material and system that produces a good color match for the person making the paint repair is the after-market paint and application procedure that will produce the best results. See Table 11–1 for some examples. Paint materials are available from paint jobbers in:

1. Base color—F/P (Factory Pak) or mix, depending on the manufacturer.
2. Pearl luster color—F/P (Factory Pak) or mix, depending on the manufacturer.

A good base color match and the proper number of applied sandwich coats (pearl luster) are critical in obtaining a good color match.

For the basecoat color:

1. Apply only the number of coats necessary to achieve full hiding.
2. Do not substitute another product color for the recommended base color coat.
3. The slightest change in the base color coat will be greatly magnified once the pearl luster coat and the clear top coat finishes are applied.
4. Follow the product manufacturer's label instructions.
5. It cannot be overstressed that tinting takes time and should be done as a last resort.

For the pearl luster coat: Make a sample sprayout panel, known in the trade as a let-down panel. Use this panel to compare the pearl luster effect of the applied refinish material buildup to that of the OEM applied material. See Figure 11–13 and the steps below for making a let-down sample panel:

1. Prepare a suitably sized test panel with the same color undercoat as used on the job. Allow it to dry, sand as required, and apply a small check hiding sticker to ensure the proper amount of base material. (Stickers are available through paint jobbers.)

FIGURE 11–13. Layout of tricoat let-down sample panel. (Courtesy of PPG Industries.)

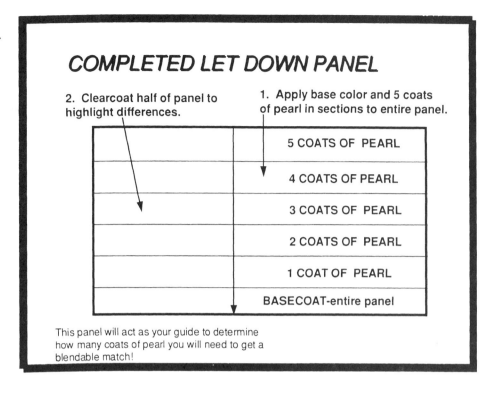

WARNING (ISOCYANATES)

It is MANDATORY that adequate respiratory protection be worn. Examples of such protection are air line respirators with full hood or half mask. Such protection should be worn during the entire painting process. Persons with respiratory problems or those allergic to isocyanates must not be exposed to isocyanate vapors or spray mists.

2. Apply the basecoat color to the complete panel to achieve full hiding and allow it to dry. Use a heat lamp to speed drying.
3. Divide and mark the panel in six equal sections as shown in Figure 11–13 to illustrate film build variations. Mark the section numbers on both sides of the panel.
4. Prepare and install five sections of premasked masking paper:
   a. Mask the "BASECOAT ONLY" section.
   b. Then mask the "BASECOAT ONLY" and #1 sections.
   c. In overlapping fashion, continue masking through section #4.
5. Starting with section #5, apply the first single medium-wet coat of pearl luster color and allow it to flash.
6. Remove the first section of masking to uncover section #4.
7. Apply another single medium-wet coat of pearl luster over the uncovered area of the panel and allow it to flash.
8. Repeat the above steps until the complete panel (except the BASECOAT ONLY section) has been sprayed with pearl luster material.
9. Place the panel under a heat lamp to dry.
10. After the panel is dry, mask off one-half of the panel lengthwise as shown.
11. Apply three single medium-wet coats of clear top coat material over the unmasked side of the panel, then remove the masking paper and place the panel under a heat lamp until dry.

The above panel illustrates the effect each additional coat of material has on the lightness/darkness of the final top coat color. A let-down panel should be used as an aid to determine the approximate number of pearl coats needed to achieve a color match by comparing the various film-build areas on the panel to an adjacent panel on the car prior to material application. Mica or pearl flake weighs more than a comparable aluminum flake. The use of agitator cups is highly recommended to keep the flakes in suspension for consistent color match.

FIGURE 11–14. Tricoat panel blend repair on two-door car. (Courtesy of PPG Industries.)

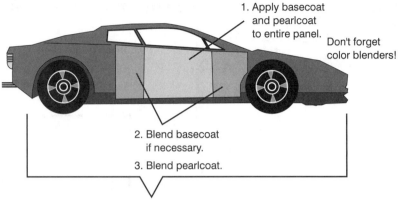

Tricoat Panel Blend Repair

Figure 11–14 shows the correct repair operations on the right door of a two-door coupe that is finished with a tricoat (pearlcoat) finish. Note that after the door is painted, the adjacent panels are blended with the basecoat color and pearlcoat color, and the entire right side of the car is clearcoated.

Try to keep the repair and blend areas as small as possible. One way to accomplish this is to use a two-gun method for the blending stage of the repair. The two-gun method can be used with both base color and pearlcoat blend edges.

- One gun contains regular reduced pearlcoat. Use the first gun to apply the pearlcoat to the main repair area. Keep the pearlcoat within the repair area.
- A second gun (keep it in the booth within reach) contains a mixture of pearlcoat and color blender to make the second gun more transparent. Use the second gun to extend your blend edges.

New Replacement Panel Refinish Procedure

The procedure described in this section is for tricoat finishes (see Figure 11–15).

PREPARATION FOR PAINTING

1. Place the new part on a workbench or part stand.
2. Wash the panel with warm water and a mild detergent.
3. Clean the entire panel with an appropriate cleaner.
4. Sand the new panel with fine sandpaper.
5. Reclean the panel with an appropriate cleaner and tack wipe the sanded area.
6. Select the proper sealer for the surface being repaired. Mix as required. Use the correct color sealer to assist in the basecoat color coverage.
7. Apply the proper sealer to the primed area and test panel.
8. Follow the paint supplier's drying instructions. If forced drying is necessary, follow the correct procedures for the materials being applied.

BASECOAT APPLICATION

The basecoat application on a new part requires that the basecoat cover the panel and adjacent blending surfaces completely to achieve an acceptable color match.

1. While the part is on the part stand, apply the basecoat and clearcoat to the inside edges. If color sealer is recommended for the inside part, then mount the part to the vehicle.
2. Select the correct color.
3. Check the paint supplier's basecoat formula. Mix the final selected color and make sure to
   a. Shake the color mix thoroughly.
   b. Strain the color mix just before filling the spray gun.
4. Select the correct spray equipment (for example, nozzle size and air cap).
5. Use the test panel or sprayout to check for accurate color matching.
   a. Apply the basecoat color to the test panel to achieve complete hiding.
   b. Apply the clearcoat over the basecoat on the test panel.

PANEL AND SECTIONAL PANEL REFINISHING

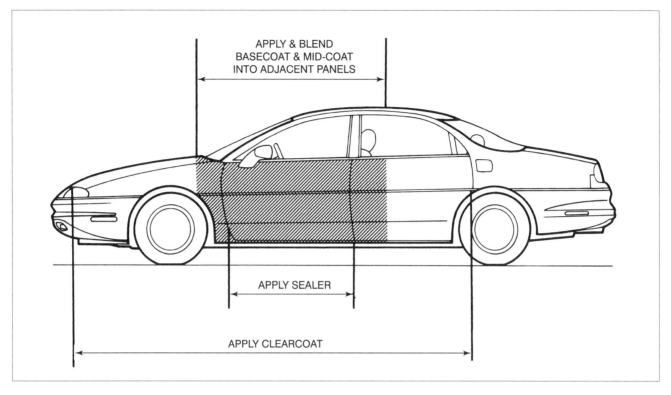

FIGURE 11–15. New replacement panel finishing procedure—tricoat finishing repair. Note the blending of panels into adjacent surfaces. (Reprinted with permission of General Motors Corporation.)

c. Compare the panel to the vehicle in sunlight to ensure a blendable color match.

d. Check a color alternate if the colors do not match.

6. Wipe if necessary.
7. Apply the basecoat color over the undercoat.
8. Apply an additional basecoat color, if necessary, to achieve hiding.
9. Blend the basecoat a short distance into the adjacent panels.
10. Allow the basecoat to dry per the paint supplier's recommendations.

MIDCOAT APPLICATION

Multicoat repairs require the application of a midcoat material over the basecoat color in the adjacent panel blend areas to achieve an acceptable color match.

1. Check the color against the vehicle to ensure a blendable final color match. Use the paint supplier's alternate tint if necessary to match the color as closely as possible.
2. Check the paint supplier's midcoat label. Mix the final selected color and make sure to:

   a. Shake the color mix thoroughly.

   b. Strain the color mix just before filling the spray gun.
3. Use the let-down panel to check for accurate color matching. Compare the panel to the vehicle in sunlight to ensure a blendable color match (see Figure 11–16).
4. Wipe if necessary.
5. Apply coats of midcoat material as required and blend each coat into the adjacent panels.
6. Allow the midcoat application to dry per the paint supplier's recommendations.

CLEARCOAT APPLICATION

Before spraying clearcoat, allow the midcoat to dry per the paint supplier's recommendations.

1. Mix the clearcoat with the required components according to the paint supplier's recommendations.
2. Strain the clearcoat.

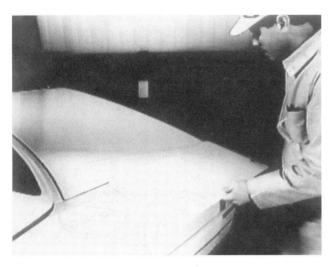

FIGURE 11–16. Comparing let down sample panel to car for color match. (Courtesy of PPG Industries.)

3. Select the correct spray equipment (for example, nozzle size and air cap).
4. Apply the clearcoat:
   a. Allow adequate time between coats.
   b. Follow the paint supplier's recommendations for air pressure at the gun.
   c. Apply the first coat of clearcoat over the new panel and approximately 3 to 4 inches beyond the midcoat.
   d. Apply the second coat of clearcoat over the entire repair area.

SECTIONAL PANEL REPAIR

Sectional panel repairs involve similar paint application as in a complete panel repair, except that the color is confined to a smaller part or section of the panel. Sectional panel repairs are possible because the repair color can be extended to natural breaklines like the following:

1. Panel crease-lines.
2. Moldings.
3. Panel edges.
4. Striping, decal, or two-tone paint breakline.

Sectional panel paint repairs are done in less time, using less paint material, and for much less labor cost than would be the case if complete panels were painted.

> **NOTE**
> Sectional panel repairs involve crease-line or reverse masking. For details concerning this masking, refer to Chapter 6.

Typical Sectional Panel Repair Procedure

The procedure outlined in this section is for tricoat finishes.

PREPARATION FOR PAINTING

1. Wash the entire area to be repaired with warm water and a mild detergent (refer to Figure 11–17).
2. Clean the entire area to be repaired with appropriate solvent cleaner.
3. Sand the repair area.
   a. Sand the primer.
   b. Sand the clearcoat area with 1200- to 2000-grade sandpaper.
4. Clean with an appropriate cleaner and tack wipe the sanded area.
5. Select the proper sealer for the surface being repaired. Mix as required. Use the correct color sealer to assist in the basecoat color coverage.
6. Apply the proper sealer to the primed area and test panel.
7. Follow the paint supplier's flash and drying time instructions. If force drying is necessary, follow the recommended procedures.

BASECOAT APPLICATION

A basecoat repair requires that the basecoat to cover the repair panel and undercoats and all blending areas completely to achieve an acceptable color match (see Figure 11–18):

1. Apply the final masking.
2. Select the correct color for the vehicle by checking the car paint code and the paint supplier's color cards.
3. Check the paint supplier's basecoat formula. Mix the final selected color and make sure to:
   a. Shake the color mix thoroughly.
   b. Strain the color mix just before filling the spray gun.

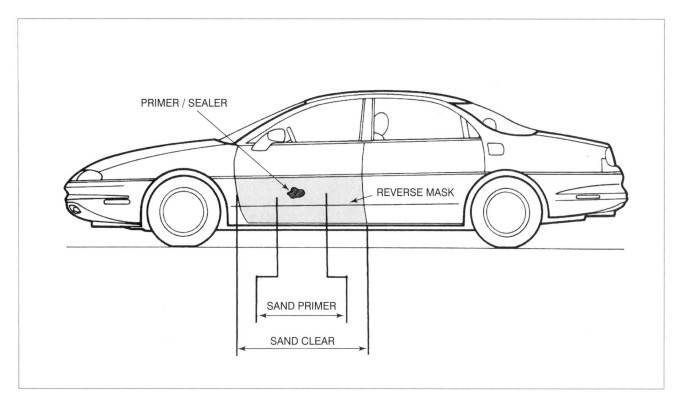

FIGURE 11–17. Sectional panel repair—surface preparation. (Reprinted with permission of General Motors Corporation.)

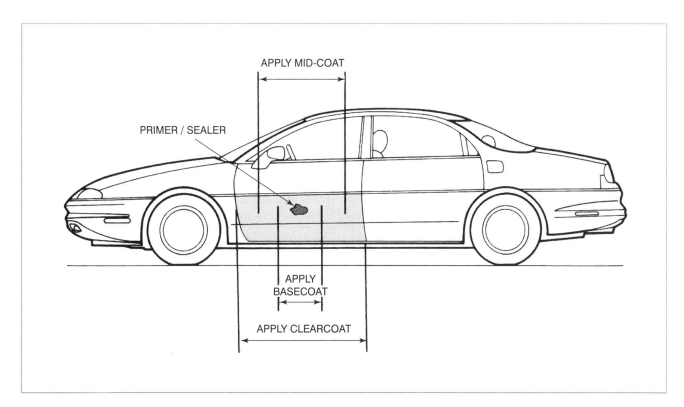

FIGURE 11–18. Sectional panel repair—tricoat finish application. (Reprinted with permission of General Motors Corporation.)

4. Select the correct spray equipment (for example, nozzle size and air cap).
5. Use the test panel or sprayout to check for accurate color matching.
   a. Apply the basecoat to achieve complete hiding on the test panel.
   b. Apply the clearcoat to the basecoat on the test panel.
   c. Compare the panel to the vehicle in sunlight to ensure a blendable color match.
   d. Check for a color alternate if the colors do not match.
6. Tack wipe the panel if necessary.
7. Apply the first coat of basecoat color over the undercoat.
8. Apply the second coat of basecoat color beyond the first coat and blend as necessary.
9. If the hiding or matching does not match your desired effect, apply a third coat to the repair area. Blend as required.
10. Allow the basecoat to dry per the paint supplier's recommendations.

MIDCOAT APPLICATION

A multicoat repair requires the application of midcoat over the basecoat to achieve an acceptable color match.

1. Identify the vehicle color. Use the paint supplier's alternate tint if necessary to match the color as closely as possible.
2. Check the paint supplier's formula label. Mix the final selected color and make sure to:
   a. Shake the color mix thoroughly.
   b. Strain the color mix just before filling the spray gun.
3. Use a let-down panel to check for accurate color matching. Compare the panel to the vehicle in sunlight to ensure a blendable color match (see Figure 11–16).
4. Tack wipe if necessary.
5. Apply coats of midcoat as required and blend each coat on adjacent surfaces.
6. Allow the midcoat to dry per the paint supplier's recommendations.

CLEARCOAT APPLICATION

Before spraying the clearcoat, allow the midcoat to dry per the paint supplier's recommendations.

1. Mix the clearcoat with the required components according to the paint supplier's specifications.
2. Strain the clearcoat.
3. Select the correct spray equipment (for example, nozzle size and air cap).
4. Apply the clearcoat at the repair spot and to the complete panel.
   a. Allow adequate time between coats.
   b. Follow the paint supplier's recommendations regarding air pressure at the gun air cap.
   c. Apply the second coat of clearcoat over the entire repair panel.
   d. Apply a third coat if polishing is required.

Acrylic Urethane Enamel Application (Single-Stage)

1. Reduce the acrylic urethane color according to the label directions. Use the proper reducer for the temperature conditions.
2. To determine the correct spray technique for a color match, proceed as follows:
   a. Spray a small test panel with several coats, allowing proper flash time between coats. Start with 40 psi at the gun.
   b. Adjust the amount of air pressure, reduction, and type of solvent to obtain a good color match.
   c. Spray the test panel to full hiding for comparison (see Figure 11–19).
   d. If a color mismatch still results, adjust the color by tinting, as explained in Chapter 14.
3. Apply sufficient coats of color for full hiding. Some colors call for two to three coats, and others call for three to four coats. Check the label.

FIGURE 11–19. Comparing sample spray out for color match to car. (Courtesy of DuPont Company.)

4. Allow the finish to dry to a dust-free condition (1 to 2 hours) before moving the car from the spray booth, or move the car to a drying room.

> **CAUTION**
>
> Painters are urged to use proper face-mask or hood-type respirators as indicated in the paint supplier's label warnings. Inhalation of isocyanate vapors and/or other by-products of spray painting may be detrimental to one's health.

Use of Clearcoat over Single-Stage Acrylic Urethane Enamel

An option available to painters is the use of a clear material with a single-stage color system. This option can be applied to either of two systems and both are highly recommended. Both systems provide increased durability and depth of color:

1. In the first system, when applying the final second or third coat of enamel, mix the final coat 50:50 with a reduced and catalyzed clearcoat compatible with the color system. While this final coat may meet minimum coating requirements, more color or color/clear mix can be applied.
2. The painter can also apply a catalyzed clear over the single-stage color, like a basecoat/clearcoat finish. Normally, 2 mils of clear (two to three coats) are needed. This system provides a deeper gloss and increased durability. Be sure to check with the paint jobber, however, that the appropriate clearcoat is used over the single-stage color.

> **NOTE**
>
> An incompatible clearcoat over single-stage color can result in several unpleasant problems, chief of which would be delamination of the clearcoat, loss of gloss, and early film degradation.

PAINT STRIPING REPLACEMENT

If a car panel with OEM striping must be repainted, new striping to match the OEM striping can be applied quickly with the following procedure:

1. Secure the proper 3-M Fine Line striping tape, or its equivalent, and the proper color from the paint jobber according to car series, model year, and type of stripe needed.
2. Clean the surface to be striped with paint finish cleaning solvent to remove any wax, grease, or other contaminants.
3. Cut a length of striping tape 2 to 3 inches longer than needed for the panel. Before cutting the tape, place a piece of masking tape or paper to the adhesive side of the tape, as shown in Figure 11–20.

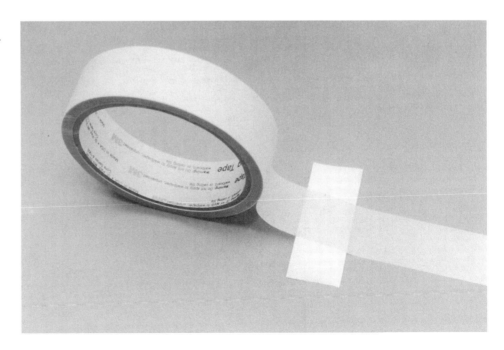

FIGURE 11–20. Securing tape on back side with paper or masking tape before cutting. (Courtesy of 3M.)

4. Align and temporarily apply the striping tape according to the OEM stripes on each side as a reference. If only one reference is available, obtain the needed reference markings from the opposite side of the car. Before sticking the tape down firmly, gunsight the tape for alignment from each end and correct any misalignment as required. Also, allow 2 to 3 inches of tape to protrude beyond the panel, as shown in Figure 11–21.

5. **After slicking down the rest of the tape firmly to the panel, remove the desired pull-out piece**(s) (see Figure 11–22) to match the OEM stripe(s). Remove the desired pullout sections at a 90° angle with a moderate, steady, continuous motion.

6. **Lightly scuff the exposed paint stripe areas through the tape** (see Figure 11–23) with a 3M Scotch-Brite pad, No. 07447, or its equivalent, to ensure good paint adhesion.

7. **Mask adjacent areas** as required. **Apply color with brush or spray equipment** as required. If color is applied by brush (see Figure 11–24), add up to 5 percent retarder to the package viscosity color. Allow the stripe(s) to dry partially (lacquer, 10 to 15 minutes; enamel, 15 to 20 minutes). Then remove the outside tape first (if a double stripe) at a 90° angle, pulling slightly away from edge. Remove the remaining tape at a 90° angle. Figure 11–25 shows the finished job.

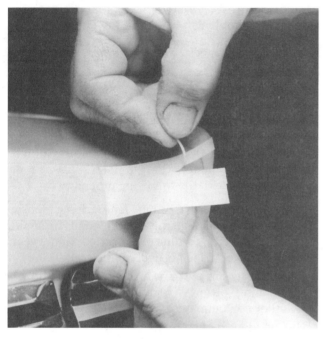

FIGURE 11–21. Striping tape protrudes at end to aid selection of proper insert. (Courtesy of 3M.)

> **NOTE**
>
> If working at the front or rear end of the stripe(s), duplicate the striping design with sections of masking tape according to the opposite side of the car. **If force-drying is required, remove the tape before applying heat lamps.**

FIGURE 11–22. Removing tape insert at 90° angle to surface. (Courtesy of 3M.)

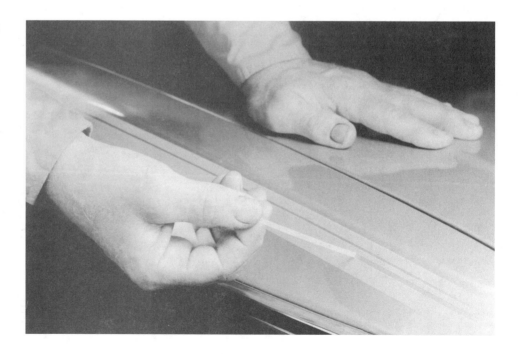

PANEL AND SECTIONAL PANEL REFINISHING

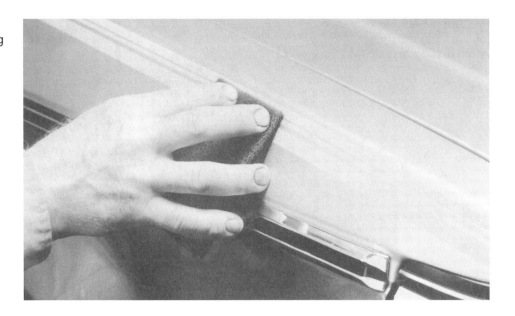

FIGURE 11–23. Scuffing paint through insert opening with 3M Scotch Brite pad. (Courtesy of 3M.)

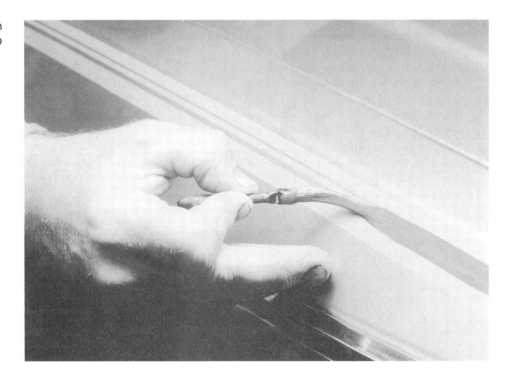

FIGURE 11–24. Application of paint stripe with touch-up brush. (Courtesy of 3M.)

REVIEW QUESTIONS

1. How can the painter in the paint shop be advised of the following on a continuing basis?
   a. Factory refinish recommendations.
   b. Latest paint supplier refinish procedures.
2. In a general manner, describe a complete panel repair.
3. In a general manner, describe a sectional panel repair.
4. What are the advantages of a sectional panel repair over a complete panel repair?
5. Do color mismatches show up more readily on panel repairs or on spot repairs? Why?
6. List several natural breaklines that make sectional panel repairs possible.
7. Outline the procedure for painting a service replacement panel.
8. Describe the procedure for applying a single-stage acrylic urethane enamel.

FIGURE 11–25. The finished job with paint stripes applied. (Courtesy of 3M.)

9. Describe the procedure for a spot/sectional panel repair of an OEM basecoat/clearcoat finish.
10. Describe the procedure for repainting an OEM basecoat/clearcoat finished panel.
11. Describe the procedure for repainting the OEM paint striping on a door panel.
12. What is meant by the side zone method of repairing a tricoat paint finish?
13. Describe the procedure for repainting a complete panel with the tricoat color system.
14. What is the principal difference between the basecoat/clearcoat and the tricoat color systems?
15. Explain the procedure for blending a panel repair on a tricoat paint finish. The damage is on the right door of a two-door coupe.
16. Explain the procedure for blending a quarter and sail panel at a roof panel on a basecoat/clearcoat finish.
17. Why is the use of an agitator cup important when applying mica pearl finishes?
18. Why is blending of colors important when making panel repairs?
19. Explain the general accepted procedure for blending single-stage paint finishes.
20. Explain the procedure for making a let-down test panel for checking a color match on a tricoat paint finish.
21. Explain the procedure for making a test panel for checking the color match on a single-stage metallic finish.

CHAPTER 12

Spot and Spot/Partial Panel Repair

INTRODUCTION

By definition, spot repairs are divided into two general categories:

1. **Single-stage color spot repairs.**
2. **Basecoat/clearcoat finish spot/partial panel repairs.**

In single-stage color spot repairs, the repairs can be confined within the limits of a given panel. As shown in Figure 12–1, the complete repair plus blending is confined to one panel. The key is having enough room to do effective blending. In Figure 12–2, however, the spot repair must be blended into adjacent panels because the repair is near those adjacent panels.

The spot/partial panel repair system has become the most popular method for repairing spot and/or partial panel sections on basecoat/clearcoat finishes. This system features blending the color at a spot location and then clearcoating the entire panel to the nearest breakline or, if no breakline exists, blending into the nearest adjacent panel (Figure 12–3).

Only through experience and spray testing can a painter judge beforehand whether to spot repair or panel repair to the nearest breakline. Any time a painter can match an adjacent panel with a color, the painter should be able to spot repair that same color, and vice versa.

All automotive colors are rated by each car company for repairability properties ranging from easy to spot repair to difficult to spot repair. Each production color falls into a rating category between these two limits. All automotive production colors are designed to be repairable. The term *repairable* means that a color can be spot or panel repaired successfully by an

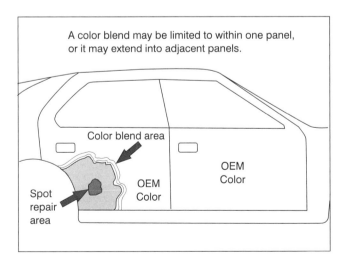

FIGURE 12–1. Blending a single-stage color within the boundaries of a panel. (Courtesy of PPG Industries.)

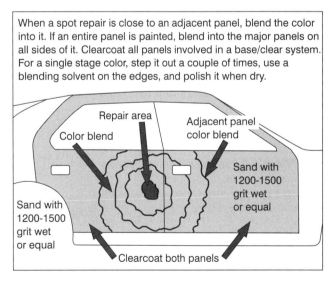

FIGURE 12–2. Blending a spot repair into adjacent panels when the repair area is near adjacent panels, as shown. (Courtesy of PPG Industries.)

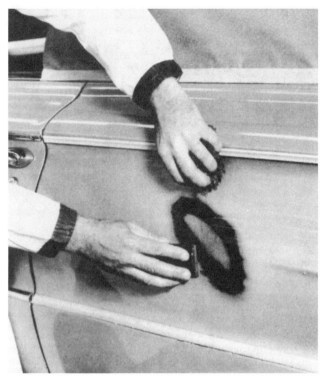

FIGURE 12–3. Featheredging technique around spot repair. (Courtesy of General Motors Corporation.)

average painter using field repair materials. Color repairability is an important factor in determining if a color is to be used in a production plant.

SUCCESSFUL SPOT AND/OR PARTIAL PANEL REPAIR

A successful spot repair has the following three repair characteristics:

1. **A commercially acceptable color match:** The color match, if not perfect, is so close that it would be acceptable to a high percentage of the motoring public.

2. **A good blend:** The repair blend should be smooth, gradual, and fully acceptable. There should be no telltale ring or pronounced shaded perimeter around the repair.

3. **Comparable gloss and surface texture:** The entire repair should have similar gloss and comparable smoothness compared to adjacent surfaces. Even the orange peel effect should be comparable.

SURFACE PREPARATION FOR SPOT REPAIRS

The identity and depth of a paint problem and type of paint system tells the painter what surface preparation is required. For every paint problem or failure, there is a proper and a required surface preparation. To know what surface preparation is needed for a specific paint problem, the painter should check the appropriate paint condition in Chapter 17. The main concern of the painter during surface preparation is to remove any affected paint as required and to prepare the surface as required before making a paint repair.

The following surface preparation operations and illustrations are designed to help an apprentice in doing top-quality work:

1. Wash the surface with water and mild detergent.
2. Clean the surface with paint finish cleaning solvent.
3. Sand and featheredge broken paint edges (see Figure 12–3). For details on featheredging, see Chapter 6.

Figure 12–4 is a typical cross-section of a sanded featheredge on metal. **A cross-section of a painted panel is a paint finish and metal panel cut straight through at right angles to the surface.** Figure 12–4 shows the following:

FIGURE 12–4. Single-stage color spot-repair cross-section: featheredging. (Courtesy of General Motors Corporation.)

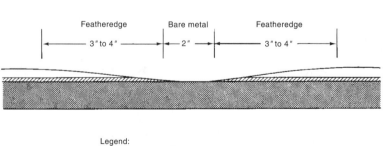

SPOT AND SPOT/PARTIAL PANEL REPAIR

a. The factory standard color, about 2.5 mils thick.
b. The factory undercoat, usually about 1.00 mil thick.
c. The metal.
d. The sanded featheredge, several inches long.

Figure 12–4 gives the painter an idea of how far back broken paint edges should be sanded. When rubbing the hand with the fingers extended across a repair area sanded correctly, the featheredge is hardly felt or it is not felt at all. The surface should be smooth to the touch. This thorough and correct featheredging is an important key to successful spot repairing.

4. Treat bare metal surfaces with two-part metal conditioner (see Chapter 6). After treatment, wipe the metal-conditioned surfaces with a water-dampened cloth to remove excess metal conditioner from the surrounding surfaces. Allow the surfaces to dry.
5. Apply the primer and primer–surfacer according to label directions (see Figure 12–5).

FIGURE 12–5. Application of primer–surfacer over spot repair. (Courtesy of General Motors Corporation.)

FIGURE 12–6. Single-stage color spot repair cross-section: primer–surfacer application. (Courtesy of General Motors Corporation.)

Figure 12–6 shows a cross-section of primer and primer–surfacer application before and after the undercoats dry.

Item 1: Straight primer is applied to bare metal.
Item 2: Freshly applied, wet primer–surfacer is shown by the upper broken line.
Item 3: Dried primer–surfacer is shown by the second broken line.
Item 4: The sanded primer–surfacer is shown by a solid line. Note how the sanded primer–surfacer bridges smoothly and uniformly across the original color coats.

6. Sand the primer–surfacer and adjacent original color per label directions.

Figure 12–7 shows what happens when an insufficient amount of primer–surfacer is present in a spot-repair situation. Insufficient primer–surfacer shows up as a dished-out patch, bull's-eye, or low spot within a spot repair. The condition is caused by an insufficient amount of primer–surfacer or by too much sanding after primer–surfacer application. Apprentices can prevent this problem by using the guide-coat filling and sanding technique.

The best precautions to follow when sanding primer–surfacer, especially when spot-repairing enamel finishes, are:

a. Sand the inside overspray edge around the center of the spot using No. 400 sandpaper, a suitable sanding block, and water. Keep away from the center area at first.
b. When sanding over the center area, check the progress of the sanding frequently. **Stop sanding at the center area as soon as the surface is smooth and free of orange peel.**

7. Finally, reclean the entire compounded area with a cloth dampened with water and very little finish cleaning solvent.

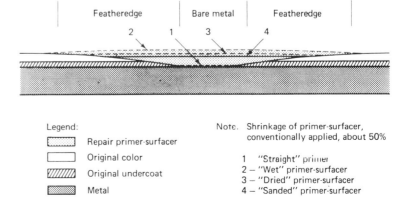

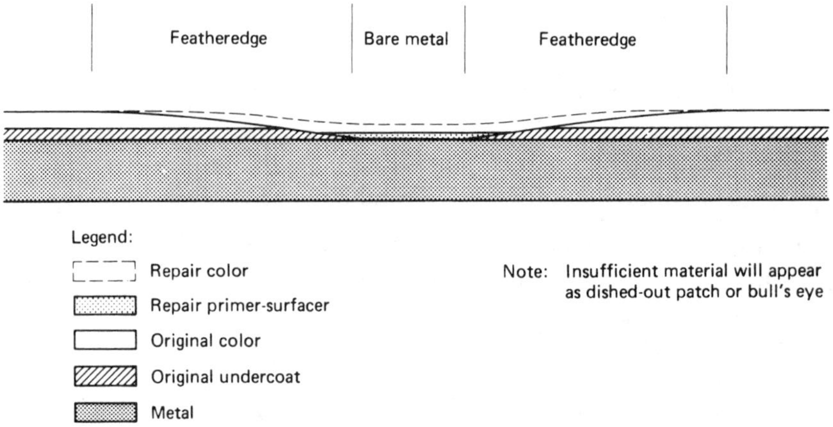

FIGURE 12–7. Single-stage color spot-repair cross-section: insufficient material. (Courtesy of General Motors Corporation.)

SINGLE-STAGE FINISH SPOT REPAIR TECHNIQUE

To achieve a successful spot repair, the painter must have several items, chief of which are:

1. The correct color code material, which may include use of a factory alternate color formula or, if necessary, a tinted color.
2. The correct solvent for the temperature conditions.
3. The correct color blender.
4. A two-spray gun setup:
   a. One gun for the color application.
   b. One gun for the color blender.
5. A proper spray technique.

Once all surface preparation operations have been done, the spray technique involves the following steps:

1. Apply color to the repair area in single coats and allow the recommended flash time between coats and blending at blend areas.
2. Achieve complete hiding with the proper number of color coats, each of which is blended as required. If damage is properly located, the repair may be blended within a given panel (Figure 12–1).
3. If damage is next to another panel, carry the color application and blending into that next panel (Figure 12–2).

The important characteristics of a successful spot repair in single-stage finishes are:

1. A blendable color.
2. Proper blending technique.
3. The artistry of the painter.

SINGLE-STAGE ACRYLIC LACQUER SPOT REPAIRS

Color Preparation

Double-check certain key items to ensure that the best equipment and color are being used, particularly when matching metallics:

1. Check the color on hand to be sure that it is the correct color code.
2. Check for correct solvents.
3. If available, set up an agitator-type paint cup, which makes spraying metallics as easy as spraying solid colors.
4. Be sure to agitate the color thoroughly by hand or on a paint shaker.
5. Prepare acrylic lacquer color with a two-spray gun system as follows:
   a. In spray gun 1, prepare a conventional reduction of acrylic lacquer color. Use slow thinner in reduction for average temperatures. Label the cup "Color."

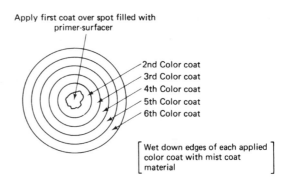

FIGURE 12–8. Blending single-stage acrylic lacquer into the old finish. (Courtesy of DuPont Company.)

SPOT AND SPOT/PARTIAL PANEL REPAIR

FIGURE 12–9. Spot-repair cross-section: single-stage color application. (Courtesy of General Motors Corporation.)

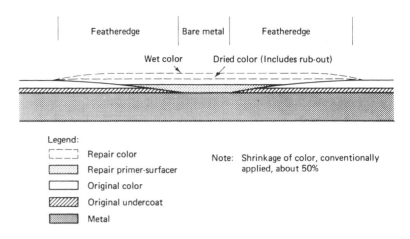

b. In spray gun 2, prepare the color blender. Label the cup "Mist Coat."

Single-Stage Color Application

Refer to Figure 12–8 and use the following procedure:

1. Set up the spray gun in the midrange adjustment (Chapter 4).
2. Spray out a small test panel with several coats and allow the proper flash time between coats. (See your local paint jobber for special black and white paper test panels.)
3. Adjust the amount of reduction, air pressure, and type of solvent to obtain a good color match. These variables in spray painting are controllable by the painter.
4. Spray test a panel to full hiding for comparison to the car.
5. If a color mismatch still results, adjust the color as explained in Chapter 14 before proceeding.
6. Apply a first coat over the primer-surfaced area (see Figure 12–9). Use a feathering technique at the beginning and end of each stroke (see Chapter 4). Apply color blender to the edges and allow it to flash.
7. Apply the second through the final coat in the same manner, extending each coat beyond the previous one as shown in Figure 12–8. Blend the edges and allow each coat to flash. Application of color is complete when full hiding is achieved.
8. Allow the repair to dry at least 1 full day but preferably one week, or force dry the repair with a heat lamp at 180° for 30 minutes.
9. Compound and polish the repair and adjacent areas as required. Figure 12–10 shows the color application and amount of shrinkage in a dried acrylic lacquer. (Shrinkage of standard color, conventionally applied, is about 50 percent.)

FIGURE 12–10. Checking test panel for color match to car. (Courtesy of DuPont Company.)

SINGLE-STAGE ACRYLIC ENAMEL SPOT REPAIRS

Spot repairing over original acrylic enamel color with acrylic enamel color is recommended by most car factories. To gain adhesion of top coats without the use of a sealer, it is necessary to clean and compound the original acrylic enamel surfaces with a fine compound by hand.

Most paint suppliers package enamel by the gallon. Quarts and pints are not factory-packaged. Since quart quantities are used the most, the painter must depend on the jobber for an accurate mix. An accurate mix is more important with solid colors than with metallics because metallic colors can be controlled for shade variation with spraying techniques.

Most paint suppliers (PPG, DuPont, Sherwin-Williams, Rinshed-Mason, and Martin-Senour) have hardeners available in the field. All these acrylic enamels, including catalyzed enamels, can be compounded lightly for gloss purposes after 24 to 48 hours if necessary. However, thorough compounding of enamel color is not recommended. When the need for thorough compounding arises due to poor blending of spot repairs, it is better for durability purposes to redo the spot repair. Simply wash off the enamel color with the appropriate reducer and do the repair again—correctly.

Spraying acrylic urethane may appear complicated to the apprentice or beginner because everything in enamel refinishing takes place more slowly. The painter must learn to adapt and be patient when working with these materials. With practice and experience, spot repairs can be made expertly as the painter becomes more familiar with the paint systems and products involved. Correct air pressure and spray technique with the correct materials are the key to top-quality workmanship.

Color Preparation

Double-check that the best equipment and color are being used, particularly when matching metallics:

1. Check the single-stage color on hand to be sure it is the correct color code.
2. Check for correct solvents for the temperature conditions.
3. If one is available, set up an agitator-type paint cup, which makes spraying metallics as easy as spraying solid colors.
4. Be sure to agitate the color thoroughly by hand or on a paint shaker.
5. Prepare acrylic enamel color with a two-spray gun system as follows:
   a. In spray gun 1, prepare a conventional reduction of acrylic enamel color with the proper reducer for the temperature. Label the cup "Color."
   b. In spray gun 2, prepare the color blender. Use the same reducer as that used in the color coat. Label the cup "Mist Coat."

Color Application

1. Set up the spray gun in the midrange adjustment (Chapter 4).
2. Spray out a small test panel with two coats of enamel color and allow 5 to 7 minutes flash time between coats. (See your local paint jobber for special black and white paper test panels.)
3. Adjust the amount of reduction, air pressure, and type of solvent to obtain a good color match. These variables of spray painting are controllable by the painter.
4. Spray two to three coats of color on the test panel for full hiding. Allow 5 to 7 minutes flash time between coats. Then compare the test panel to the car for a color match. If the color matches satisfactorily, proceed with step 6. If the color does not match, proceed with step 5.
5. If a color mismatch still results, adjust the color as explained in Chapter 14 before proceeding.
6. Apply a first coat over the primer-surfaced area (see Figure 12–9). Use a feathering technique at the beginning and end of each stroke (see Chapter 4). Apply color blender to the edges and allow it to flash.
7. Apply a second and, if necessary, a third coat in the same manner, extending each coat beyond the previous one as shown in Figure 12–9. Allow flash time between coats. Blend the edges as required.

SPOT/PARTIAL PANEL REPAIRS ON MULTICOAT FINISHES

A painter can do the following about color matching:

1. Get the correct color code from the car.
2. Maintain color-mixing equipment.
3. Use color variance decks.
4. Use the proper light source to evaluate color.
5. Use the proper air pressure, solvent selection, spray equipment, etc., as necessary.
6. Tint the color if necessary to make it blendable.
7. Blend the color as necessary.

The above steps can provide the correct color for any and all basecoat color repairs. The problem with basecoat/clearcoat finishes, however, is not in the color; it is in the clearcoat. Clearcoats are designed to be the toughest and most durable finishes in the automotive trade. There is no way to make spot repairs flawlessly on clearcoats within a panel so that it is 100 percent acceptable.

Most car companies, and particularly factory warranties, require that **if no clean breakline or body line exists for stopping the application of clearcoat, extend the application to the nearest panel edge.** However, in a nonwarranty situation, the blending of a clearcoat can be done as explained in Figure 12–11.

SPOT AND SPOT/PARTIAL PANEL REPAIR

FIGURE 12–11. Blending clearcoat of quarter panel into roof panel. (Courtesy of PPG Industries.)

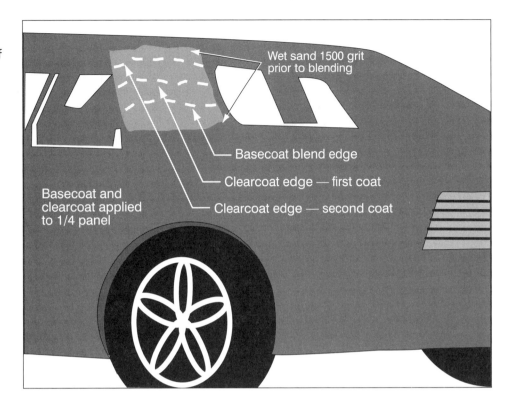

Use the following procedure for blending the clearcoat of a quarter panel into the roof panel:

1. Choose as small an area as possible to minimize the blend area.
2. Prepare the area for the blend. A light sanding with 1200- to 1500-grit paper is recommended.
3. Spray the basecoat to hiding and step out the edges where necessary.
4. Apply the first coat of clear. Stop the edge in the designated blend area. Dust the edge with DX 830 Blending Solvent.
5. Apply the second coat of clear. Extend the edge just beyond the first clear blend edge, about 1 to 3 inches. Extending the blend edges too far may lead to accidental removal of some of the clear when the edge is polished because the film build may be too thin.
6. Dust the edge with DX 830 Blending Solvent. You may lower your air pressure somewhat but too low an air pressure will not atomize the blending solvent properly. If you lower the pressure appreciably to work in a tight area, turn the fluid nozzle in to cut back on fluid delivery.
7. Apply a third coat of clear, if necessary, as already outlined.
8. When it is dry, polish the edge lightly by hand or machine. Use a foam or similar pad and a finessing type polish so you do not buff the clear edge and leave a line. (The only way to fix the line is to apply more clear!)

Blending a Spot Repair on a Basecoat/Clearcoat Finish

Figure 12–12 shows a spot repair with basecoat/clearcoat color on a quarter panel. The repair is finished with clearcoat.

1. Prepare the entire quarter panel for painting by cleaning and sanding with 1500 to 2000 sandpaper.
2. Determine the correct basecoat color from the paint code on the car.
3. Spray out a sample panel to the proper hiding and apply two coats of clearcoat to the panel. Allow the proper flash time between coats and allow proper drying before applying clearcoats.
4. Compare the sample color to the car under proper lighting conditions. Tint or adjust the color as required to make it blendable.
5. Apply the first coat to the repair area with a feathering application at the blend areas.
6. Apply a second coat of basecoat over the complete affected panel and extend the application into the adjacent panel at the top of the quarter

Blending Spot Repair

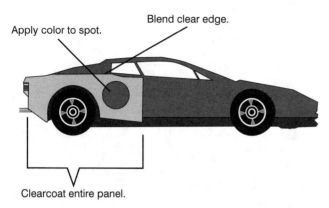

FIGURE 12–12. Blending a spot repair in quarter panel lower area. (Courtesy of PPG Industries.)

panel for several inches as shown. Allow it to flash. Featheroff all application of color at the blend area edges.

7. If necessary, apply a third coat of color over the entire repair area and extend the application beyond the previous coat.

Clearcoat Application

Before applying clearcoats, allow the basecoat color to dry according to the paint supplier's recommendations.

1. Mix the clearcoat with the required components per the label directions.
2. Strain the clearcoat.
3. Use the recommended spray equipment and safety gear.
4. Apply the first coat of clear to the repair panel and extend the application beyond the blend area, where it is feathered off.
5. Apply the second coat of clearcoat to the entire affected panel and featheroff the application at the top of the quarter panel.
6. Apply a third coat of clear to all the repair areas if polishing is required. Carefully featheroff the blend areas with blending solvent.

Spot Repair of Door on Tricoat-Finished Car

1. Repair the damage in the middle of the door as shown in Figure 12–13 and carry repair operations through the sanding of the undercoats in preparation for color application.
2. Prepare the entire side of the car for final clearcoating by washing; sanding with 1500, 2000, or 2500 sandpaper; and recleaning all panels with a solvent cleaner.
3. Obtain the correct basecoat and midcoat color codes from the car body number plate or service parts identification label with help from a paint jobber. Secure the correct paint system components from the jobber.
4. Find the uncleared and unpearled section of the basecoat (groundcoat) color on the car. Check under the door sill plate, on the trunk floor pan, on the trunk lid inner panel, in the engine compartment sides, on radiator braces, etc.
5. Spray out a sample of uncleared basecoat color on a test panel and compare to the car OEM basecoat color. If necessary, make a final adjustment to the basecoat color to make it blendable. The basecoat must be blendable or better to make the repair possible. If the basecoat color is not blendable, you will not be able to achieve a color match. The true color comes from the basecoat.
6. You are now ready to make a let-down panel as explained in Chapter 11 (see Figure 11–7). The purpose of a let-down panel is to predetermine accurately the number of pearlcoats needed to duplicate the OEM finish.
7. Check the let-down panel against the car under proper lighting conditions to determine which test panel matches the car best.

By holding the let-down panel next to the OEM color in a properly lighted area, the painter can decide which panel of the five most closely matches the OEM color. If one of the panels matches the OEM finish, that particular panel answers the question about the number of pearlcoats necessary to make an acceptable repair.

If the let-down panel indicates that the proper match lies between coats of pearl (between three and four coats, for example), the best way to approximate this "half-coat" is to use a color blender in your chosen repair system. Your next coat after 3 will be a 1:1 mix of ready-to-spray pearlcoat and DBU 500 or DBC 500 color blender (based on the basecoat system you are using) in a separate paint cup. Other paint companies have equivalent clear color blender products. Because these color blenders are essentially basecoat products without pigment, you will have created the half-coat you need to obtain an acceptable level of pearl effect for a blendable match. By using this method, you have made your pearlcoat more transparent.

To produce a half-coat of pearlcoat, prepare the materials as follows: 1 pint of reduced DBU pearlcoat plus 1 pint of reduced DBU 500 Color Blender

SPOT AND SPOT/PARTIAL PANEL REPAIR

FIGURE 12–13. Spot repair on door of tricoat-finished car. (Courtesy of PPG Industries.)

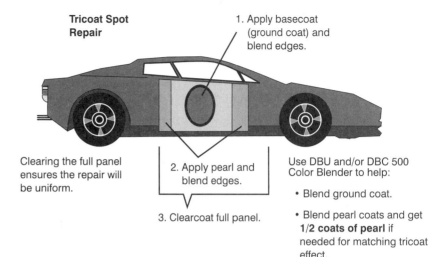

(reduced at the same ratio, with the same reducers of the DBU pearlcoat) equals 2 pints of half-coat of pearlcoat. If you are using DBC basecoat, use DBC 500 Color Blender in the same way. If necessary, the 1:1 ratio can be changed (more or less color blender added) to match whatever pearlcoat effect you need. The "half-and-half" method is a convenient starting point.

The following basecoat, midcoat, and clearcoat application procedures are part of the spot repair of a door on a tricoat finished car started on page 192.

BASECOAT APPLICATION

1. Apply the first coat of the basecoat color over the repair area and blend the color at the edges.
2. Apply a second coat of the basecoat color beyond the first coat and blend the edges as necessary.
3. If necessary, apply a third coat beyond the second coat and blend the edges as required.
4. Allow the basecoat to dry per the paint supplier's recommendations before applying the midcoat color.

> **CAUTION**
>
> Try to keep the repair as small as possible; these repairs can easily become time-consuming.

> **CAUTION**
>
> Blend the base color with approved color blenders, based on your chosen system. You can add up to two parts of reduced color blender to every one part of your ready-to-spray basecoat color (2:1 ratio).

MIDCOAT APPLICATION

1. Mix the final selected midcoat color thoroughly.
2. Strain the color mix just before filling the spray gun.
3. Use the let-down panel to determine the number of midcoats needed to achieve the color match. Use the two-gun system of applying midcoats and blending material.
4. Apply the number of midcoats as determined by the let-down panel and blend each coat over the new basecoat color on adjacent surfaces as required. Allow the proper flash time between midcoat applications.

> **CAUTION**
>
> Blend midcoats with a second gun and prepared with a mixture of two or more parts of color blender to one part of ready-to-spray midcoat material. You may wish to use more color blender than the 2:1 ratio. **The danger is that any extra coats of full strength pearlcoat applied over OEM pearl at the blend will show up as a dark halo or line!**

CLEARCOAT APPLICATION

Before applying the clearcoats, allow the midcoat to dry per the paint supplier's recommendations.

1. Mix the clear with the required components to the paint supplier's recommendations.
2. Select the correct spray equipment (for example, nozzle size and air cap).

3. Apply the clearcoat over the newly applied midcoat application on the repair panel and on complete adjacent panels.

   a. Allow adequate time between coats.

   b. Apply the second coat of clear over the repair panel and over the complete adjacent panels.

   c. Apply a third coat of clear over the entire repair area if polishing will be required.

REVIEW QUESTIONS

1. Explain the procedure for making a spot repair on a single-stage color finish.
2. Outline the procedure for spot repairing a given panel on a basecoat/clearcoat finish.
3. Why is compounding of adjacent surfaces required when spot repairing original single-stage enamel finishes?
4. When making a spot repair on a basecoat/clearcoat finish, how much or how little of the panel may be clearcoated?
5. What is color blender and how does it work?
6. Explain the procedure for blending a quarter sail panel to the roof panel joint.
7. Explain the procedure for making a let-down sample panel to check the number of pearlcoats required to match an OEM pearlcoat finish.
8. Explain the procedure for matching a midcoat color on a tricoat finished vehicle.
9. How can a painter achieve a color match when the color shade needed is between two spray-out panels, one having three coats and the other having four coats of pearl finish?
10. What does the term *paint repairability* mean?
11. There are three requirements for making a successful spot repair:
    a. Color match.
    b. Blend.
    c. Gloss and surface texture.
    Explain the minimum level of acceptability for each.
12. Name several paint problems a painter can correct without repainting.

CHAPTER 13

Compounding and Polishing

INTRODUCTION

In the earlier days of refinishing, most repairs were done with lacquers, and all work required compounding and polishing to complete the job. Since then, lacquers have been replaced with new technologies. With the introduction of basecoat/clearcoat finishes, new technologies in compounding and polishing also became required to keep pace with new polishing challenges. But even with all these changes in paint technologies, the reasons for using compounds and polishes remain very much the same.

The purpose of compounding and polishing is revealed in the following:

1. To restore gloss and beauty to aged finishes.
2. To increase gloss and smoothness on new finishes.
3. To prepare finishes during refinishing for blending operations and adhesion.
4. To repair minor refinish problems such as the following without repainting:
   a. Dirt nibs and orange peel.
   b. Sandscratches.
   c. General overspray, etc.
5. To repair finishes affected by environmental damage such as:
   a. Acid rain.
   b. Iron rust spotting.
   c. Bird droppings, etc.

COMPOUNDING IN REFINISHING

Compounding, as used in the refinish trade, involves the use of many sharp abrasives that come in many sizes and forms. The main job of compounding is to cut or level a coarse surface and make it smoother. Compounding removes normal sandscratches, dry spray, and minor surface imperfections. Compounding is also used over smooth, painted surfaces during repainting operations because the cutting and shearing action of compounds abrades the surface (makes it rougher) to aid the adhesion of topcoats.

Before basecoat/clearcoat finishes, compounds were classed as fine, medium, and coarse. With the introduction of basecoat/clearcoat finishes, finer brands of compounds became required. Also, with the development of fine sandpapers (ranging from 800- to 2500-grit size) required by the basecoat/clearcoat finishes, rubbing compounds have also been developed to keep pace with the sandpaper trade.

The cutting action of compounds parallels the cutting action of sandpaper. When a 1500-grit sandpaper makes sandscratches, a rubbing compound with equivalent 1500-grit abrasive can remove those scratches. It is necessary to use the proper compound to remove specific sandscratches.

Paint technicians must understand the meaning of *micron* and *mil* because they must work within factory and/or paint company specifications when making paint repairs to keep up the durability of a paint finish and/or for warranty purposes. A mil is 1 one-thousandth of an inch. Automotive paint thicknesses are measured in mils. **One mil is about the thickness of aluminum foil.** See Chapter 9 for paint finish thicknesses. A micron is one-twentieth of a mil. It takes 20 microns to equal 1 mil. The size of sandscratches is measured in microns.

When making repairs on basecoat/clearcoat topcoats, no more than $\frac{1}{2}$ mil of clearcoat should be removed to maintain film durability. Paint technicians should use a paint thickness gauge to guide these repairs. If more than $\frac{1}{2}$ mil of clearcoat is removed, additional clearcoat needs to be added.

TABLE 13–1 Products for 3M Perfect-It II Paint Finishing System

Part No.	Description	Items/Unit	Units/Case
3M™ GLAZES/WAXES			**QUANTITY/CASE**
06050	3M™ Perfect-It™ foam polishing pad glaze/dark	12 oz.	6
06051	3M™ Perfect-It™ foam polishing pad glaze/light	12 oz.	6
05995	3M™ Perfect-It™ foam polishing pad glaze/light	1 qt.	12 qts.
05996	3M™ Perfect-It™ foam polishing pad glaze/dark	1 qt.	12 qts.
05997	3M™ Perfect-It™ hand glaze	1 qt.	12 qts.
06054	3M™ premium liquid wax	12 oz.	6
06005	3M™ premium liquid wax	1 qt.	12 qts.
06006	3M™ premium liquid wax	1 gal.	4 gals.
06055	3M™ premium paste wax	14 oz.	6
06009	3M™ Perfect-It™ wax and glaze wipe	125	4 boxes
3M™ SUPERBUFF™ BUFFING PADS		**PADS/BAG**	**QUANTITY/CASE**
05700	3M™ Superbuff™ buffing pad	1	24
05701	3M™ Superbuff™ 2 Plus 2 pad	1	24
3M™ HOOKIT™ SBS™ COMPOUNDING PAD/BACKUP PAD		**PADS/BAG**	**QUANTITY/CASE**
05711	3M™ Hookit™ SBS™ Plus compounding pad	1	24
05717	3M™ Hookit™ SBS™ backup pad	1	1
3M™ PERFECT-IT™ FOAM POLISHING PAD/BACKUP PAD		**PADS/BAG**	**QUANTITY/CASE**
05725	3M™ Perfect-It™ foam polishing pad	2	12
05718	3M™ Perfect-It™ backup pad—for use on 3M™ Superbuff™ adaptor #05710	1	1
3M™ PERFECT-IT™ II RUBBING COMPOUND			**QUANTITY/CASE**
05973	3M™ Perfect-It™ II rubbing compound	1 qt.	12 qts.
05974	3M™ Perfect-It™ II rubbing compound	1 gal.	4 gals.
3M™ IMPERIAL™ WETORDRY™ COLOR SANDING PAPER SHEETS		**SHEETS/SLEEVE**	**SLEEVES/CASE**
02022	Micro fine 1200	50	10
02023	Micro fine 1500	50	10
02044	Micro fine 2000	50	10

Courtesy of 3M Automotive Aftermarket Division.

POLISHING

Polishing is the application of a product, a substance, made especially for the purpose of creating gloss. Most of these products are of a wax or chemical compound construction and include special fine abrasives. The purpose of adding abrasives is that the single polish can clean and polish in one operation. Polishes give protection to a paint finish as a clear glossy sealer, which can be noted by the beading action of water on horizontal surfaces. Good polishes are very slippery and deflect water, dirt and the elements. Thus, they protect the finish.

Custom polish jobs involve thorough chemical cleaning of the surface, compounding the complete surface, and applying a special sealant type polish. These polish jobs last about one year. A reapplication of the sealant polish itself, after cleaning, usually lasts another year.

All compounds and polishes use a special solvent base composed of chemicals, water, and/or special solvents. Each polish company has its own patented manufacturing system. Polishes that do not contain abrasives are so stated on the label. Polishes with no abrasives are known as sealants.

Table 13–1 is a summary of products used in the 3M Perfect-It™ II paint finishing system available through paint jobbers in the refinish trade. Other compound and polish companies have products that may be equivalent to 3M's. However, the 3M system has been approved by GM, Ford, and Chrysler, and it is available to body and paint shops nationwide. For further information regarding the availability of approved compounding and polishing systems, check with your local paint jobber.

COMPOUNDING AND POLISHING

Hand Rubbing Procedure

Hand compounding is not easy. Before starting, prepare the compound according to the label directions.

1. Apply the compound sparingly to a rubbing ball made of water dampened and wrung-out clean, soft cloth. Flannel cloth is excellent for this purpose. The amount of compound on the cloth is determined by the size of the area worked. Do not apply hand compound directly to the painted surfaces.
2. Rub with straight back-and-forth motions in one direction. Use both hands on the rubbing ball and apply moderate pressure. Continue rubbing until the desired smoothness and gloss are achieved. Work small areas at a time. Results do not happen instantly. It may take several minutes to do one spot only several inches square. As the gloss begins to appear, ease hand pressure and polishing action until a final gloss is attained. Repeat each application as necessary.
3. With a clean cloth, wipe the surface clean by removing all traces of rubbing compound.
4. Use a suitable soft-bristle brush to aid in removing compound from gap spacings, crevices, and tight corners. These areas on a car cannot be reached with a machine polisher, which is why hand compounding becomes necessary.

Cleaning the Wool Pad During Use

1. Go to a suitable location in the building, or outside the building if weather permits.
2. Lay the polisher on the floor with the pad up, grip the handle, and turn on the polisher.
3. Apply the bonnet cleaning tool or suitable tool to the right side of the pad as the polisher is running, and move the tool toward the center.
4. Work the tool from the center to the right edge and back in a straight line. Repeat this movement of the tool back and forth several times until the pad is clean.

Machine Compounding Procedure

1. Prepare the car by applying masking tape ($\frac{3}{4}$ inch) to the crease-lines, raised panel edges, and sharp corners that will be contacted by the rubbing pad to prevent cut-throughs.
2. Prepare the rubbing compound by following the label directions. Usually, it is necessary to add water and agitate the mixture thoroughly.
3. Prepare the polisher with a recommended, clean, properly tufted, carpet-type pad. Have pad cleaner handy for occasional use during compounding to keep the pad clean.
4. Wear a suitable apron or shop coat to protect clothing. Use glasses and a dust mask.
5. Apply rubbing compound to the surface with a suitable brush (2- to 4-inch) from a special container or from a suitable plastic squeeze bottle. **Never apply compound to the wheel. Avoid the use of too much compound.**
6. Use the polisher as follows:
   a. **After the compound has been applied, press the rubbing pad to the compound, in the flat, and generally distribute compound to the area to be rubbed.** Never press on the polisher. Let the polisher's own weight do the job. Always do a small area at a time, such as 2 feet by 2 feet square.
   b. Now turn on the polisher. **Start with left-to-right and right-to-left strokes. Lift the right half of the pad up slightly when moving to the right, and lift the left half of the pad up when moving to the left.** These strokes are equivalent to the buffing strokes.
   c. Overlap each left-to-right and right-to-left stroke by 50 percent.
   d. After 2 square feet have been covered going from left to right, change directions to move fore and aft. Use the upper half of the pad on the surface by raising the lower half to move toward the operator. Use the lower half of the pad on the surface by raising the upper half to move away from the operator.
   e. Overlap each fore-and-aft stroke by 50 percent.
   f. Two times over with left-to-right strokes, and two times over with fore-and-aft strokes constitutes a complete compounding cycle.
7. If too much compound remains at the end of the compounding cycle, too much compound was used at the start.
8. If compound disappears sooner than two or three times over the surface, you are not using enough compound.
9. As each area is compounded, move to a new area. Overlap each area 3 to 4 inches.
10. Check the surface closely during operation to determine the progress of the desired cutting action. Reapply compound to the surface as necessary.

TABLE 13–2 Compounding Problems, Causes, and Remedies

Problem and Cause	Remedy
1. Cut-throughs at panel edges and corners. *Cause:* Excessive compounding without protection.	1. Clean and brush touch-up as required. *Prevention:* Tape sharp edges.
2. Although polisher and pad are okay, compounding is slow and gloss is poor. *Cause:* Substandard compound.	2. Contact another paint jobber and change to a different brand.
3. Compounding process slow due to excessive orange peel and overspray. *Cause:* Poor solvents and poor spray technique.	3. Water-sand surface with No. 1500 or 2000 sandpaper before compounding.
4. Rub-through in middle of panel; primer shows. *Cause:* Apparently insufficient color applied or excessive compounding.	4. Spot- or panel-repair as required.
5. Pad clogs up fast; poor results. *Cause:* Hand compound used with power polisher.	5. Clean or change pad as required: use proper compound.

NOTE

The use of water for wet sanding with No. 1500 or 2000 sandpaper before compounding is determined by the condition of the surface, that is, the amount of dirt nibs, overspray, and/or orange peel present. The objective of rub-out and polish operations is to match adjacent surfaces in terms of gloss and surface smoothness.

11. Remove masking tape from all crease-lines and panel edges. Rub out and polish these surfaces by hand. Also, work on those areas that were inaccessible with the polishing wheel.

For a list of problems that may arise during the compounding process, see Table 13–2.

To complete a compounding and polishing job, wash the car completely, including all crevices, corners, chrome, glass, and exterior plastic parts. Using a chamois or equivalent cloth, wipe all surfaces, windows, and chrome parts. When doing a complete paint job, clean all windows inside and outside and vacuum the interior trim and floor carpets.

3M PERFECT-IT™ II PAINT FINISHING SYSTEM

The 3M PERFECT-IT™ II paint finishing system is covered at this time because it can make paint repairs on basecoat/clearcoat finishes without the need for repainting. The system is easy to follow and places emphasis on making repairs on basecoat/clearcoat finishes while removing less than $\frac{1}{2}$ mil of clearcoat. For this purpose, a paint thickness gauge is required. If more than $\frac{1}{2}$ mil of clearcoat is removed, the addition of more clearcoat is required. Paint problems that can be repaired with the PERFECT-IT II paint finishing system are wheel or swirl marks, low gloss, oxidation, minor scratches, overspray, orange peel, dust nibs, paint runs and sags, and the effects of acid rain. Table 13–3 lists the basic tools and equipment for use with the 3M Perfect-It II paint finishing system.

CAUTION

When making the following repairs on basecoat/clearcoat finishes to remove paint problems without repainting:

1. Use the **"less is best method."** For example, when sanding is necessary, always sand with the finest grit sandpaper. Proceed to the next coarsest grits only when necessary.
2. Remember: **Do not cause any more problem than already exists.** Follow the recommendations and steps. Using too coarse a sandpaper too early causes deep scratches, a result that defeats the purpose of the repair.

COMPOUNDING AND POLISHING

TABLE 13-3 Basic Tools and Equipment Recommended for Perfect-It™ II Paint Finishing System

Item	Part Number[a]
Micro fine sandpaper—1200	02022
Micro fine sandpaper—1500	02023
Micro fine sandpaper—2000	02044
Sponge rubber sanding pad	
Machine polisher (direct drive—rotary)	
Machine Polisher (dual action [D.A.])	
Compounding and polishing pads:	
3M Hookit backup pad	0517
	0518
Wool compounding pad	05711
Foam polishing pad	05700
Foam polishing pad	5725
Rubbing compound	5973
Foam polish pad glaze	5995—White
	5996—Dark gray
Safety equipment (dust respirator, goggles, and gloves)	
Paint thickness gauge	

[a]All components that do not show a part number are available through your local paint jobber.

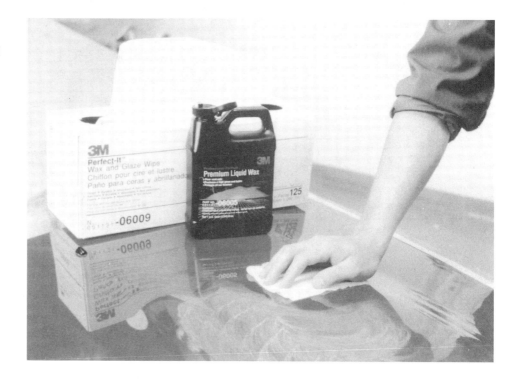

FIGURE 13-1. Hand glazing/waxing using wax and glaze wipes with liquid wax. (Courtesy of 3M Automotive Aftermarket Division.)

1. Clean the panel with soap and water.
2. Clean the panel with wax and grease remover.
3. Check and record the thickness of the paint at the spots to be repaired with a paint thickness gauge. (See Chapter 9 for details about the use of a paint thickness gauge.)
4. At the spot to be repaired (which represents damage), apply polish (PN 5995) using a foam polishing pad (PN 5725) on a rotary polisher.
5. Inspect the area that you polished (Figure 13-1). If the imperfection is removed, complete the rest of the vehicle in same manner. If it is

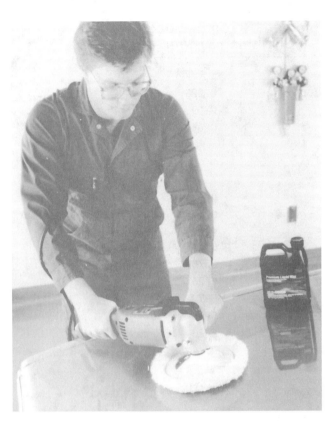

FIGURE 13–2. Compounding can remove between 1200 and 1500 scratches while leaving minimal swirl marks. (Courtesy of 3M Automotive Aftermarket Division.)

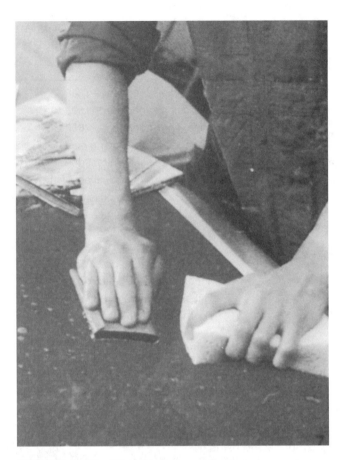

FIGURE 13–3. Wet-sanding can remove imperfections such as dust nibs and paint runs, and can level orange peel. (Courtesy of 3M Automotive Aftermarket Division.)

not, move to the next more aggressive step, which is compounding.

6. Compound the area with a wool or foam compounding pad (PN 05711) and compound (PN 5973) and inspect the surface (Figure 13–2). If the surface is okay, apply polish as in step 4 to complete the job.
7. Check and record the thickness of the paint at the spot under repair.
8. If the condition is still present, wet-sand the area with 2000-grit sandpaper and check your progress with a rubber squeegee (Figure 13–3).
9. Check and record the thickness of the paint at the spot under repair.
10. If progress is too slow, move to the next more aggressive sandpaper grit, such as 1500, and repeat wet-sanding the repair area. Inspect the area for the problem condition and recheck the paint thickness with the gauge.
11. If progress is still too slow and if the paint defect remains, wet-sand the repair area with 1200-grit sandpaper. Inspect the area for the problem condition and recheck the paint thickness with the gauge.
12. Using rubbing compound and a compounding pad, as in step 6, compound the area to remove sandscratches.
13. Polish the area to remove compounding swirl marks as in step 4 using a foam polishing pad (5725) on a rotary polisher and with PN 5995 or 5996 glaze (Figure 13–4).
14. Clean the area with 3M Perfect-It detailing cloth (PN 39016 or 6016) to remove dust or haze. Then inspect the surface for final appearance under proper lighting conditions, such as natural sunlight.
15. Black, red, dark blue, or other dark colors have a tendency to show swirl marks more than light colors. If they are present, use a random action/dual action polisher or sander and attach a foam polishing/finishing pad. Apply polish (PN 5995 or 5996) to remove the swirl marks and to complete the job (Figure 13–5).
16. Apply a protective coating of wax or sealant as desired.

COMPOUNDING AND POLISHING

FIGURE 13–4. Polishing the car surface using a conventional rotary polisher and foam polishing pad glaze. (Courtesy of 3M Automotive Aftermarket Division.)

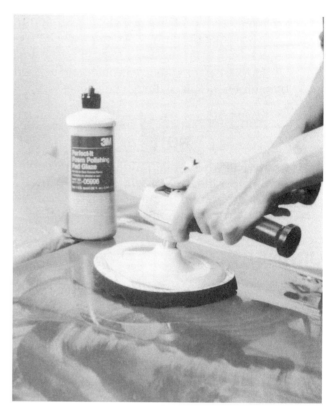

FIGURE 13–5. Removing swirl marks by using a random action/dual action polisher or sander and foam polishing finishing pad and polish. (Courtesy of 3M Automotive Aftermarket Division.)

The following is a list of conditions that can happen to any paint technician during the application of topcoats to a car. The paint technician can take steps to reduce some of the risk of these defects.

RUNS AND SAGS

Check your gun technique first, then the speed of the solvents used, gun setup, etc. If you are trying a new clear- or topcoat, make sure you practice before applying it to a production vehicle.

DIRT

Most dirt in any paint job comes from three sources:

1. **The vehicle:** Do a little more prep cleaning before painting.
   a. Blow off the door jambs, cracks, crevices, wheel wells, etc., one more time before you put the vehicle into a booth to start painting.
   b. Remask if the masking paper or media is tattered or loaded with dry overspray from previous paint applications.
   c. Use DX 103 Multiprep as an antistatic spray prior to applying topcoat. This spray removes the static charge inherent in the painting process. Static attracts airborne dust and dirt particles.
2. **The technician:**
   a. Wear disposable painter's coveralls or special paint suits. They are usually coated with a slick facing that does not hold dust. Many are also solvent-resistant.
   b. Stay away from cloth coveralls. They are dust carriers.
   c. Use a blowgun to dust yourself off before you enter the booth.
   d. Wear a hat or head sock to keep hair where it belongs, on your head, not in your paint work.
3. **The spray booth:** If the booth is maintained properly, this source of dirt can be minimized.
   a. Change booth intake and exhaust filters on a regular schedule based on use. If the filters look plugged, they very well could be. Check them often.

b. Make sure fan blades are clean and motors are working properly. Check drive belts frequently.
c. Make sure door seals are in place and in good condition. Replace them when necessary.

> **NOTE**
>
> Not only will a well-maintained booth keep the paint job cleaner, it will also help the products dry to and to cure better.

ORANGE PEEL

This defect is usually affected by spray gun technique, solvent or hardener choice, gun setup, surface preparation, and a host of other factors.

1. A step-by-step review of each possibility is necessary to determine what needs to be corrected.
2. This issue comes up most often when the orange peel **does not match the vehicle being refinished.** Some areas have texture; some do not.
3. The buffing process will remove the orange peel, but it is better to determine ways to reduce it by checking the processes and equipment being used.

FISHEYES

This problem comes from either oil-based contamination not being removed from the surface before painting or by airborne contamination landing on the panels during the painting process. Some oil-based products common to a body shop are polishing compounds that contain certain silicones, mechanical lubricants (almost *any* oil-based lubricant), automotive waxes and dressing for interiors or exteriors, etc.

To prevent fisheyes:

1. Wash every car before it enters the shop.
2. Use wax and grease removers properly.
3. Use disposable wipes.
4. Do not use shop towels or laundered rags for cleanup.
5. Keep any possible source of contamination out of the prep and painting areas.
6. Have a designated polishing and compounding area away from the paint department if at all possible.

The following procedure often works well for eliminating fisheyes:

1. Wash the surface thoroughly with clean wipes and wax and grease remover.
2. Wet-sand the affected area with suitable sandpaper and above remover.
3. Rewash the affected area with clean wipes and wax and grease remover.

When removing defects, the following steps and products are involved in the process:

Step 1: First make sure the surface is clean. Soap and water removes water soluble particles, dust, etc. Wax and grease cleaners like DX 330 (PPG) are mild and safe as wet-sanding agents. DX 390 is low VOC cleaner.

Step 2: If sanding is required on basecoat/clearcoat finishes, start sanding with 2000-grit sandpaper. Wet-sanding has historically been a preferred method by most collision centers. If the defect is not removed, graduate sandpaper and use the next coarsest grit, which is 1500, etc.

Step 3: If less than $\frac{1}{2}$ mil of clearcoat has been removed as indicated by a paint thickness gauge, the next step is to compound and then to polish the area to complete the repair.

Step 4: If the problem condition is still present, additional sanding must be done in preparation for replacing the topcoats as required.

Some special sandpapers for dual-action sanders (DAs) have been developed in the past few years. They can be effective but their use takes some practice. They seem to work best on large flat panels such as hoods and deck lids. To use the sandpapers effectively, the sander must be a quality tool with very smooth operation characteristics. Several tool companies make such sanders. Contact your local jobber for recommendations and more information.

For **dry sanding,** use sandpaper in the following order: 800-grit DA, 1000-grit DA, and 1200-grit DA. Sandpaper is available in several brands of abrasives. **The key is to remove as little paint film as possible.** Other tools for removing defects are available and can be effective:

1. Waterbug mechanical sander.
2. Barrel sanding blocks for isolated defect removal.
3. Vixen file.
4. Razor blade.

When **buffing,** a painter must make the following choices:

1. Air versus electric buffers.
2. Thermoplastic versus thermoset topcoats.

Thermoset (urethanes) require a slower speed, about 1500 rpm. The buffer types shown above have some characteristics that should be considered. **Air buffers** are easy to control and are lightweight, and their power supply (air) is readily available. However, they are noisy and, depending on the individual tool, they may be hard to control at low speeds. **Electric buffers** have a steady power supply, and they are available in variable speed models with fair speed control. However, their cords tangle easily, they are heavy and noisy, and they can create sparks.

Whichever type of buffer you choose, make sure it has variable speed because many of today's compounds work better at slightly different rpms. It is difficult to use one speed for everything. Buffing procedures have changed as new compounds were developed to polish thermoset urethane topcoats:

1. Now buffers are used in the 1200–1500 rpm range versus the 1800–2000 rpm or more used for lacquer type compounds.
2. Slower speeds are generally more effective overall with today's compounds.

Buffing pads have changed considerably because of the changes in today's compounds and polishes.

1. First, a quick-change backing pad is needed on the buffer.
2. A wool-blend compounding pad is excellent for compounding.
3. A type of sponge pad may be used for compounding and polishing.
4. A special design sponge pad is excellent for fine finish polishing and applying final glazes.

The use of a dual-action (DA) polisher at the right speed (rpm) is sometimes helpful after using a rotary polisher for obtaining the best results in polishing. The use of a quick-change fastening system means quick pad changes during buffing operations.

For best results when buffing:

1. **Buff "off" the edge of a panel or body line, not "into" it.**
2. **Keep the speed slow, and keep moving.**
3. **Use light to medium pressure, only what is necessary.**
4. **Keep the polish on the surface. Don't let it dry too much.**
5. **Buff about two square feet at a time.** Do not try to do a complete hood at once.

The above tips are good buffing practices that can make the job much easier.

A few final words about compounding and polishing:

1. **Take the time to clean before you paint.**
2. **When making a repair, remove as little of the paint as possible, only enough to correct the condition.**
3. **When you must compound, make sure you use the correct cutting material for the size of the scratch.**
4. **When you decide to polish, make sure you use the right glazing product for your paint.**

DAMAGE FROM ACID RAIN AND INDUSTRIAL FALLOUT

Can acid rain and industrial fallout damage an automotive finish? They definitely can, so it is important to know the causes and the proper steps required to repair the damage.

Acid rain is created when sulfur dioxide or nitrogen oxides are released into the atmosphere and combine with water and the ozone to create either sulfuric or nitric acid. It is estimated that the United States alone produces 30 million tons of sulfur dioxide and 25 million tons of nitrogen oxides yearly. Over two-thirds of the sulfur is emitted from power plants burning coal, oil, or gas. Iron and copper smelters, automobile exhaust, and natural sources like volcanoes, wetlands, and forest fires account for most of the remaining pollutants. Once released into the ozone, these acids are readily dissolved into cloud droplets that, if low enough in pH, can cause significant damage.

The standard for measuring acid rain is the pH scale. It runs from 0 to 14, with 7 being neutral, or equal to distilled water. A pH reading of 4 is ten times more acidic than a solution of acid and water with a pH of 5, and 100 times more acidic than a pH of 6.

The level of acid rain varies greatly around the country. For example, South Carolina is reported to be one of the most acidic states in the nation. In Los Angeles, fog has been measured to have the acidic strength of lemon juice. Rainfall in the northeast is extremely corrosive to car paints and finishes. For example, the average pH of rainfall in New Jersey is an acidic 4.3. GM now has clauses in some of its new car warranties that exempt them from liabilities involving paint damage in high pH areas.

Identifying Acid Rain Damage

How can you tell if a vehicle finish has been damaged by acid rain? Damage generally occurs to the paint pigments, with lead-based pigments the most susceptible. Typically, the damage looks like water droplets that have dried on the paint and caused discoloration. Sometimes the damage appears as a white ring with a clear, dull center. Severe cases show pitting.

Discoloration varies depending on the color. For example, the rain damage to a yellow finish may appear as a white or dark brown spot. Medium blue may have a whitening look. White may be discolored pink, and medium red may be discolored purple.

Single-stage metallic finishes can become damaged because an acidic solution reacts with the aluminum particles and etches away at the finish. Freshly painted cars are more easily damaged than aged finishes. Lacquers and uncatalyzed enamel finishes are most susceptible to damage, followed closely by catalyzed enamel. Clearcoated finishes add a layer of protection against acid rain, so later model vehicles with basecoat/clearcoat are less susceptible to damage. A clearcoat protects the paint pigments from discoloration, but it is still possible for acid rain to create a peripheral etch or ring on the clearcoat.

Repairing Acid Rain Damage

Restoring a finish damaged by acid rain is generally not too difficult. The procedure varies, however, depending on the level and depth of the damage. The following steps outline repair procedures according to the level of damage as illustrated in Figure 13–6. You can stop at the stage in which the problem has been corrected. Remember that polishing or compounding removes part of the original finish and thereby reduces the overall life of the finish.

SURFACE LEVEL DAMAGE ONLY (A)

1. Wash with soap and water.
2. Clean with wax and grease remover.
3. Neutralize the area by washing with baking soda solution (1 tablespoon baking soda to 1 quart of water) and rinse thoroughly.

DAMAGE EMBEDDED IN SURFACE COAT (B)

1. Follow the cleaning and neutralizing steps listed above.
2. Hand polish the problem area. Inspect and continue if necessary.
3. Buff with a polishing pad. Inspect frequently and remove as little of the original finish as possible to cure the problem.
4. Use rubbing compound. Inspect and continue if necessary.
5. Wet-sand with 1500- or 2000-grit sandpaper and compound. If damage is still visible, repeat with 1200-grit sandpaper. Do not use a coarser grit than 1000.

DAMAGE THROUGH TO UNDERCOAT (C)

1. Follow cleaning and neutralizing steps listed for surface level damage above.
2. Sand with 400- to 600-grit sandpaper.
3. Reclean and neutralize prior to priming and repainting.

Identifying Damage from Industrial Fallout

Generally, damage from industrial fallout is caused when small, airborne particles of iron fall on and stick to a vehicle's surface. The iron can eventually eat through the paint, causing the base metal to rust. The damage is sometimes easier to feel than to see. If you sweep your hand across the suspected damage, you will likely detect a gritty or bumpy feel. Rust-colored spots may be visible on light-colored vehicles.

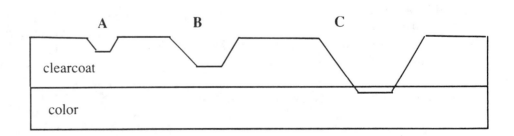

FIGURE 13–6. How acid rain affects paint finishes: (a) light damage, (b) moderate damage, (c) severe damage. (Courtesy of PPG Industries.)

COMPOUNDING AND POLISHING

Repairing Industrial Fallout Damage

The steps for repairing damage caused by industrial fallout are similar to those for repairing acid rain damage, but with the following exception: After washing the car, treat the repair area with a fallout remover, a chemical treatment product made especially for treating industrial fallout damage. **Do not buff the damaged area before removing the fallout because buffing will drive the particles into the paint surface.** Also, if the particles break loose and become lodged in the buffing pad, they can cause deep gouges.

One company that makes fallout remover is ZEP Industries. To obtain this or an equivalent material, check with your local paint jobber.

REVIEW QUESTIONS

1. List four reasons why compounds and polishes are used.
2. What is the main job of a compound?
3. What is the main job of a car polish?
4. What is a practical way of applying rubbing compound on a surface before turning on the machine?
5. How does a painter determine which rubbing compound works best for speed and quality of finish?
6. Describe a procedure for repairing moderate orange peel and minor sags and dirt nibs without repainting.
7. Describe how rubbing compounds work as cutting agents.
8. Describe the procedure for repairing fallout damage consisting of rust specks in paint.
9. Describe the procedure for hand rubbing.
10. Describe the procedure for repairing acid rain damage.
11. Describe the machine rubbing procedure.
12. What safety equipment should a painter use when compounding and polishing?
13. What is the recommended rpm rating for polishers used on basecoat/clearcoat finishes?
14. What grades of abrasives are used in the 3M Perfect-It II paint finishing system?
15. What type of bonnet is best for polishing with glaze materials?
16. When compounding, how can a technician avoid cut-throughs at high panel edges, sharp crease-lines, and corners?
17. What is the procedure for cleaning a wool compounding pad during use?
18. The depth of a sandscratch is measured in microns. How many microns equal 1 mil of paint?
19. What type of bonnet is best for machine compounding?
20. How much clearcoat can be removed during a polishing repair while maintaining the integrity of a paint finish without repainting?

CHAPTER 14

Color-Matching Fundamentals and Techniques

INTRODUCTION

This chapter is designed for painters and paint shops that do not have automatic color-matching equipment. Essentially, the chapter covers the achievement of successful color matches manually, which also involves the human eye and certain basic color-matching principles.

When a color-match problem is expected or encountered, the painter should spray out one or more test panels to full hiding for comparison to the car. By adjusting the variables under his or her control on additional test panels, the painter can often achieve a good color match. When a good color match cannot be achieved under these circumstances, tinting is required. **Tinting most definitely should be done as a last resort.**

The basics on color matching and tinting as covered in this chapter should aid both the novice and the experienced painter in achieving color matches. Best results are achieved by working closely with paint formulas for specific colors being adjusted. If formulas are not available at a paint shop, check with the paint jobber of the paint material being used for answers to specific questions and specific tints required. If possible, take a painted car part to the paint jobber.

As paint technicians, painters need to know more about color and adjustment then ever before because of the ever increasing advances in automotive paint technologies and application methods. To adjust a color by tinting, the painter should have a fundamental knowledge of basic color theory. He or she should understand the basics as well as the tints themselves before picking up tinting colors in an attempt to "bring a color in." It is only through practice and experience, two important ingredients, that a painter becomes qualified as an expert in the art of tinting.

VARIABLES AFFECTING COLOR MATCH

Because of their construction and behavior, metallic colors are sensitive to the solvents with which they are reduced and the air pressure with which they are applied. A **variable** is a part of the spray painting environment (such as temperature, humidity, and ventilation) or a part of the spray painting process (such as amount of reduction, evaporation rate of solvents, air pressure, and type of equipment). Table 14–1 is a chart of variables and their effects on the shade of a metallic color. As each variable changes during application, metallic finishes are affected accordingly.

Light penetrates the surface of the cured paint film, reflects off the aluminum flakes, and then passes around the pigment particles of varying densities, thus producing the ultimate color shade. Figure 14–1 portrays a basecoat/clearcoat finish that produces the **ultimate color shade.**

Figure 14–2 portrays a normal orientation of metallics and pigments. The uniform dispersion and density of aluminum flakes and the pigment particles produce a **standard metallic color shade.**

Figure 14–3 portrays a flat orientation of metallics. To produce a **light metallic color shade** requires an accumulation of aluminum flake dispersed nearly horizontally at the top of the paint film, thereby obscuring most of the pigment particles beneath.

Figure 14–4 portrays a perpendicular orientation of metallics. The dense flotation of pigment particles near the surface of the paint combined with aluminum flake dispersed nearly perpendicularly results in a **dark metallic color shade.**

TABLE 14–1 Variables Affecting the Color of Metallics

Variable	To Make Color Lighter	To Make Color Darker
SHOP CONDITIONS:		
a. Temperature	Increase	Decrease
b. Humidity	Decrease	Increase
c. Ventilation	Increase	Decrease
SPRAY EQUIPMENT AND ADJUSTMENTS:		
a. Fluid tip	Use smaller size	Use larger size
b. Air cap	Use air cap with greater number of openings	Use air cap with lesser number of openings
c. Fluid adjustment valve	Reduce volume of material flow	Increase volume of material flow
d. Spreader adjustment valve	Increase fan width	Decrease fan width
e. Air pressure (at gun)	Increase air pressure	Decrease air pressure
THINNER USAGE:		
a. Type of thinner	Use faster evaporating thinner	Use slower evaporating thinner
b. Reduction of color	Increase volume of thinner	Decrease volume of thinner
c. Use of retarder	(Do not use retarder)	Add proportional amount of retarder to thinner
SPRAYING TECHNIQUES:		
a. Gun distance	Increase distance	Decrease distance
b. Gun speed	Increase speed	Decrease speed
c. Flash time between coats	Allow more flash time	Allow less flash time
d. Mist coat	(Will not lighten color)	The wetter the mist coat, the darker the color

Courtesy of General Motors Corporation.

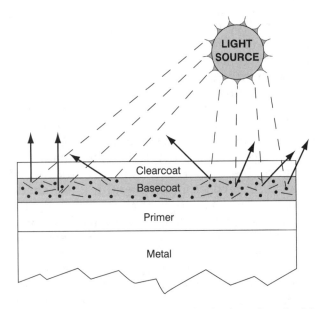

FIGURE 14–1. Ultimate color shade. (Reprinted with permission of General Motors Corporation.)

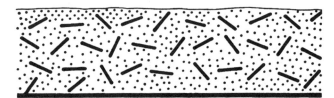

FIGURE 14–2. Normal orientation of metallics and pigments. (Reprinted with permission of General Motors Corporation.)

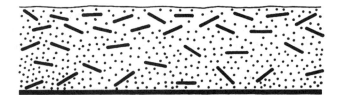

FIGURE 14–3. Flat orientation of metallics. (Reprinted with permission of General Motors Corporation.)

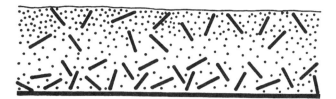

FIGURE 14–4. Perpendicular orientation of metallics. (Reprinted with permission of General Motors Corporation.)

OTHER CAUSES OF COLOR MISMATCHES

It would be untrue to state that all color mismatches are because of the paint technician's spraying techniques or his or her failure to follow product label directions, although many are. But there may be any number of other reasons why the color did not match the car, including:

- Balling of the metallic flake
- Bending of the metallic flake
- Shearing and wearing of the metallic flake
- Fading of the color pigments
- Unstable tints
- Thin paint
- Wrong color
- Underbaked finish (at OEM level)
- Overbaked finish (at OEM level)
- Ultraviolet breakdown
- Portions of vehicle painted at different plants or other sources
- Poor jobber mix
- Unbalanced thinner (true, latent, and diluent)

Factory-applied materials can be duplicated using standard repair techniques. In many cases, however, application alone may not correct all mismatches. No matter what the reason or who is to blame, the problem becomes the painter's. Every painter should be familiar with the following two tinting terms:

Face: The direct observation of a color at a right angle.
Pitch or **side tone:** The observation of a color from an acute angle, possibly from a 45° angle or less.

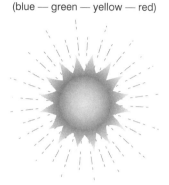

FIGURE 14–5. Picture of the sun.

LIGHTING CONDITIONS AND COLOR INSPECTION

Figure 14–5 is a picture of the sun. All color, as the eye sees it, originates from the radiant energy of the sun. Color is made possible by the reflection of light from all that is seen. The white light of the sun is composed of all the colors of the spectrum, including red, orange, yellow, green, blue, indigo, and violet. All these colors are available for reflection. The best way to evaluate a color match is to make the evaluation under the best possible daylight conditions.

Figure 14–6 is a graph depicting the intensity at which all colors are seen clearly in natural daylight. Natural daylight is variable. Lighting studies indicate that color analysts, artists, finishers, and the like, prefer the light from a natural, moderately overcast north sky. The color of this light is shown in Figure 14–6. Note that all colors are present in the light and there is a little more blue than in the others, which accounts for the bluish characteristics of the north sky daylight. Natural daylight then represents the ideal with which light sources are compared to determine how well the artificial source can be substituted for the natural one.

Figure 14–7 shows that a standard fluorescent light (with only blue, yellow, and red) contains only a portion of the visible light that is available for reflection. Figure 14–8 is a graph of average fluorescent daylight. This graph shows the abrupt changes of the fluorescent simulation versus the smooth flow of the natural daylight graph curve. Violet, green, and red show up much more under this type of light. Because of the emitting of high-energy peaks of certain colors, a person's judgment is misled.

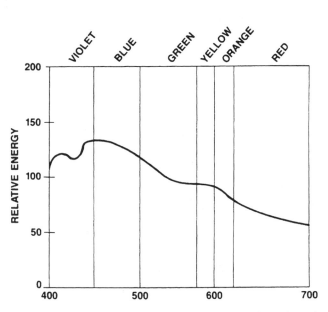

FIGURE 14–6. Graph of natural daylight. (Reprinted with permission of General Motors Corporation.)

Figure 14–9 is a picture of incandescent light, which contains only a portion of the visible light for reflectance (yellow and red). Figure 14–10 shows a graph of ordinary incandescent light. Note how low this light is in violet, blue, and green for reflectance.

Figure 14–11 shows that the color intensity of the Macbeth filtered daylight-type light is close to natural daylight. The disadvantage of this type of light is that it will not show the amount of metallic sparkle in the color that the sunlight will. Therefore, the amount of metallic sparkle in a paint finish can only be matched in bright sunlight.

FLUORESCENT LIGHT
(BLUE — YELLOW — RED)

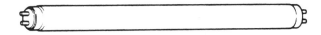

CONTAINS ONLY A PORTION OF THE VISIBLE LIGHT THAT IS AVAILABLE FOR REFLECTION.

FIGURE 14–7. Fluorescent light (blue, yellow, red).

INCANDESCENT LIGHT
(YELLOW — RED)

CONTAINS ONLY A PORTION OF THE VISIBLE LIGHT THAT IS AVAILABLE FOR REFLECTION.

FIGURE 14–9. Incandescent light (yellow, red).

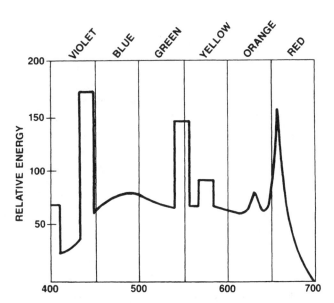

FIGURE 14–8. Graph of average fluorescent daylight. (Reprinted with permission of General Motors Corporation.)

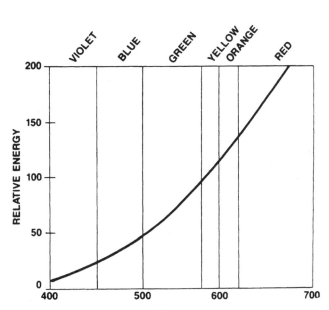

FIGURE 14–10. Graph of ordinary incandescent light. (Reprinted with permission of General Motors Corporation.)

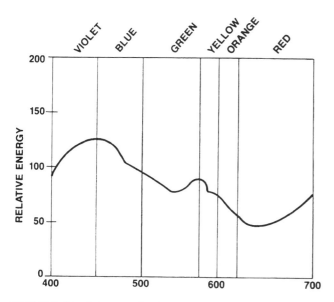

FIGURE 14–11. Graph of Macbeth filtered daylight, Spectralight®. (Reprinted with permission of General Motors Corporation.)

CONCEPT OF BLACK AND WHITE SPRAY-OUT PANELS

To be sure that there is adequate hiding for the evaluation of a color match, the paint companies are making available black and white test spray-out panels to the after-market paint refinisher. These panels can be used to spray out a color prior to applying it on a car to check for a color match against an undamaged area. The black and white concept is to apply color over both areas of the test card or panel until the black and white areas are no longer distinguishable. This application ensures a proper film build and proper hiding of applied color. You can secure test cards at your local paint jobber.

THEORY OF COLOR

To perform color adjustment by tinting, the painter should have a fundamental knowledge of basic color theory. The painter should understand the basics as well as the tints themselves before attempting tinting operations. Tinting should be done most definitely as a last resort.

Pigments can reflect and absorb light rays. When a person sees blue, the pigment absorbs all the light rays with the exception of blue ones, which it reflects. When a person sees red, the pigment absorbs all light rays except red ones, which it reflects. The eye is like a tiny radio receiver and picks up these waves of light as they are reflected.

A particular color does not appear the same under different types of light. Figure 14–12 shows how the same color appears differently under sunlight and under an incandescent light. The reason is that the incandescent light has less wavelength reflectance because there are fewer light components to reflect.

FIGURE 14–12. How we interpret color. (Reprinted with permission of General Motors Corporation.)

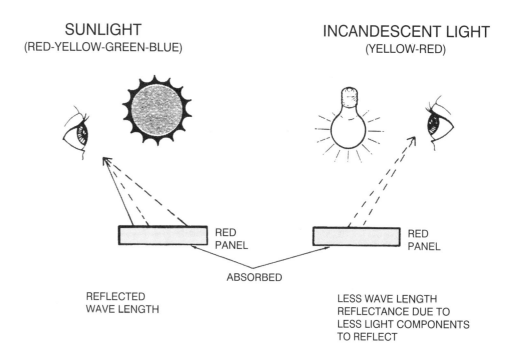

Metamerism is a phenomenon whereby the spectral reflectance curves in a color match under one light source but do not match under a second light source. Metameric matches can occur when paint materials are formulated with different dyes or pigments.

HOW WE SEE COLOR AND WHY

To communicate clearly about color matching, we need to use a common language to describe color. Every color has three dimensions. The painter must understand each of them if he or she wants to adjust a color successfully. The three dimensions of a color are *hue*, *value*, and *saturation*.

Hue

Hue is the excitation of the sense of sight created by beams of light that allow us to distinguish one color shade from another. Hue is a specific shade of a given family of color. An example of this is flame red. Flame red is a specific hue within the red color range. Hue is often referred to as **color** or **cast**. Color is adjusted clockwise or counterclockwise, never across the axis of the color wheel. Refer to Figure 14–13.

1. **Primary colors:**
   Red
   Yellow
   Blue
2. **Secondary colors** (mixture of two primary colors):
   Orange
   Green
   Purple or violet
3. **Intermediate colors** (mixture of an adjacent primary and secondary color):
   Red-orange
   Yellow-orange
   Blue-green
   Blue-violet or blue-purple
   Red-purple or red-violet

Using this same method we can have an infinite amount of colors. At present, paint manufacturers have worked with up to 38,000 different color hues.

Value

Value is the lightness or darkness of a color. White and black are noncolors. **White is at the top of the neutral axis and black is at the bottom of the axis. Adding black or white does not change color hue. It changes value and saturation.**

Value:	Lightness or darkness of a color
White:	Highest in value
Black:	Lowest in value
Metallic:	Gray

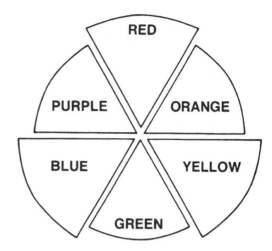

FIGURE 14–13. Primary and secondary colors. (Reprinted with permission of General Motors Corporation.)

Saturation

All colors are most saturated with color intensity at the outer rim of the color wheel. Another term used to describe the third dimension of a color is **chroma**. As (1) black, (2) white, or (3) metallic are added to a color, it becomes (1) desaturated, (2) lighter, or (3) grayer. See the BRYG color wheel in Figure 14–14. At position 1, the metallic color is a light blue metallic color. At position 2, it is a medium blue metallic color. At position 3, it is a rich dark blue metallic color.

As white is mixed with a color, it becomes desaturated toward the light side. As black is added to a color, it becomes desaturated toward the dark side. The addition of metallic to a color desaturates it toward the gray side.

Remember, if a color is rich in hue, it is saturated and it is located at the outer rim of the color wheel. As a color is desaturated, it is close to the center of the color wheel, or neutral axis.

The three dimensions of a color can be summarized as follows:

1. It is a color.
2. It is either light or dark.
3. It is either rich or muddy.

THE BRYG COLOR WHEEL

The BRYG color wheel is used to represent the position of each color family with regard to hue and saturation dimensions. The letters BRYG stand for blue, red, yellow, and green. **Hue dimension**

COLOR-MATCHING FUNDAMENTALS AND TECHNIQUES

FIGURE 14–14. The BRYG color wheel. (Reprinted with permission of General Motors Corporation.)

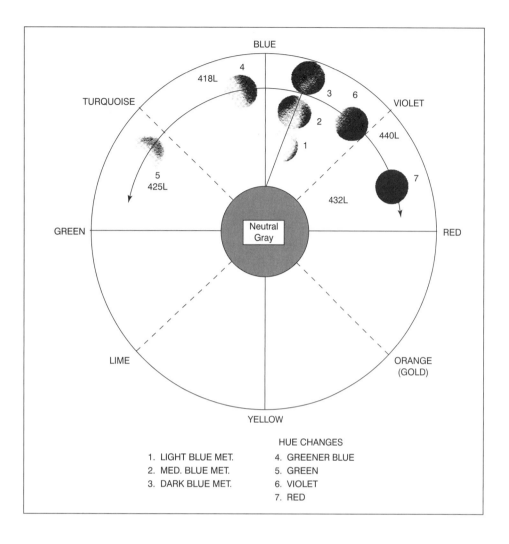

changes rotate around the wheel. Saturation dimension changes move from inside the wheel out. The center of the wheel represents the neutral gray area. The farther away from the center is the color, the more saturated the color becomes. For example, a slightly red blue metallic color would be plotted somewhere along the blue line, as shown in Figure 14–14. Depending on its saturation, it would be plotted closer to the neutral gray center or farther out to the edge. Position 1 would be considered a light blue metallic, while 2 would be considered a medium blue metallic, and 3 would be considered a dark blue metallic.

Saturation changes can be adjusted by tinting with aluminum flake to desaturate the color or adding more of the primary blue to saturate the color. Hue changes can be adjusted by adding a greener blue (4) or green (5) or a redder blue violet (6) or red (7).

The color wheel is used in the same manner throughout the color families. BRYG on the color wheel is used in both metallic and solid color families.

Summary of the Dimensions of Color

Value of a color:	Light or dark
Hue of a color:	Is it bluer, greener, yellower, or redder?
Saturation/desaturation:	Refers to purity; is the color cleaner, brighter, or dirtier (grayer)?

MUNSELL COLOR TREE CONCEPT

The Munsell color tree (see Figure 14–15) was developed by Albert H. Munsell, an accomplished printer and art instructor, to create order in the world of color. He recognized that there was no generally accepted way of describing colors exactly. People made vague comparisons to colors in nature, such as sky blue, lemon yellow, or ruby red. There were endless arguments over colors and color names. Munsell identified

FIGURE 14–15. The Munsell color tree. (Reprinted with permission of General Motors Corporation.)

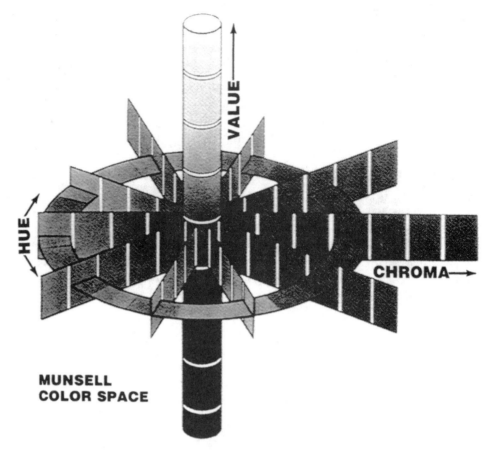

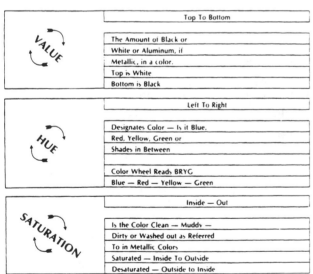

FIGURE 14–16. Color dimension summary. (Reprinted with permission of General Motors Corporation.)

the three basic qualities of color (**hue, value,** and **chroma**) and arranged them on numerical scales. The three qualities of color can be considered the three dimensions of a space in which colors are arranged in an orderly way. All perceptible colors relative to each other are located by means of their **hue, value,** and **saturation** or **chroma** on the color tree.

Figure 14–16 is a basic road map on color matching that isolates and summarizes the three dimensions of a color with tips on how to control each dimension in color matching. The tips are summarized at the right of the figure. The sequence of adjustment is important too: value, first; hue, second; and saturation, third. When each dimension is adjusted as directed in this chapter, the result is an acceptable, if not perfect, color match.

HOW TO SOLVE COLOR-MATCHING PROBLEMS

I. **Learn to describe color correctly and in the proper sequence.**
   A. **Lighter or darker: value evaluation is first.**
      1. Side-angle viewing
      2. Direct viewing
   B. **Cast or hue: color evaluation is second.**
      1. Redder
      2. Greener
      3. Bluer
      4. Yellower

C. **Brighter or grayer: saturation evaluation is third.**
II. **Learn to compare the original paint to the color being sprayed.** This comparison is done by asking these three questions:
   A. First, "Is the car's finish lighter or darker compared to the color being sprayed?"
   B. Then ask yourself; "Is the car's finish redder or greener or bluer or yellower compared to the color being sprayed?"
   C. Then ask yourself, "Is the car's finish brighter (in color intensity) or grayer in color compared to the color being sprayed?"

When you can answer the three questions, you have a starting point for tinting the repair formula to match the car. If it is a good match already, you are ready to paint.

III. **General tinting procedure**
   A. The first step in tinting a color is to adjust for lightness or darkness. See Table 14–1 and steps IV A and B below.
   B. The next step is to adjust the cast or hue of a color.
   C. The last step is to adjust for brightness or grayness when steps IIIA and B are okay.
IV. **Adjust the color for lightness or darkness as follows:**
   A. **Methods to lighten a metallic color.**
      1. After spraying a wet coat, follow with a coat at half-trigger, or with reduced fluid flow, while maintaining air pressure.
      2. Raise the air pressure at the gun (5 psi increments).
      3. Add more thinner. Use a faster drying thinner than what is in the mix.
      4. Let the reduced color set for 10 minutes in the spray cup, pour off the top into a clean container, stir the remaining material, and spray.
      5. Add additional metallic to the color.
   B. **Methods to darken a metallic color.**
      1. Apply double-wet coats. Reduce the gun fan size slightly.
      2. Lower the air pressure at the gun (5 psi increments).
      3. Add 2 ounces of retarder per reduced cup.
      4. Let the reduced color set for 10 minutes in the spray cup, pour the top portion off into a clean spray cup, stir, attach gun, and spray.
      5. Add the predominant dark color to the mix (from the formula).
   C. Spray the test panel to full hiding after each step in steps IV A and B and check against the car for a color match.
V. **After adjusting the color for lightness or darkness and the cast is off, tinting for the correct cast is required.** The steps involved in tinting for the correct cast are as follows:
   A. Decide how the original finish is off as compared to the material sprayed on the test panel. For example, the original finish may be redder than the test panel. A color can be described in being off in cast only as being redder, yellower, greener, or bluer.
   B. See the proper color formula to determine which tinting color to use, or use a suitable company tinting guide as found in this chapter. Also, refer to Table 14–2 for a lead to a color.

TABLE 14–2 If a Color Is "Too Something in Cast," Do the Following to Change the Cast

Color	Add	To Change A Cast of
Blue	Green	Red
Blue	Red	Green
Green	Yellow	Blue
Green	Blue	Yellow
Red	Yellow	Blue
Red	Blue	Yellow
Gold	Green	Red
Gold	Red	Green
Maroon	Yellow	Blue
Maroon	Blue	Yellow
Bronze	Yellow	Red
Bronze	Red	Yellow
Orange	Yellow	Red
Orange	Red	Yellow
Yellow	Green	Red
Yellow	Red	Green
White	Yellow	Blue
White	Blue	Yellow
Beige	Green	Red
Beige	Red	Green
Purple	Green	Red
Purple	Red	Green
Aqua	Blue	Green
Gray	Blue	Yellow
Gray	Yellow	Blue

Courtesy of the Sherwin-Williams Company.

C. Each color can vary in cast in only two directions. See Table 14–3 on color shifts.
D. Once the color necessary to adjust the cast correctly is determined, the amount must then be calculated.
 1. The first "hit" should be determined from the chart. Use the least amount necessary of the particular tint to change the color effectively. See Table 14–4.
 2. Mix the color thoroughly. Trigger the gun to clear the color passages. Spray the test panel to full hiding, allowing proper flash time between coats. Allow to dry and check with the OE panel. Add additional tint in the specified increments and repeat the process until the color match is acceptable.
 3. Record on the back side of the test panel the color names and color amounts so the tinting can be duplicated for more of the same color when the need arises.

VI. **Final adjustment (saturation)** The final adjustment is made only after the color is correct in lightness or darkness and hue (cast). Emphasis in this color-matching chapter is primarily on metallic colors.
 A. **If saturation (increase in color intensity) is required,** make the color more saturated by adding the primary family color noted in the formula. Arrive at the proper saturation through a test panel comparison.
 B. **If desaturation (making the color grayer) is required,** add the proper aluminum pigment and/or other tint noted in the formula. Again, arrive at the proper desaturation through spraying out a test panel for comparison.
 C. **If color is slightly gray or dirty,** spray on a wet color coat, followed by an additional coat applied from a slightly greater distance at half-trigger and the same air pressure to lay the metallics flatter at the surface. If this application is a basecoat color, apply clear per label directions, allow to dry, and check against the original finish.
 D. **To make a color more gray** and if the formula is not available, add a small amount of white mixed with a very small amount of black. Spray out a test panel, let it dry, and check for color match to OEM panel.
 E. **To lighten the pitch depth of a color,** add white. Use the proper white in the proper

TABLE 14–3 Color Shifts

- Colors either greener or redder in cast
  1. Blues
  2. Yellows
  3. Golds
  4. Purples
  5. Beiges
  6. Browns
- Colors either yellower or bluer in cast
  1. Greens
  2. Maroons
  3. Whites and off-whites
  4. Blacks
  5. Grays
- Colors yellower or redder in cast
  1. Bronzes
  2. Oranges
  3. Reds
- Colors bluer or greener in cast
  1. Aqua
  2. Turquoise

Remember

A color can be described as being off in cast or hue only as being redder, yellower, greener, or bluer. Each color can vary in only two directions. To decide what tints should be used to adjust a color, refer to the formula mix for that color.

Courtesy of General Motors Corporation.

TABLE 14–4 Tinting Measurements

- Trace: 0–1 gram
  - 25 drops = approximately 1 gram
  - $\frac{1}{8}$ teaspoon = approximately $\frac{3}{4}$ gram
- Small: 1–4 grams
  - $\frac{1}{4}$ teaspoon = approximately 1.5 grams
  - $\frac{1}{2}$ thinner cap = approximately 4 grams
- Medium: 4–10 grams
  - $\frac{1}{2}$ teaspoon = approximately 3 grams
  - 1 teaspoon = approximately 6 grams
  - 1 thinner cap = approximately 8 grams
- Large: 10–20 grams
  - $1\frac{1}{4}$ thinner caps = approximately 10 grams
  - $2\frac{1}{2}$ teaspoons = approximately 15 grams
  - 3 teaspoons = approximately 18 grams
  - 1 tablespoon = approximately 19 grams

NOTE

The above chart is a guide for determining an equivalent amount measurement when the indicated amount of tint to add to a color is given in grams.

Courtesy of General Motors Corporation.

volume. If in doubt, check with the paint jobber. The following are general guidelines for 1 pint of reduced paint:

1. For light metallics, use $\frac{1}{2}$ teaspoon white.
2. For medium metallics, use $\frac{1}{4}$ teaspoon white.
3. For dark metallics, use $\frac{1}{8}$ teaspoon white.

F. **A flattening agent** can be used in a metallic color to increase metallic sparkle. As much as 15 percent to 20 percent may be used.

G. **For additional information on saturation,** check with the paint jobber. Also, refer to the paint supplier's tinting guide.

H. **When color is acceptable for blending,** blend the repair color into the panel to achieve an acceptable color match. Allow to dry. If this application is a basecoat color, apply two to three coats of clearcoat per label directions over the complete panel. Allow to dry. If the application is an acrylic lacquer, rub out and polish as required.

IMPORTANT

It is impossible to make a color brighter without throwing off the two previous corrections, which are value (lightness or darkness) and hue (color).

UNDERSTANDING PAINT FORMULA INGREDIENTS

All major and popular paint formulas are usually part of a paint shop if it is equipped with a paint mixing setup from any particular paint company. If a paint shop is without a paint mixing setup, questions on paint formulas can be answered by the paint jobber. There are two important things to remember when examining paint formulas:

1. Various tints of certain percentages make up the formula of a color.
2. The mixing formula is an important tool to the refinisher when tinting of a color is required. Each formula provides the identity and specific amount of each component that makes up the formula.

The hue of pure tints can change drastically when mixed or "let down." All paint manufacturers make available to the after-market a list of their formula mix tints on color chip cards. These cards show the tint colors in their pure state and explain how each reacts when "let down." The formula mix tints include the following information:

1. Name of the tint
2. Product identity number
3. Strength of the tint

A rule of thumb when examining the formula mix ingredients is that, if the quantity of a given tint in the formula is minimal, it most probably is very high in strength. Adding a small quantity of the tint to a mix will make a noticeable color difference. Become familiar with the strength of each tint. **Basecoat color tints are among the strongest available.** Extreme care must be exercised in their use.

HOW TO USE A COLOR TINTING GUIDE

The first pages of a tinting guide generally list those procedures that should have been tried before you decide to tint the color.

1. Gun technique
2. Color mixed properly
3. Color reduced properly
4. Proper blending techniques

Once you have determined that you need to tint:

1. Find the color family closest to the car color you are working with.
2. Under good lighting conditions (preferably daylight), compare the dried refinish color and the car to be refinished.
3. Determine the color problem. If the color requires a hue adjustment, does it need red or green, blue or yellow? If one of the hue directions is off, add trace amounts of the tinting color noted in the formula to correct it. If a formula is not immediately available, the painter can be guided by the tinting guides in this chapter. Use only trace amounts of the tinting colors noted.
4. If the hue direction appears to be okay, then evaluate lightness and saturation in the same manner. If lightness is the problem, add trace amounts of white or black. If saturation is the problem, make the color more saturated by

adding the primary family color noted. If desaturation is needed, add aluminum or another tint noted in the formula mix.

5. Once you have determined what color to tint with, be careful not to add too much. Add small amounts and check the match after each "hit" under the same lighting conditions, at full hiding after it is reasonably dry. Remember, wet materials will often dry down to a different shade. You can always add more tinting color. But if you have added too much, you may find it impossible to tint your original color back. When this occurs, it often requires starting over with fresh color.

Keep an accurate record of each color tinted and the amount of tinting color used. Record this information on the back of each spray test panel so that more of the same color can be made quickly when the need arises.

HINTS FOR COLOR MATCHING

Before tinting a color, be certain that a color variation is not due to:

1. Incomplete agitation
2. Incorrect reduction
3. Incorrect air pressure
4. Insufficient number of coats
5. Poor spray technique

Painters should get the idea out of their minds that they can achieve a perfect panel-to-panel color match because there is no such thing. There are too many variables, both in the refinish situation and at the OEM level that can cause variations in a color match. That is why blending into adjacent panels with the same color has become the most sensible option. Go for **commercial acceptability** in a color match. **This is a color match that, although not perfect, would be acceptable to a very high percentage of the motoring public.**

A car should be viewed from the following angles to check a color match:

1. 45° to 60° (side tone)
2. Head on (face)

The head-on appearance is the least important. A painter cannot blend a color with a bad side tone and lose it properly.

If you have tried painting the car with several products prior to deciding to tint, sand the material off. Many colors applied over a color tend to oversaturate side tones. Always spray out a test panel with the repair color and check it against the car if problems with a color match are suspected.

Use only half of the repair color; keep the other half untouched. A record should be kept of what is done to the original formula when tints are added.

Before attempting to tint any color, all variables should be tried, including more paint or even a different batch of paint. **It cannot be overstressed that tinting takes time and should be done as a last resort.**

Tinting Colors Used in Solid Color Families

Solid colors are those colors that contain no metallic flakes. They are harder to tint to match as well as to blend to match. To tint a solid color, no variables can be modified to blend and vary color effect. A solid color has to be tinted within the confines of a color hue and to near exactness. Figure 14–17 portrays a color wheel showing the most popular colors used when the tinting of solid colors is required.

Tinting Colors Used in Metallic Color Families

Metallic colors are paints that can be tinted to within what is commonly known as a "ball-park" match, knowing that gun techniques can often compensate for any minor color mismatches. To lighten a metallic color, add more metallic flakes (see Figure 14–18). The addition of white to a metallic color can result in distorted side-tone effects. It might match head on, but the side tone may be off-color and generally on the light side. The side tone may not be a mismatch from the metallic standpoint, but from the color itself and the lack of metallic sparkle.

COLOR PLOTTING

Figure 14–19 shows a color plotting example. A color wheel is used to represent the position of each color with regard to hue and saturation. Hue dimension changes rotate around the wheel, while saturation changes move from inside the wheel out. The center of the wheel represents the neutral gray area. The

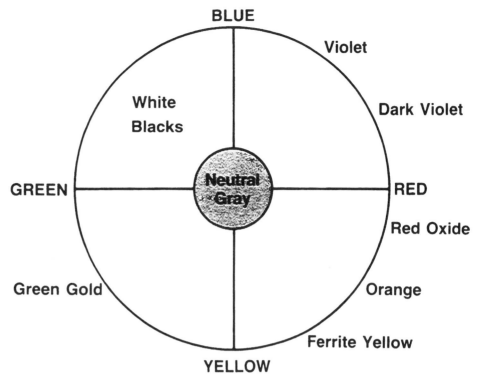

FIGURE 14–17. Tinting colors used in solid color families. (Reprinted with permission of General Motors Corporation.)

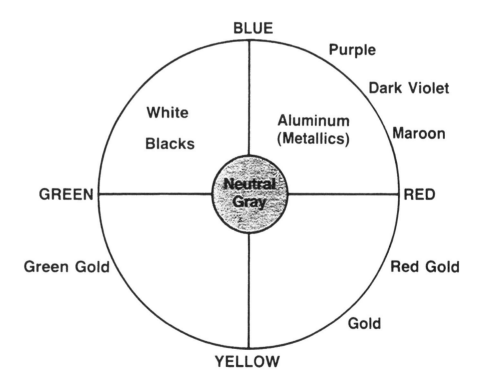

FIGURE 14–18. Tinting colors used in metallic color families. (Reprinted with permission of General Motors Corporation.)

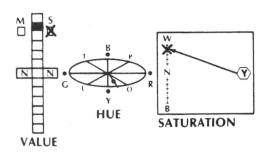

COLOR: LT. YELLOW CODE: 50			
CODE	MOVEMENT	SCALE	TOTAL
402	WHITE	336	336
452	YELLOW	385	49
451	LT. YELLOW	398.5	13.5
457	YELLOW/RED	407.5	9
405	BLACK	419	11.5
465	BINDER	790.5	371.5
485	BALANCER	975	184.5

FIGURE 14–19. Color plotting example. (Reprinted with permission of General Motors Corporation.)

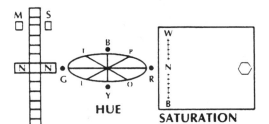

COLOR		CODE	
CODE	MOVEMENT	SCALE	TOTAL

FIGURE 14–20. Color plotting chart. (Reprinted with permission of General Motors Corporation.)

further from the center is the color, the more saturated the color will become.

The color being plotted is a light yellow, code 50. There is no aluminum in the formula; therefore, it is a solid color. A large quantity of white and light tints is called out; therefore, it is high in value. The hue would fall between the yellow and red dimension of the color wheel, closer to the yellow area. The color is highly desaturated, almost on the white side.

Figure 14–20 is a blank color plotting chart and Figure 14–21 is a blank color evaluation sheet. They can be reproduced in any size and number by a print shop for use by paint jobbers, paint shops, vocational training schools, and individuals.

REVIEW QUESTIONS

1. What is the best way to check out a refinish color when color-matching problems are suspected?
2. Before tinting a color, the painter must be certain that the color variation is not due to what four common items over which he or she has control?
3. Name three variables over which a painter does not have control. (It is assumed, in this case, that the painter does not own the shop.)
4. Name five common causes of color mismatches.
5. Inspection of a paint finish is done in proper lighting conditions from what two angles?
6. Explain how a person sees color.
7. What are the three primary colors?
8. Name three secondary colors.
9. Describe the value of a color.
10. Describe the hue of a color.
11. Describe saturation of a color.
12. Describe the BRYG wheel.
13. Describe the Munsell color tree.
14. When tinting a color, what is the first quality that should be adjusted?
15. When tinting a color, what is the second quality that should be adjusted?
16. What four major colors will adjust the hue of any color mismatch?
17. Pair each description number with the proper color dimension that it describes:

    Hue _____
    Value _____
    Saturation _____

    1. The strength or intensity of a color
    2. A specific shade of a given family of color
    3. The lightness or darkness of a color

FIGURE 14–21. Suggested record for spray out test panel color evaluation. (Reprinted with permission of General Motors Corporation.)

COLOR EVALUATION

Color Code _____ Color Name _____ WA # _____

Compared to Vehicle Color, Sprayed Material is: _____

LIGHTNESS:

☐ Light   ☐ OK   ☐ Dark

SATURATION:

☐ Desaturated   ☐ OK   ☐ Saturated

HUE:

☐ Blue   ☐ Green   ☐ Yellow   ☐ Red   ☐ OK

TINT ADDED*	AMOUNT**
_____	_____
_____	_____
_____	_____
_____	_____
_____	_____
_____	_____

AIR PRESSURE AT GUN _____

AMOUNT OF THINNED MATERIAL _____

SOLVENT USED _____

VISCOSITY READING _____

*MUST BE CLEAR COATED TO GAIN COLOR MATCH IF BC/CC MATERIAL
**AMOUNT OF TINTS USED MAY VARY DEPENDING ON BUILD DATE OF VEHICLE

Date: _____   Signed _____

CHAPTER 15

Painting Interior Plastic Parts

INTRODUCTION

In the auto industry, plastics are divided into two categories according to their characteristics and refinish requirements:

1. **Thermoset plastics are hard or rigid in construction,** and they are **not affected by high heat.**
2. **Thermoplastic plastics are flexible;** they can bend or flex and recover shape quickly. They **are affected by high heat.**

To determine if a plastic is of the thermoset or thermoplastic type, perform a **hot wire test** as follows:

1. Heat a steel nail (held in pliers) to red hot.
2. Touch the heated wire to the back side of the plastic being tested.
    a. If the wire has no effect on the plastic, the plastic is of the **thermoset type.**
    b. If the plastic softens and melts when contacted by the hot nail, the plastic is of the **thermoplastic type.**

There is no set rule governing where each type of plastic may be used in the manufacture of a car. However, **each category of plastics requires specific refinish procedures.**

Chapter 15 explains how to identify and refinish **rigid plastic parts.**

Chapter 16 explains how to identify and refinish **flexible plastic parts.**

HOW PARTS ARE SERVICED

Most automotive interior plastic parts are serviced in one color or in a prime coat to keep the inventory of service parts to a minimum. When these parts are replaced in service and require a different color, they must be repainted. Plastic parts on cars in service need repainting for many reasons. The most frequent reasons are collision damage involving parts replacement and fading due to excessive exposure to the elements.

A painter must be able to identify each major type of plastic and be familiar with approved paint systems and materials for each plastic to do refinish work properly and with maximum durability. Painters become familiar with plastic parts best by working with them over time. Figure 15–1 shows how several different plastics are used on car interiors.

HOW TRIM PARTS ARE COLOR KEYED

The painting of all interior body components is color keyed by the trim combination number on the body number plate (see Chapter 2 for location and description). The first two numbers of the trim combination indicate the basic color. Sometimes a basic color is made up of two colors. The third and/or fourth numbers or letters, when used, indicate the trim type and trim design, which is usually one of the following:

1. Cloth and vinyl
2. All vinyl
3. All leather

FIGURE 15–1. Typical use of different plastics on car interiors. Polypropylene (PP), acrylonitrile-butadiene-styrene (ABS), and polyvinyl chloride (PVC) are leading plastics used in the industry. (Courtesy of Oldsmobile Division, General Motors.)

An example trim combination number is **19 N**. The **19** can mean that the trim color is black. The letter **N** can indicate an all-vinyl interior.

Before using any color on the interior of a car, be sure that the color complies with the trim number as shown on the body number plate and on the paint supplier color chart. Also note if the color chart calls for colors to be low gloss or semigloss. Color charts sometimes show color chips of trim colors to help the painter identify interior colors more quickly and accurately.

FEDERAL AND FACTORY PAINT STANDARDS

Painting of interior autobody parts is quite different from painting the exterior of a car for several reasons:

1. A federal regulation requires that all components in a driver's vision area be painted with a low-gloss or flat finish to prevent sun glare or sun reflection into the driver's eyes. The parts affected are:
   a. Instrument panel upper section cover
   b. Instrument panel at front of cover
   c. Radio front speaker grilles
   d. Windshield pillars and other parts at the front end
2. *Note:* The same principle applies to components around the back window that could reflect sunlight indirectly through the inside rearview mirror. These parts are:
   a. Back window side and lower finishing moldings
   b. Radio speaker grilles
   c. Rear window de-fogger grilles
   d. Rear package shelf trim foundations
3. All other interior painted parts are usually finished with a semigloss finish, which is about a 25- or 30-degree gloss finish, unless otherwise specified, in a color and gloss matching fashion. Normally, when an interior acrylic lacquer is

applied in a conventional manner with a medium solvent, the finish is close to semigloss. Interior semigloss parts are as follows:

a. Door, headlining, and side wall trim, including parts attached
b. Arm rests
c. Seat cushions and seat backs, including parts attached
d. Head rests
e. Station wagon and hatch-back-style rear load floors.

TESTING QUALITY AND DURABILITY OF PAINT PRODUCTS

Each painter is the paint engineer in his or her particular paint shop. Use only proven and fully tested products in the paint shop. Following this commonsense rule helps paint shops avoid substandard paint work due to faulty products. The painter usually has enough problems when he or she works with approved materials. As a reminder to painters, the following sections discuss the key properties built into each refinish material as used on the interiors of cars.

Wear and Abrasion Resistance

A painted trim part should look well, clean well, and be normally wear resistant. Normal wear resistance means continued durability when exposed to normal service use and continued cleaning without paint failure.

One common failure of a poor paint finish is called crocking. **Crocking** is the wearing off or transfer of paint pigments from the paint to a person's clothing. The condition is caused by poor paint and by a combination of rubbing, pressing, heat, and perspiration. Applying color over a wax surface also contributes to the problem. The condition happens, for example, when a person takes a trip on a hot day and sits on a newly painted vinyl seat cushion. If color transfers onto the clothing, staining or discoloring it, the paint is faulty.

Crack Resistance: Flex Test

Good service paint finishes are built to withstand cracking in all types of weather, just as original finishes do. The test for this paint property consists of the following steps:

1. Subject the painted flexible trim part to a temperature of 5° to 20°F below freezing for about 30 minutes.
2. Flex the part by bending it in half, if possible, or stretch the part the normal distance that it should stretch.
3. If the paint finish cracks on being flexed in cold temperatures, the paint finish is faulty.

Color Fade Resistance

Painted trim parts should show a similar performance to an original car paint finish with respect to fade resistance. One of the greatest enemies of a paint finish is the ultraviolet portion of sunlight. A good paint product has an ultraviolet screener added that cuts off ultraviolet light at the surface so that paint finishes last longer. With excessive exposure, both service paint and original finish should fade uniformly. Uniform fading is a sign of good service paint. If service paint fades excessively before the original color, it is faulty paint.

Proper Adhesion

Equally important to the appearance and color of a finish is its adhesion. A paint finish should not be removable with masking tape. In fact, one test for adhesion is with the use of masking tape. Perform the adhesion test at all edges and at the center of the panel being tested. **Perform the test when the paint is fully dried.**

1. Apply a good grade of masking tape to each area being tested. Apply tape to an area 4 inches wide and 6 inches long, with sufficient tape raised at the edge for a good grip.
2. Pull to remove the tape in one fast motion at right angles to the surface, while the panel is supported as required.
3. **If no paint is removed with the masking tape, the adhesion is good.**

If tiny bits of paint are removed where the finish was damaged, chipped, or cut, this is normal. Test labs usually make tape-pull tests over an "X" that has been cut into the paint finish on the panel. However, if large sections of paint, one inch square or larger, are peeled up from the surface, the paint adhesion is poor.

Proper OEM Appearance Standards

Service paint finishes should match, look like, and wear like the original paint finish. Good appearance and durability qualities are built into a good paint finish. Therefore, whenever a paint finish is factory approved, the painter knows that the product can be used worry free. The factors contributing to good appearance in a repair situation are:

1. Color match and good blend in the case of a spot repair
2. Matching gloss level
3. Matching orange peel level

PAINTING INTERIOR PLASTIC PARTS

The most practical way to paint interior plastic parts is to paint the parts before installation. By following this simple procedure, the painter saves on masking time and does the painting with ease because the part is completely accessible.

> **IMPORTANT**
>
> Before painting a plastic car part, identify the plastic. **Identifying the plastic is important because it leads to the proper refinish procedure.** Failure to use the proper products and procedures will result in a film failure.

Generally, **many interior plastic parts are of rigid plastic construction** and some parts are of flexible plastic construction. Also, **most of the exterior plastic parts are of flexible plastic construction,** and occasionally rigid plastics may be used. There is no set rule on where rigid and flexible plastics are used.

The best way to identify the plastic of an interior or exterior car part is to observe the back of the part. Here you will find capitalized identification letters for plastic identification. (See Tables 15–1 and 16–1.) These codes consist of two or more capitalized letters, such as ABS, ABS/PVC, PP, etc.

Two different paint systems are required for painting the plastic parts on a car:

1. Chapter 15 features the rigid plastic parts painting procedure, which is used for painting all the plastics that appear in Table 15–1.
2. Chapter 16 features the flexible plastic parts painting procedure, which is used for painting all the plastics that appear in Table 16–1, including all vinyl plastics. (See Chapter 16.)

The purpose of the following tables and figure is to simplify the painting of rigid plastic parts:

1. Table 15–1 identifies the plastics used on cars.
2. Table 15–2 lists products and paint systems for painting rigid plastic parts.
3. Figure 15–2 outlines the refinish steps and products for painting rigid plastic parts as recommended by R-M and Diamont.

Here is a detailed procedure explaining how to paint rigid plastic parts using the R-M/Diamont system. DuPont and PPG have equivalent refinish systems.

Procedure

PLASTIC PREPARATION

1. **Prep work:** Wash virgin (unpainted) plastic substrate with soap (mild detergent) and warm water to remove grease, wax mold release agents, and other contaminants. Then dry with paper towels and or lint-free cloth.
   For OEM primed and basecoat/clearcoat painted plastic substrates, proceed to step 2.
2. Wash with R-M 902 Plastic Cleaner to remove wax, silicone, and other contaminants.
3. Scuff with Scotchbrite #7447 (red) very fine pad. Rinse with R-M 902 Plastic Cleaner. Then dry with paper towels and use tack cloth prior to actual painting.

APPLICATION

4. **Adhesion promoter:** DS30 Diamond Transparent Sealer—2:1:30 percent by volume. Mix 2 parts DS30 Transparent Sealer with 1 part PH12 Hardener and 30 percent UR40 Universal Reducer. Prepare DS30/PH12/UR40 according to label instructions. Apply 1 thin even coat at 50 to 65 psi at the spray gun. Allow 15 to 30 minutes drying time before any spray application of Diamont Basecoat Color, HS Solo Color, or DP20 Primer.
5. **Primer:** If filling is required, DP20 Primer can be used. A flex agent for the primer is not required over rigid plastic parts. DP20 Primer for rigid parts: DP20/PH12/UR40 at 4:1:1 by volume. Spray two or three coats of DP20 Primer over DS30 at 55 to 65 psi. Let dry for 1.5 hours at 68°F or force dry at 140°F for 30 minutes. Hardeners and reducers may vary according to actual shop conditions. Allow the panel to cool to room temperature before wet sanding with 600-grit sandpaper. Wash and rinse with R-M 902 Plastic Cleaner and let dry.
6. **Topcoats:** Spray two or three coats of Diamont Basecoat or HS Solo Color until hiding has been achieved. Set the air pressure at 45 to 55 psi.

TABLE 15-1 Identification of Rigid Plastic Parts

Code	Family Name	Common Trade Name	Typical Application
ABS	Acrylonitrile butadiene styrene	ABS, Cycolac, Lustran, Kralastic	"A" pillars, consoles, grilles
ABS/PC	ABS/PC alloy	Pulse, Proloy, Bayblend	Doors, instrument panels
ABS/PVC	ABS/PV alloy	Proloy, Pulse, Lustran, Cyclovin	Door panels, grilles, trim
BMC	Bulk molding compound	BMC	Fender extensions
EMA	Ehtylene methyl acrylate/Ionomer	Surlyn, EMA, Ionomer	Bumper guards and pads
METTON	Metton	Metton	Grilles, kick panels, running boards
MPPO	Modified polyphenylene oxide	MPPO	Spoiler assembly
MPPO/ABS	Modified polyphenylene oxide/ABS alloy	Triex, Elimid	Brakelight housing
PA	Polyamid	Zytel, Vydyne, PA, Minion	Fenders, quarter panels
PET	Thermoplastic polyester	Rynite	Trim
PBT/PPO	PBT/PPO alloy	Germax	Claddings
PBTP	Polybutylene terepthalate	PBT, PBTP, Pocan, Valox	Wheel covers, fenders, grilles
PBTP/EEBC	Polybutylene terepthalate/EEBC alloy	Bexloy, "M," PBTP/EEBC	Fascia, rocker panels, moldings
PC	Polycarbonate	Lexan, Merlon, Calibre, Makrolon, PC	Taillight lenses, IP, trim, valance panels
PC/ABS	PC/ABS alloy	Germax, Bay Blends, Pulse	Doors
PPO	Polyphenylene oxide	Azdel, Hostalen, Marlex, Prfax, Noryl, GTX, PPO	Interior trim, door panel, splash shields, steering column shroud
PPO/PA	Polyphenylene polyamid	PPO/PA, GTX 910	Fenders, quarter panels
PR/FV	Fiberglass reinforced plastic	Fiberglass, FV, PR/FV	Body panels
PS	Polystyrene	Lustrex, Styron, PS	Door panels, dash panels
RTM	Resin transfer molding compound	RTM	Body panels
SMC	Sheet molding compound	SMC	Body panels
TMC	Transfer molding compound	TMC	Grilles
UP	Unsaturated polyester (thermosetting)	SMC, BMC, TMC, ZMC, IMC, XSMC, UP	Grille opening panel, liftgates, flareside fenders, fender extensions

Plastic Refinish Precaution It is important to remove all surface contaminants from any substrate prior to painting. In the case of plastics, the added obstacle of mold release agents can contaminate the surface. Not removing these contaminants can result in the failure of the refinish system.

Some contaminants are water soluble and some are not. It is important to wash the part with soap and warm water and wipe it with a cleaning solvent. The solvent recommended for use in these procedures is R-M 902 Plastic Prep Cleaner.

Prebake Notice Since many rigid and flexible plastic parts may exude mold release agents when baked, prebake any part that will be baked during the painting process. Always clean the surface with 902 Plastic Prep Cleaner before and after baking to ensure all mold release agents have been removed.

Allow 5 minutes flash time between coats. Allow 15 to 20 minutes final flash prior to clearcoat application. Flex agent is not required for the basecoat or HS Solo on rigid plastic parts.

7. **Clearcoat:** Follow label instructions for preparations and application of Diamont Clears. Spray two or three coats at 50 to 60 psi, with 5 minutes flash between coats. Allow 10 to 15 minutes final flash before baking or force drying DC88 or DC89. DC76 and DC92 are immediate bake clears.

Diamont Rigid Clears	DC76	DC76/DH15, 17 or 19	3:1 (use 35–45 psi for immediate bake)
	DC88	DC88/DH44 or 45	4:1
	DC89	DC89/DH44 or 45	4:1
	DC92	DC92/DH42 or 46/UR Series	3:1:20–70 percent (immediate bake)

TABLE 15–2 Products for Painting Rigid Plastic Parts

Product Description	R-M Diamont	BASF Glasurit	DuPont	PPG
Adhesion promoter	DS30 +PH12 +UR40	934-0	2322S	DPX801
Primer	DP20 +PH12 +UR40	934-200 934-100	1120S 1140S 1125S	K36 K200 NCP250 NCP270
Topcoats				
Basecoat color	Diamont/ Supreme BC	Glasurit 54 Line	Chroma Premier Chroma Base	Deltron DBU Deltron 2000 DBC
Clearcoats	Diamont Limco	Glasurit 54 Line	Chroma Premier	DCU line DCD line DAU line
Single-stage color	HS Solo/ Supreme SS Limco 123	Glasurit 21 Line	"E" Quality Chroma One	Deltron DAU Concept DCC
Gloss flattening agents				
Urethane type	See paint jobber.	923-55	See paint jobber.	DX685
Nonurethane type	850	521-300	4528S	DX265

Note: By the time this book is distributed, product availability may change. For the latest in refinishing on any product, always check with your local paint jobber.

8. Allow to cure overnight or bake at 140°F for 30 minutes. Before compounding or repair is attempted, allow to cool for 30 minutes.
9. Local VOC regulations may restrict the use or suggested reductions of these refinish products. Check local regulations to determine what products can be used.

> **NOTE**
>
> DuPont, PPG, and many other paint companies have refinish products and systems equivalent to the above.

PRODUCTS FOR PAINTING BODY INTERIOR VINYL TRIM PARTS

The following sections discuss paint products and systems that are car factory– and/or paint supplier–recommended for repair of vinyl interior paint finishes. Each product and its proper use is explained. For the availability of each, check with your local paint jobber. If the nearest jobber does not handle a specific item, check with the next nearest paint jobber.

Vinyl Cleaner (or Paint Finish Cleaning Solvent)

Vinyl cleaner consists of a blend of solvents designed to clean dirt and foreign matter from vinyl painted surfaces. Always use this cleaner or a paint finish cleaning solvent before repainting automotive vinyl surfaces, and follow the cleaning procedure below:

1. Apply the cleaner on the area to be cleaned and allow to set for 30 seconds to 1 minute, depending on the soiled condition.
2. Wash the surface with the cleaner and a water-dampened cloth as required. Wipe the surface dry with a clean cloth.
3. Use a soft-bristled brush to loosen deeply embedded dirt and to speed the cleaning action.
4. Repeat the cleaning operations if necessary.

PAINTING INTERIOR PLASTIC PARTS

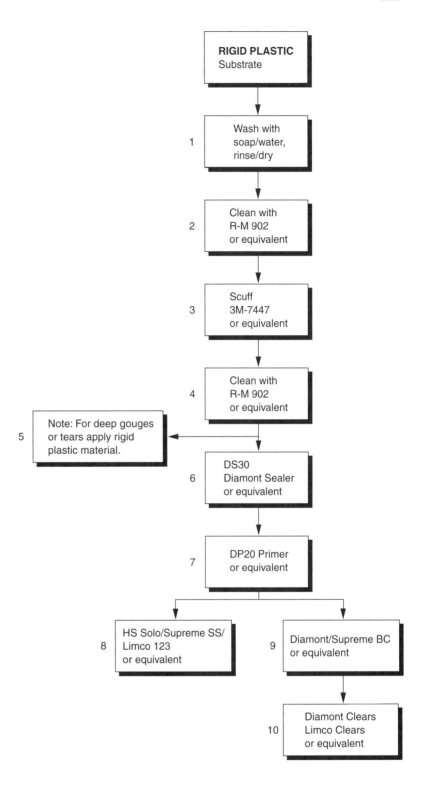

FIGURE 15–2. Step-by-step outline for repainting rigid plastic parts. (Courtesy of BASF Corporation.)

Vinyl Conditioner

The purpose of vinyl conditioner is to prepare a vinyl surface to be repainted. Vinyl conditioner consists of a blend of strong solvents that penetrate and soften original vinyl finishes and cause repair vinyl paint to fuse better into vinyl surfaces. Use the following vinyl conditioning procedure:

1. Apply vinyl conditioner generously to the surface to be painted with a clean, lint-free cloth. Wipe the conditioner onto the surface with a one-directional wiping motion.

2. Allow the conditioner to remain on the surface for 30 to 60 seconds. Wipe it off with a clean, lint-free cloth while the surface is still wet.

> **CAUTION**
>
> Always wipe in one direction by lifting the cloth on the return stroke and overlapping as required. Rotate the cloth as wiping progresses. Do not rub back and forth.

3. Repeat these operations as necessary to condition the surface. The surface can be repainted with vinyl color immediately after conditioning is complete.

Interior Vinyl Spray Colors

Vinyl spray colors are specially formulated for painting interior vinyl plastic trim parts. In addition to matching interior colors, factory-approved vinyl colors must match two levels of gloss: low gloss and semigloss. Leading paint supplier color charts show how available interior vinyl colors are color controlled to match interior trim codes for repair of all Chrysler, Ford, and General Motors cars for current and past model years.

For the availability of vinyl trim colors, check local paint jobbers and paint supplier color charts. Many paint suppliers have vinyl colors on the market for the repair of automotive interiors. For best results in painting vinyl parts, follow the directions of the car factory or vinyl paint supplier.

Interior Clear Vinyl Finishes

Clear vinyl finishes are available in two gloss levels for painting interior trim parts: low gloss and semigloss. Instrument panel trim parts can be painted with semigloss vinyl paint and then finished with a double coat of low-gloss vinyl clear to meet car factory specifications. However, most instrument panel–specified vinyl colors are available in the low-gloss finish. Apply clear vinyl of the proper gloss level to instrument panel parts, seat cushions, seat backs, head rests, and arm rests using the following application procedure:

1. Allow the vinyl color to dry 20 to 25 minutes.
2. Agitate the clear vinyl thoroughly.
3. Apply two conventional wet coats of clear vinyl. Allow flash time between coats.
4. Allow the coats to dry 2 to 4 hours before installing the part or putting the car into service. Allow to dry overnight before using seat cushions.

Vinyl and Flexible (Soft) ABS Plastic Parts Finishing

The outer cover material of flexible instrument panel cover assemblies is made of ABS plastic. The same is true of many padded door trim assemblies. The soft cushion padding under ABS covers is urethane foam plastic.

The most widely used flexible vinyls (polyvinyl chloride) are coated fabrics used in seat trim, some door trim assemblies, headlinings, and sun visors. Most (all-vinyl interior) head rests are covered with flexible vinyls. Examples of hard vinyls are door and front seat back assist handles, coat hooks and exterior molding, and bumper inserts.

The paint system for vinyl and flexible ABS plastic involves the use of interior vinyl color and a clear vinyl top coat.

> **NOTE**
>
> No primer or primer–sealer is required.

Use the following painting procedure:

1. Wash the surface thoroughly with vinyl cleaner or with paint finish cleaning solvent.
2. Prepare the vinyl surface for paint with vinyl conditioner according to the directions on the label or as given in this chapter.
3. Using the proper vinyl color as recommended by the car factory, apply interior color according to the label instructions. Apply color in wet coats for complete hiding. Allow sufficient time between coats.
4. Before the final color coat flashes completely, apply one wet double coat of clear vinyl for proper gloss level according to the color chart.
    a. Use a nonglare clear vinyl as described in this chapter for instrument panel and rear window area parts.
    b. Use a semigloss clear vinyl for all other areas.

5. Allow the parts to dry as required before installing them.

> **NOTE**
>
> Apply only enough color for proper hiding and wear to avoid washout of the grain effect.

Gloss Flattening Compound

Gloss flattening compound is available in two types through leading paint suppliers and jobbers (see Table 15–2). One type is for nonurethane finishes and the other is for urethane finishes. These compounds can be used to reduce gloss on interior repair colors to match adjacent trim parts. Use the following procedure for gloss flattening compounds:

1. Prepare the compound according to the label directions by agitating it thoroughly.
2. After color-coating the part and determining that flattening compound is required, add the compound in small amounts, such as 5 percent at a time, to reduced color and stir thoroughly.

> **CAUTION**
>
> Some paint materials can take only a certain amount of flattening agent. After that, the paint material loses its durability.

3. Spray out a small sample panel and allow it to dry several minutes. Check for the proper gloss level with adjacent parts. Adjust the color with additional flattening compound as required.
4. Apply the final one or two color coats to the part as required.

LUGGAGE COMPARTMENT FINISHING

When the interior surfaces of rear compartments must be painted, the accepted practice is to finish the job with body color. Before doing so, however, all necessary surface preparation operations, such as metal conditioning, if necessary, and primer or primer–sealer application, should be done.

REVIEW QUESTIONS

1. As a general rule, in how many colors are paintable plastic parts serviced?
2. How are interior colors of a car identified?
3. At what gloss level are driver's vision-area trim parts painted?
4. At what gloss level are interior side wall trim parts painted?
5. What is crocking?
6. Describe the flex test that painted interior flexible plastic parts must pass for the plastic paint finish to be acceptable.
7. What chemical ingredient is added to interior and exterior paint finishes to make them color fade-resistant?
8. Describe the adhesion test that paint finishes on interior plastic parts must pass to be commercially acceptable.
9. Explain the identification process for interior plastic parts.
10. Describe the procedure and materials required for painting rigid or hard plastic parts.
11. Explain how to use a vinyl conditioner.
12. What is the purpose of a clear vinyl finish? In what gloss levels is it available?
13. How are interior vinyl trim colors controlled to match the car interiors?
14. Explain how R-M (No. 891) Flex Agent is mixed with color and what interior parts can be painted with this system.
15. What is the procedure for applying polypropylene primer?

CHAPTER 16

Painting Flexible Plastic Parts

This chapter covers (1) painting flexible exterior plastic parts according to factory and/or paint supplier recommendations and (2) painting vinyl tops according to factory and/or paint supplier recommendations.

PAINTING FLEXIBLE PLASTIC PARTS

This part of the chapter deals with how flexible exterior plastic parts are designed, how they are painted at the factory, and how they should be painted in the refinish trade.

Design of Flexible Exterior Plastic Parts

Certain car parts are made of flexible plastic to prevent damage to the car or parts whenever these parts are normally flexed under low-speed driving conditions, such as 5 to 10 mph. This incidental contact occurs frequently in heavy city traffic conditions when two cars lightly touch or bump, bumper to bumper. The condition also occurs commonly in parking lots when car bumpers lightly contact a barrier post or a wall during parking.

Location of Flexible Plastic Parts

The use of paintable plastic panels in car manufacturing has increased over the years. The first use of flexible plastics on cars was limited to a few body panels that needed a flexing action. Examples of early design plastic parts are:

1. **Filler panels:** located between the bumper and the car body.
2. **Extension panels:** located mostly at the rear of rear quarter panels.
3. **Front end panels** (also known as fascia): many complete front end assemblies are made of flexible plastics.
4. **Spoilers and valance panels:** these panels are made of various plastics.
5. **Bumper covers:** many car designs require plastic covers on bumpers that are painted to match the body color and thus enhance the beauty of the car.

This list represents only the beginning of plastic material usage. Many cars on the market today, like the GM Saturn, have all major exterior car panels of flexible plastic construction. Ford, Chrysler, and import companies also have special car lines that have flexible plastic exterior car panels.

Elastomeric Paint Finishes

Most original paint finishes on flexible exterior plastic parts are baked elastomeric enamels. The word *elastomeric* has the same meaning as *flexible*, *rubberlike*, and *elastic*. These words mean that something can be bent, creased, and stretched, and then it can recover its original size and shape without cracking or injury of any type. Flexible plastics are designed to do the same thing without cracking or breaking. Also, elastomeric paint finishes are designed to do the same thing while firmly attached to flexible plastics.

An elastomer is a manufactured compound with flexible and elastic properties. The resin system of elastomeric enamels and lacquers is made of these elastomeric compounds.

Requirements of Flexible Exterior Plastic Paint Repairs

Flexible plastic paint repairs must pass the following tests:

1. **Color match:** The color match between painted flexible parts and the car should be commercially acceptable.
2. **Gloss:** The gloss of the repair panel should match adjacent panels reasonably well.
3. **Crack resistance:** Repair paint finishes should be as resistant to cracking as original finishes in all types of weather (see Chapter 15).
4. **Proper adhesion:** Once applied and thoroughly dried, the paint finish should not be removable with masking tape (see Chapter 15).
5. **Proper OEM appearance standards:** Service paint finishes should match, look like, and wear like original paint finishes. Good appearance and durability qualities are built into a good paint finish.

How Flexible Plastic Parts Are Serviced

Most automotive exterior plastic parts are serviced in one color or in a prime coat to keep the inventory of service parts to a minimum. When these parts are replaced in service and require a different color, they must be painted. Plastic parts on cars in service need repainting for many reasons.

Selection of the Correct Paint Repair System

There are many good elastomeric paint finishes on the market. Table 16–1 shows several of the best systems available. All are approved by most car factories. There are other equally good products available on the market. Unfortunately, there are many substandard products on the market that have a tendency to crack when flexed in cold or even normal temperatures. The painter, as the paint engineer in his or her shop, makes an important decision when

TABLE 16–1 Paint Systems and Products for Painting Flexible Plastic Parts

Product Description	R-M Diamont	BASF Glasurit	DuPont	PPG
Flexible additive to make paint flexible	891	521–111	350S 792S 793S 9250S	369 950
Adhesion promoter	864	934–0	2322S	DPX-801
Primer/surfacer/sealer	DP20/25	285–22/ and 929 series hardener	1120S 1125S 9250S and 1135S	DPW-1844 and DPW-1821
Acrylic lacquer (single-stage)	Mix reduced color with flexible additive per label directions.			
Acrylic enamel (single-stage)	Mix reduced color with flexible additive per label directions.			
Acrylic urethane (single stage)	Mix reduced color with flexible additive per label directions.			
Basecoat color system	Do *not* mix basecoat colors with flexible additive.			
Basecoat clears (Add flex additive to clears per label directions.)	DC76/DH15 DC88/DH44 DC89/DH44 DC92/DH42	923–54 923–85 923–55 (See paint jobber.)	380S 580S 780S 7500S 7600S 7800S	(Flex 950) and DCA-468 DC-1100 DC-1260 DC-1275 DC-1290
Interior vinyl trim colors (per jobber directions)	Trim code and color code and 891	Trim code and color code and 521–111	Trim code and vinyl mix	Trim code and UCV number
Vinyl clear				
Nonglare	—	—	—	UCV-69
Semigloss	—	—	—	UCV-71

Note: By the time this book is distributed, product availability may change. For the latest in refinishing on any product, always check with your local paint jobber.

selecting a particular paint system. Equally important to selecting a high-quality product is the application of that product.

Painting Flexible Plastic Parts

Painting flexible exterior plastic parts can be done with a paint system selected from Table 16–1. All plastics painted at the factory are paint repairable in service. It is best to paint full sections of panels to the nearest breakline, because thorough sanding and cleaning are required to promote adhesion of the repair paint system. Thorough sanding will remove all gloss.

> **CAUTION**
>
> **Spot repair** of flexible plastic parts **is not recommended.**

The following tables and figure simplify the painting of flexible plastic parts:

1. Table 16–1 lists special products and paint systems for painting flexible plastic parts.
2. Table 16–2 identifies the plastics used on cars. The proper identification will help the painter in using the proper refinish procedure.

TABLE 16–2 Identifying Flexible Plastic Parts

Code	Family Name	Common Trade Name	Typical Application
EEBC	Ether/Ester Blocked Co-Polymer	EEBC	Bumpers
EEBC/PBTP	EEBC/Polybutylene Terepthalate	EEBC, PBTP, Bexloy	Bumpers, rocker panels
EMPP	Ethylene Modified Polypropylene	EMPP	Bumper covers
EPDM	Ethylene/Propylene Diene Monomer	EPDM, Nordel, Vistalon	Bumpers
EPM	Ethylene/Propylene Co-Polymer	EPM	Fenders
MPU	Foam Polyurethane	MPU	Spoilers
PE	Polyethylene	Alathon, Dylan, Lupolen, Marlex	—
PP	Polypropylene (blends)	Noryl, Azdel, Marlox, Dylon, Pravex	Inner fenders, spoilers, kick panels
PP/EPDM	PP/EPDM alloy	PP/EPDM	Spoilers, grilles
PUR	Polyurethane	Colonels, PUR, PU	Fascias, bumpers
PUR/PC	PUR/PC alloy	Texin	Bumpers
PVC[a]	Polyvinyl Chloride	Apex, Geon, Vinylite	Body moldings, wire insulation, steering wheels
RIM	Reaction Injected Molded Polyurethane	Rim, Bayflex	Front fascias, modular windows
RRIM	Reinforced Reaction Injected Molded PUR	RRIM	Fascias, body panels, body trims
TPE[b]	Thermo Polyethylene	TPE, Hytrel, Bexloy-V	Fascias, bumpers, claddings
TPO[b]	Thermopolyolefin	Polytrope, Renflex, Santoprene, Telcar, Vistaflex, ETA, Apex, TPO	Bumpers, end caps, rubber strips, sight shields, claddings, interior B post
TPP[b]	Thermo-Polypropylene	TPP	Bumpers
TPU	Thermopolyurethane, Polyester	TPU, Hytrel, Texin, Estane	Bumpers, body side moldings, fenders, fascias

[a]Coated fabrics, like seat trim and vinyl tops, are classed as PVC
[b]Light scuffing with 3M-7447.

FIGURE 16–1. Step-by-step outline for repainting flexible plastic parts. (Courtesy of BASF Corporation.)

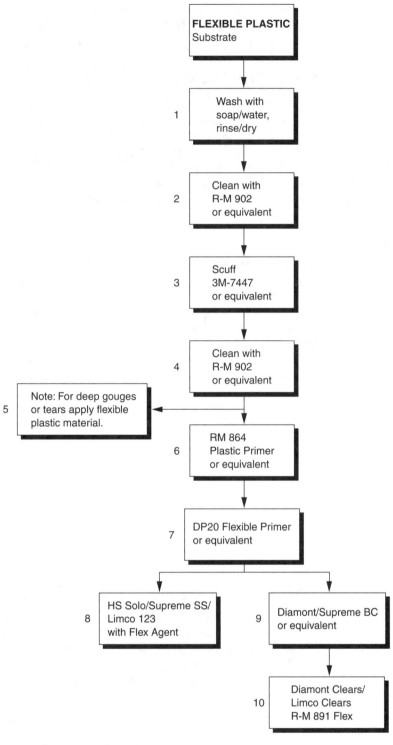

3. Figure 16–1 outlines the refinish steps and products for painting flexible plastic parts as recommended by R-M and Diamont.

Here is a detailed procedure that explains how to paint flexible plastic parts using the R-M/Diamont system. DuPont and PPG have equivalent refinish systems. (For products and procedures used in painting hard or rigid plastics, see Chapter 15.)

Procedure

PLASTIC PREPARATION

1. **Prep Work:** Wash virgin (unpainted) plastic substrate with soap (mild detergent) and warm water to remove grease, wax mold release agents, and other contaminants. Then dry with paper towels and or lint-free cloth. For OEM

primed and basecoat/clearcoat painted plastic substrates, proceed to step 2. For all other materials, proceed to step 4.

2. Wash with R-M 902 plastic cleaner to remove wax, silicone, and other contaminants.
3. Scuff with Scotchbrite #7447 (red) very fine pad.
4. Rinse with R-M 902 plastic cleaner. Then dry with paper towels and use tack cloth prior to actual painting.

APPLICATION

5. **Adhesion Promoter:** R-M 864 Clear Polypropylene Plastic Primer.

   Spray R-M 864 according to label instructions (RFU—easy to use). Apply one medium wet coat at 40 psi at the spray gun. Allow 15 to 30 minutes dry time before any spray application of Diamont Basecoat Color, HS Solo Color, or DP20 Flexible Primer.

6. **Flexible Primer:** When priming a flexible part, the primer surfacer must have a flex agent added to the mix ratio. Regular primer surfacer (without a flex agent) is too hard and will not flex.

   DP20 Primer for Flexible parts: DP20/PH12/UR Reducer/DF25 at 3:1:1:1 by volume.

   Spray two or three coats of Flexible DP20 Primer over R-M 864 at 55 to 65 psi. Let dry for 1.5 hours at 68°F or force dry at 140°F for 30 minutes. Hardeners and reducers may vary according to actual shop conditions. Allow the panel to cool to room temperature before wet sanding with 600-grit sandpaper. Wash and rinse with R-M 902 Plastic Cleaner and let dry.

7. **Topcoats:** Spray two or three coats of Diamont Basecoat or HS Solo Color until hiding has been achieved. Set the air pressure at 45 to 55 psi. Allow 5 minutes flash time between coats. Allow 15 to 20 minutes final flash prior to clearcoat application.

8. **Clearcoats:** Follow label instructions for preparations and application of Diamont Clears. Spray two or three coats at 50 to 60 psi with 5 minutes flash between coats. Allow 10 to 15 minutes final flash before baking or force-drying DC88 or DC89. DC76 and DC92 are immediate bake clears. To make the clear flexible, use R-M 891 Flex Agent. Follow label instructions.

Diamont Flexible Clears	DC76	DC76/DH15, 17, or 19/891	3:1:1 (use 35–45 psi—immediate bake)
	DC88	DC88/DH44 or 45/891	4:1:1
	DC89	DC89/DH44 or 45/891	4:1:1
	DC92	DC92/DH42 or 46/891/UR Series	3:1:1 + 20 percent (immediate bake)

9. Allow to cure overnight or bake at 140°F for 30 minutes. Before compounding or repair is attempted, allow to cool for 30 minutes.

REPAIR OF BUMPER COVER BEFORE PAINTING

When the smooth, painted surface of a bumper cover is cut, gouged, or damaged, the cover can be repaired successfully with the following procedure, which is a 3M system of plastic repair. First the gouged section must be filled and leveled with flexible parts repair material (3M part number 05900 or 05901), Parts A and B. The repair procedure is explained and illustrated below. (A flexible parts putty [3M part number 05903] may be needed.)

1. Clean the repair area with soap and water, and then clean the complete repair surface with paint finish cleaning solvent. Mask as required.
2. Sand away the damaged area to create a tapered, smooth featheredge (Figure 16–2). First taper the damage with a 36-grit disc. Finish the rough featheredge with a 180A-grit disc (Figure 16–3).
3. If backup aid is needed, clean the back side of the plastic and apply a suitable tape, like 3M's Auto Body Repair Tape No. 06935, or its equivalent (Figure 16–4).
4. Mix equal parts of Flexible Parts Repair Material (FPRM) 3M No. 05900 or 05901, Parts A and B, according to label instructions (Figure 16–5).
5. Apply the material to the repair area with a squeegee (Figure 16–6). First, apply a tight skin coat to the surface. Then build the FPRM slightly higher than the undamaged areas. Allow the mixture to cure 30 minutes at 60° to 80°F until firm.
6. After curing, rough featheredge the plastic with grade 180A (Figure 16–7).
7. Apply Flexible Parts Putty, 3M No. 05903, as needed to fill pinholes and sandscratches (Figure 16–8). Allow the putty to dry 15 to 30 minutes.
8. Sand the putty with grade 240A on a soft hand block (Figure 16–9). Then, using grade 320 (by

FIGURE 16–2. Tapering damaged area. (Courtesy of 3M Automotive Aftermarket Division.)

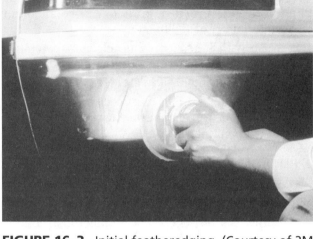

FIGURE 16–3. Initial featheredging. (Courtesy of 3M Automotive Aftermarket Division.)

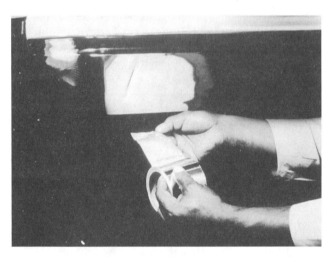

FIGURE 16–4. Backup material. (Courtesy of 3M Automotive Aftermarket Division.)

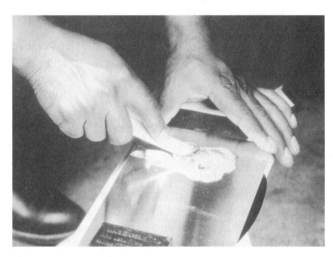

FIGURE 16–5. Mixing FPRM A and B. (Courtesy of 3M Automotive Aftermarket Division.)

machine) or 400 (by hand), sand the complete OEM finish on the flexible plastic part to remove 80 percent to 90 percent of all gloss to the nearest breakline.

9. Wipe and blow off the repair panel (Figure 16–10). Mask to paint the complete panel to the nearest breakline.
10. Apply a double wet coat of Flexible Parts Coating, 3M No. 05905, or R-M's HP-100 to the repair area (Figure 16–11). Allow 10 to 15 minutes of flash time. Apply a second double coat to the entire panel section. Allow to dry 45 minutes to 1 hour and then sand to desired smoothness with 320A or 400 sandpaper (Figure 16–12).
11. Clean spray gun immediately after use, first with water, then with lacquer thinner.

> **NOTE**
>
> If the flexible primer–surfacer–sealer product recommended in step 10 is not available, prepare and use a primer–surfacer compatible with a flex agent as recommended by the paint jobber. Mix the primer–surfacer with a suitable flex agent (see Table 16–1) according to the label directions. Apply two to three coats to repair the area and allow it to dry. Sand to desired smoothness with No. 400 sandpaper. If using a PPG system, apply PPG DPX-844 to the repair area and to the complete panel prior to topcoating with color.

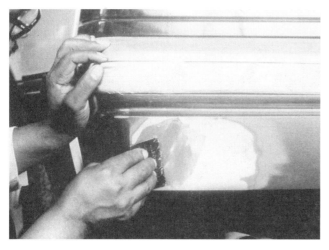

FIGURE 16–6. Applying FPRM. (Courtesy of 3M Automotive Aftermarket Division.)

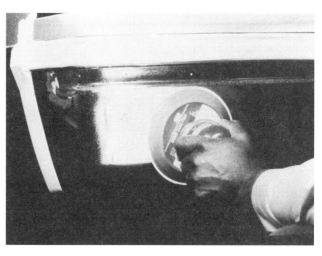

FIGURE 16–7. Final featheredging. (Courtesy of 3M Automotive Aftermarket Division.)

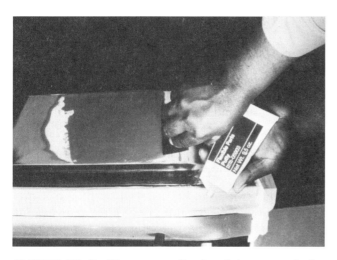

FIGURE 16–8. FP putty application (gloves needed). (Courtesy of 3M Automotive Aftermarket Division.)

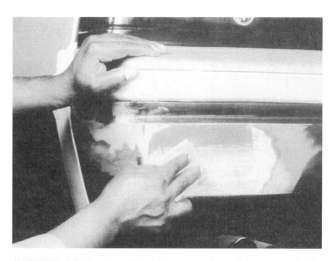

FIGURE 16–9. Sanding the putty (no gloves needed). (Courtesy of 3M Automotive Aftermarket Division.)

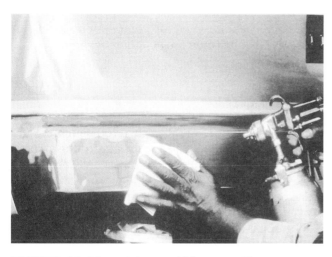

FIGURE 16–10. Wiping and blowing off repair area (no gloves needed). (Courtesy of 3M Automotive Aftermarket Division.)

FIGURE 16–11. Application of the primer–surfacer–sealer (gloves needed). (Courtesy of 3M Automotive Aftermarket Division.)

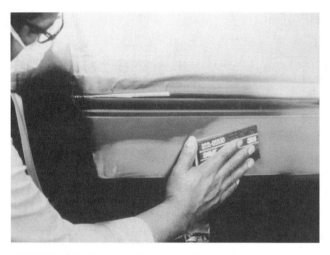

FIGURE 16–12. Sanding the primer–surfacer–sealer (no gloves needed). (Courtesy of 3M Automotive Aftermarket Division.)

FIGURE 16–13. Final color application (gloves needed). (Courtesy of 3M Automotive Aftermarket Division.)

12. Prepare color for application as follows (Table 16–1; Figure 16–13):
    a. Mix flexible additive with standard acrylic lacquer or enamel according to the label directions.
    b. Do not use a flex agent in basecoat colors.
    c. If a color match problem is expected, spray out and color match a test panel as explained in Chapter 14 before applying color on car.
    d. If you are working with a basecoat color, apply clearcoat that includes a flex agent according to label directions.
13. Complete the job by cleaning the spray gun and removing the masking.

PAINTING VINYL TOPS

This part of the chapter deals with how to paint vinyl tops in the automotive refinishing trade according to factory and/or paint supplier recommendations.

How Vinyl Tops Are Color Coded

For color identification, factory-installed vinyl tops are color coded on the body number plate (see Chapter 2). Black-and-white tops are self-explanatory for color identification. Production vinyl tops are often color matched to the body color. However, if a vinyl top was installed by a dealer or trim shop after production, the color of the top can be matched in vinyl color paint material by the paint jobber for repair or repainting purposes.

How to Order Vinyl Top Colors

To order vinyl top color for a car, you must have two pieces of information:

1. Make and model year of the car
2. Code for the vinyl top color (if included on the body number plate)

Black and white can be ordered by phone. However, if the vinyl top color is not known, the paint jobber can determine the vinyl color for the car by examining the vinyl top and checking a color chart book.

Major Vinyl Top Paint Systems

Three of the leading vinyl top paint systems available in the refinishing trade are supplied by:

1. PPG Industries
2. DuPont
3. Rinshed-Mason (part of BASF Corporation)

Usually, one or more of these vinyl top paint systems are available through the local paint jobber. Many additional paint suppliers produce vinyl top colors of equivalent quality to those listed in Table 16–3. Some vinyl top paint systems on the market have proved not to be of factory standard quality. The painter must exercise good judgment in determining which paint system to use. The paint jobber who provides the best-quality materials and the best service to a paint shop is usually the paint jobber doing the most business.

PAINTING FLEXIBLE PLASTIC PARTS

TABLE 16–3 Products for Painting Vinyl Tops

Product	PPG	DuPont	Rinshed-Mason
Paint cleaning solvent	DX-330	3919-S and 3939-S[a]	No. 900
Vinyl prep conditioner	UK-405	—	—
Vinyl colors for vinyl tops	Exterior color code and UCV number	Exterior color code and vinyl mix number	Standard color code acrylic lacquer or enamel and 891 and proper solvent

[a]Use on GM cars painted with LDL-type acrylic lacquers.

Vinyl Top Paint System Color Availability

Black and white vinyl top colors are usually available in factory-packaged form ready to use from PPG and DuPont. These colors can be obtained immediately. However, all other vinyl top colors are produced to match production vinyl colors by a mixing machine formula as each color is ordered. Although there may be some basic difference among paint suppliers' products, the suppliers use essentially genuine vinyl colors for vinyl tops.

CAUTION

Vinyl colors come ready to spray and should not be reduced.

Rinshed-Mason Vinyl Top Paint System

The Rinshed-Mason vinyl top paint system has been used successfully in the trade for several years and is factory-approved. The system makes use of Urethane Flex Agent (No. 891), which is added to R-M acrylic lacquer or enamel color. This combination makes an excellent vinyl top paint. No color matching is required because the top color already matches the car color. No special primer is required. Once a simple color preparation is completed, the material is as easy to apply as any vinyl top colors. The R-M system produces a flexible, hard film with excellent durability.

Vinyl Top Painting Procedure (for All Paint Systems)

1. Wash the old top with household bleach, a brush, and plenty of water. Rinse the top and car thoroughly in clean water. Figure 16–14 shows a vinyl top before painting.
2. Mask off moldings and the car as required (see Figure 16–15).
3. Clean the top thoroughly with paint finish cleaning solvent (see Figure 16–16).
4. Prepare the vinyl top paint material by mixing thoroughly and by straining into a paint cup.

NOTE

If using the Rinshed-Mason system, prepare the color as follows:

---

a. Mix:
   1 pint R-M acrylic lacquer or acrylic enamel color
   1 pint No. 891 Flex Agent
   $\frac{1}{2}$ pint PNT-90 lacquer thinner or reducer
b. Agitate thoroughly and strain into a paint cup. This preparation will do a complete vinyl top. Follow the label directions. The material has a limited pot life.

FIGURE 16–14. View of vinyl top with problems before painting. (Courtesy of BASF Corporation.)

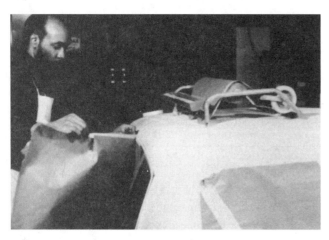

FIGURE 16–15. Masking off car before painting vinyl top. (Courtesy of BASF Corporation.)

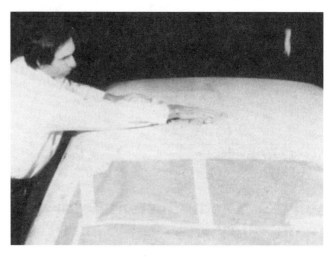

FIGURE 16–16. Washing and precleaning vinyl top. (Courtesy of BASF Corporation.)

5. Blow out all gap spacings and crevices around the top and tack-wipe the top as required.
6. Use a NIOSH-approved respirator when applying the color. With a small spray fan, apply a small, wet banding coat along all the edges around the complete top.
7. Using a full, wide-open spray pattern, apply a first full-wet coat of vinyl color as follows:
   a. Use precautions as required to prevent dripping.
   b. Control the hose by positioning it over the shoulders and back.
   c. Start the color application along the near side and proceed to the center of the top.
   d. On the opposite side of the car, start at the center and maintain the wet application of color to the near side.
   e. Keep the application wet with a full and uniform 50 percent to 75 percent stroke overlap.
   f. Keep the spray gun as perpendicular to the surface as possible.
   g. Apply color to the sail and rear quarter areas as required.
8. Apply a second full-wet coat, starting as in step 7, for complete hiding and uniformity of wetness.

NOTE

Any remaining streaks can be removed with blending solvent applied as required immediately after painting the top.

---

9. After proper drying per label instructions, remove the masking. Allow to dry as required before putting the car into service.

REVIEW QUESTIONS

1. In how many colors are paintable flexible plastic parts serviced?
2. How are plastic parts identified when determining the repair procedure?
3. How are colors for painting flexible plastic parts identified?
4. Describe the location of a flexible plastic filler panel.
5. Describe the R-M paint materials and procedure for painting flexible plastic parts (see Figure 16–1).
6. Describe the heat test to determine if a plastic is thermoplastic (soft) or thermosetting (hard).
7. Explain the materials and procedure required for painting vinyl and (soft) plastic parts.
8. Describe the materials and procedure for repainting a polyurethane plastic bumper cover after a deep cut has been repaired.
9. What does *thermoplastic* mean?
10. What does *thermosetting* mean?
11. What is the best way to identify the type of plastic on a loose part?
12. Describe the materials and procedure for refinishing a new replacement bumper cover.
13. Whom may a painter contact if he or she is unsure of specific regulations in effect in a given area?
14. Describe the 3M system for repairing a cut and damaged bumper cover with flexible plastic repair material before paint operations are done.

15. What can happen when conventional car colors with no flex agent are applied to flexible plastic parts?
16. Where are the color codes for vinyl tops found on cars?
17. Explain how to order vinyl top colors for a car.
18. Describe the procedure for painting a vinyl top on a car.

CHAPTER 17

Paint Conditions and Remedies

INTRODUCTION

The purpose of this chapter is to familiarize the painter with the identity and description of the more common paint conditions and problems that a painter encounters in the refinish trade. Painters are expected to repair these problems and conditions in a top-quality fashion.

Each paint condition and problem is described and, where helpful, is illustrated. Understanding the cause of a paint problem or condition is vital to the selection of correct repair procedures. Essentially, each paint condition and problem has its own specific repair requirements. The identity of any particular paint problem tells the painter the required repair procedure that must be followed to make a quality repair. The more a painter learns about refinish products, repair systems, and the common problems that are encountered, the more qualified a painter becomes.

A remedy for each paint condition and a prevention system for avoiding paint problems are included where applicable in this chapter. The type of finish affected is shown in the headings in parentheses.

ACID AND ALKALI SPOTTING (LACQUER AND ENAMEL)

Acid and alkali spotting (see Figure 17–1) has affected automotive finishes for many years. A more thorough coverage of the subject appears in Chapter 13 under the topic "acid rain." This condition can occur on solid or metallic colors in both the acrylic lacquer and enamel finishes, and on basecoat/clearcoat finishes. The condition appears as stained spots on the finish and affects mostly flat or horizontal surfaces. On standard red metallics,

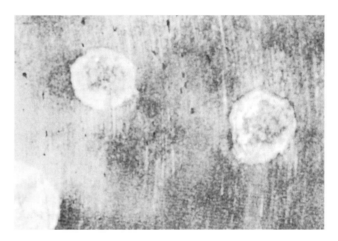

FIGURE 17–1. Acid and alkali spotting. (Courtesy of General Motors Corporation.)

spots are darker red; on blue metallics, spots are darker blue; and so on. On solid colors, discoloration spots may be of any color, depending on the specific contaminants.

Cause. On metallic colors, the condition apparently is caused by fall-out or moisture in the form of rainwater containing acid or alkaline materials that attack the aluminum flake or clearcoat. On solid colors, specific pigments may be affected by specific materials from fall-out.

Remedy. In mild cases, rub out and polish. In severe cases, sand to remove the condition *completely* and color-coat as required.

Prevention. To prevent recurrence of the condition, apply clear urethane enamel to the affected surfaces in accordance with factory specifications. Also, keep the paint finish clean and polished.

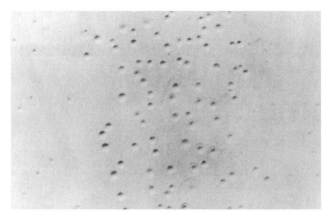

FIGURE 17–2. Blistering. (Courtesy of General Motors Corporation.)

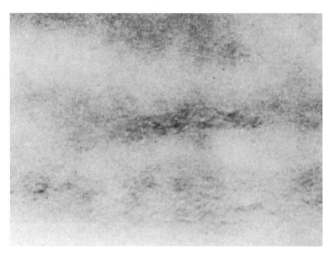

FIGURE 17–3. Blushing. (Courtesy of General Motors Corporation.)

BLISTERING (LACQUER AND ENAMEL)

Blistering is the appearance of several small, dome-shaped blisters in the paint finish. They can range in size from $\frac{1}{16}$ to $\frac{1}{4}$ inch or larger. Blisters are usually grouped together, as shown in Figure 17–2.

Cause. Blistering is usually caused by moisture becoming trapped within the paint film and then expanding between the metal and the undercoat, or between the undercoat and the color coat, causing different-size blisters to form. The depth of the condition is determined by cutting off some of the blisters and inspecting them and the spots from which they were removed. Inspection determines whether the finish blisters off the undercoats, or if the color and undercoat come off the metal. The painter can make a determination if he or she knows the color of the factory and service undercoats.

Remedy. In minor cases, blisters may be sanded out, resurfaced, sanded, and color-coated. In severe cases, the finish must be removed down to the metal before refinishing.

BLUSHING (LACQUER)

Blushing is an off-color, milky, or dull mist formation on the surface of freshly applied color (see Figure 17–3).

Cause. Blushing is caused by a mixture of water with fresh paint as it is applied, or when highly humid air condenses on the surface during spray painting and water mixes with the paint material, causing a kick-out in the paint binder.

Remedy. In most cases, spraying a coat of slow-evaporating solvent with 10 percent to 20 percent retarder immediately over the affected area will dissolve the blushed condition and restore the normal appearance of the finish. *If blushed color dries*, add retarder to the reduced material and apply a final extra-wet color coat as required.

Prevention. Keep the paint and car at room temperature. Use good-quality solvent. Use the required amount of retarder when spraying in high humidity and warm temperatures.

BULL'S-EYE (LACQUER AND ENAMEL)

A bull's eye is a spotted, ringed outline or low area in the color coat (see Figure 17–4). It often gives the illusion of a different color, depending on the reflection of light and shadows in the area. The primer might show.

Cause. A bull's-eye is the result of poor featheredging and/or poor spot-repair technique. Contributing causes are poor primer–surfacer sanding technique and/or undercoat shrinkage after the color was applied.

Remedy. In very minor cases, sand with No. 500 or 600 fine sandpaper, rub out, and polish. In severe cases, on standard colors, sand and featheredge the area correctly; build up the surface with primer–surfacer as required; and spot-blend color to match the adjacent surfaces.

PAINT CONDITIONS AND REMEDIES

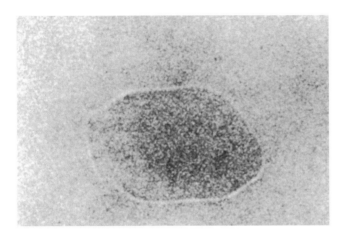

FIGURE 17–4. Bull's eye. (Courtesy of General Motors Corporation.)

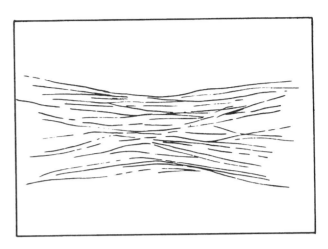

FIGURE 17–5. Checking and cracking. (Courtesy of General Motors Corporation.)

Prevention. Use the guide coat sanding system (see Chapter 8), particularly when sanding small spots on enamels. Also, use the correct spot repair technique.

CHALKING (LACQUER)

Chalking appears as a loss of gloss and a powdery surface on standard colors. Eventually, the surface becomes flat and powdery.

Cause. Chalking is the natural breakdown of paint film by prolonged weathering. Ultraviolet light from the sun, together with moisture, are the greatest enemies of a paint finish. Breakdown of any paint binder system starts at the outer surface.

Remedy. Remove a light chalking condition simply by polishing the car to restore the original gloss. If chalking is severe, compound and polish the surface to restore gloss and appearance. If chalking returns abnormally quickly, re-color-coat the car as required, or compound color and apply suitable urethane clear coating.

Prevention. Agitate the paint thoroughly and use balanced solvents when applying color. Keep the car on a regular cleaning and polishing schedule. Use clear urethane coating.

CHECKING AND CRACKING (LACQUER AND ENAMEL)

This condition is also known as line checking and/or cracking. The condition appears as a series of long, straight lines, usually traveling with the direction of a panel (see Figure 17–5). Sometimes lines are curved, going in various directions. Depending on the thickness of the color and the severity of the condition, the lines may be quite short or as long as 18 inches.

Cause. The condition is usually due to excessively thick color coats or application of a new color over an old color that was cracked and not completely removed.

Remedy. Remove the cracked color coat from the affected area to the original undercoat and recolor the coat as required. On standard acrylic lacquer colored cars, original undercoats are not affected by line checking.

Prevention. Painters should know how much total paint thickness is applied to every paint job. This determination can be made with a paint thickness gauge, which is described in Chapter 9.

MICRO-CHECKING (LACQUER)

When observed normally, micro-checking appears as severe dulling of the paint finish, with little or no gloss. However, when examined with a magnifying glass, 5× to 7× power (see Figure 17–6), the condition appears as a high volume of very small cracks in the paint surface. The cracks do not touch one another.

Cause. Micro-checking is the beginning of paint film breakdown. The condition apparently is caused by the ultraviolet rays of the sun.

Remedy. Completely remove the affected color coat and re-color-coat as required.

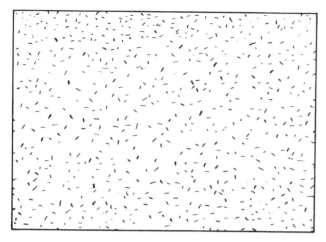

FIGURE 17-6. Micro-checking (magnified seven times). (Courtesy of General Motors Corporation.)

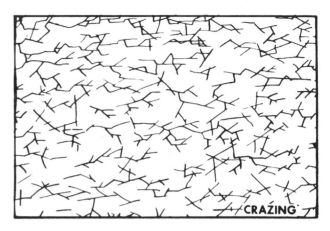

FIGURE 17-7. Crazing. (Courtesy of General Motors Corporation.)

Prevention. If primer–surfacer is required in a repair situation over a finish with micro-checking or very fine crazing, use a water-base primer–surfacer like R-M's 806 Barrier Coat or HP-100 Nova-Prime, which should solve the checking problem.

CRAZING (LACQUER)

Crazing usually appears as a fine spider-web type of cracking in the color coat (see Figure 17–7). The cracks may vary from very fine (requiring a magnifying glass to see) to relatively coarse. The crack lines connect to each other. Crazing sometimes occurs immediately after repairs are attempted.

Cause. Crazing occurs when excessive stresses, which occasionally may be set up in a paint color film during the time it cures, are suddenly released. The condition may occur when repairing acrylic lacquer that was used in a previous repair and has gone through a curing process for a period of time.

Prevention. Before repairs are attempted, test the acrylic lacquer color (to be repaired with acrylic lacquer) as follows:

1. Apply a *drop of repair lacquer thinner* to the surface to be repaired.
2. Allow the thinner to evaporate.
3. Inspect the *paint surface within the thinner ring* for crazing.

A lack of crazing indicates that the paint surface can be color-coated or blended into normally. The appearance of crazing within the thinner ring indicates the color cannot be repaired directly with the acrylic lacquer system.

1. **Option 1:** If repairing with acrylic lacquer, use a waterbase primer–surfacer like R-M's 806 Barrier Coat or HP-100 Nova-Prime according to label directions. After sanding, acrylic lacquer can be applied with no problem. This option solves the stripping problem.
2. **Option 2:** To avoid removal of acrylic lacquer that would craze with a lacquer repair, make the panel repair with catalyzed acrylic urethane finish.

Remedy. If a panel surface to be repaired already has a severe crazing condition, remove the complete affected color coat from the factory undercoat and color-coat as required.

DIRT IN THE FINISH (LACQUER AND ENAMEL)

Freshly painted surfaces with dirt in the finish show an uneven grittiness from flying dirt particles, foreign matter, and lint that have landed on or in the paint during spray painting (see Figure 17–8).

Cause. Generally, the condition is caused by dirt coming from one or more of the following sources:

1. The car.
2. The paint.
3. The atomizing air.
4. The working area.
5. Poor spraying techniques.

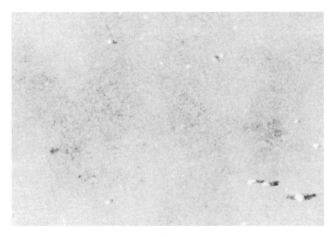

FIGURE 17–8. Dirt in the finish. (Courtesy of General Motors Corporation.)

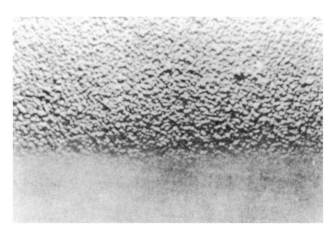

FIGURE 17–9. Dry spray. (Courtesy of General Motors Corporation.)

6. Lint on clothing.
7. Static electricity.
8. Improper operation of the paint booth.

> **NOTE**
> For an in-depth look, see the special Binks summary on the causes of dirt in a finish at the end of this chapter.

Remedy. To repair the surface in minor cases, compound and polish as required. In severe cases, water-sand lightly with No. 500, 600, or finer sandpaper; then compound and polish.

DRY SPRAY (LACQUER AND ENAMEL)

This condition is easily detected by a certain uniform, fine grittiness and dullness (see Figure 17–9). It usually forms a linear pattern or follows the direction of spray gun travel.

Cause. The condition is usually caused by holding the spray gun at an angle or too far from the surface. The condition can also be caused by insufficient solvent, excessive air pressure, a dirty spray gun, or spraying in a draft.

Remedy. In minor cases, rub out and polish as required. In severe cases, water-sand with No. 400 and/or 500 sandpaper and color-coat as required.

ETCHING (LACQUER)

Etching is a severe form of water spotting in which the entire paint surface within the periphery of each spot is etched or eaten away. The condition may appear as small or large water spotted areas and usually appears on the flat or horizontal surfaces. Etching penetrates much more deeply into the finish than does water spotting.

Cause. The condition may be caused by bird droppings, insects, or contaminants, in which case a strong chemical deposit is allowed to react with the finish for a prolonged period of time.

Remedy. If the condition is mild, sand to remove the condition and color-coat as required. If the condition is severe, sand to remove it and apply undercoats and color coats as required.

Prevention. The best prevention is to keep the paint surface clean and polished. A double coating of clear urethane enamel offers excellent protection.

FISHEYES (LACQUER AND ENAMEL)

Fisheyes are small, round, craterlike openings that appear in the finish immediately after it has been applied (see Figure 17–10).

Cause. Application of color coats over a surface contaminated with silicones.

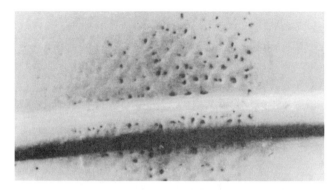

FIGURE 17–10. Fisheyes. (Courtesy of DuPont Company.)

Prevention

1. Clean the surface with wax and a silicone-removing agent such as Prep-Sol, Pre-Kleano, Acryli-Clean, or equivalent.
2. Sand the surface as required.
3. Reclean the surface with a silicone-removing agent.
4. Proceed with the colorcoat application

If the preceding prevention steps are not successful and fisheyes appear with the application of the first coat, add fisheye eliminator, known as Fish-Eye-Preventor, Fish-Eye-Eliminator, or the equivalent, to the reduced color according to paint supplier directions, and continue color-coating immediately.

Remedy. To repair a paint surface with a dried fisheye condition, sand the surface smooth and color coat as required, incorporating the previous prevention steps.

LIFTING OF ENAMELS

This condition can appear as raising and swelling of the wet film or peeling when the paint surface is dry. It also looks like puckering and crinkling (see Figure 17–11).

Cause. Lifting is caused by solvents in a refinish paint attacking a previously painted surface. The following are additional causes:

1. Improper drying of previous coating
2. Sandwiching enamel between two lacquer finishes
3. Recoating improperly cured enamel
4. Spraying over unclean, incompatible surfaces

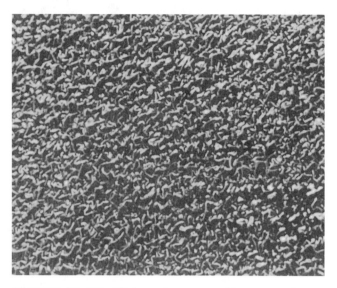

FIGURE 17–11. Lifting of enamels. (Courtesy of DuPont Company.)

Remedy. Remove the lifted paint finish and refinish as required.

Prevention

1. Clean old surfaces thoroughly.
2. Allow all subcoats full drying time.
3. Seal old finishes.
4. Avoid the use of acrylic lacquers over uncured air-dried enamels.
5. Avoid sandwiching an enamel coat between two lacquer coats.

MOTTLING (LACQUER AND ENAMEL)

Mottling appears as dark, shaded, or off-color spots and streaks in the paint finish (see Figure 17–12). This condition occurs primarily in metallic finishes. A moderate amount of mottling is to be expected in metallic finishes as a standard condition.

Cause. Metallic colors are made from a combination of different pigments and aluminum flakes that have different weights and particle sizes. When the film is sprayed extra wet, the paint ingredients have a natural tendency to separate and float into groups. Under normal conditions, this tendency is small in magnitude and cannot be seen by the naked eye. However, certain conditions aggravate this tendency to a point where the separation of the pigments and

PAINT CONDITIONS AND REMEDIES

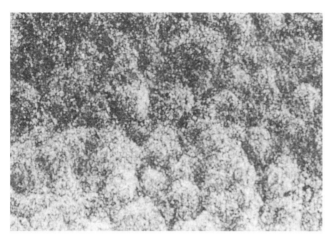

FIGURE 17–12. Mottling. (Courtesy of General Motors Corporation.)

metallic flakes becomes visible. The conditions are caused by using solvents that dry too slowly; allowing pigment particles to migrate; applying color on a cold surface or in a cold room; or applying color coats too wet.

Remedy. In minor cases, no correction is required. In severe cases, perform the following operation:

1. Allow the applied color to flash thoroughly.
2. Apply additional color to the affected area in a *fog-coat* fashion, as follows:
   a. Apply color from an 18- to 20-inch distance with a full spray fan and continual swirling movement until the condition disappears.
   b. Apply a blending solvent as required to bring the repair to the proper gloss.

OFF-COLOR (LACQUER AND ENAMEL)

The color is off shade or does not match.

Cause. There are reasons why a repair color does not match a car. Three conditions that can cause off-color at the factory are balling, bending, or shearing of the metallic flakes. Other causes at the factory are overbaking or underbaking of the finish and breakdown of the ultraviolet screener. Off-color due to field repair problems include poor jobber mix, unstable tints, improper spray techniques, and the spray painting variables (temperature, humidity, and ventilation). There are additional causes.

Remedy

1. Determine the specific color-matching problem (see Chapter 14).
2. Make the necessary color adjustments through spray technique or tinting to achieve the best color match on a spray-out test panel.
3. Make the same acceptable color-match repair on the car.

EXCESSIVE ORANGE PEEL (LACQUER AND ENAMEL)

Orange peel is a natural occurrence in refinishing in which the resultant finish has uneven formations on the surface that look like the skin of an orange (see Figure 17–13). A certain amount of orange peel occurs in normal refinishing and is acceptable. When a surface becomes extra coarse or rough with orange peel and becomes a distraction, it is excessive.

Cause. Excessive orange peel is actually a defect of flow or leveling. This condition is brought about by any one or a combination of the following:

1. Using the wrong type or a poor grade of solvent
2. Insufficient reduction of color
3. Too high an air pressure
4. Improper adjustment of the spray gun
5. Poor spray gun technique: holding the gun too far from or too close to the surface
6. Spraying in a draft
7. Coats applied too dry and thin
8. Cold shop or metal temperatures

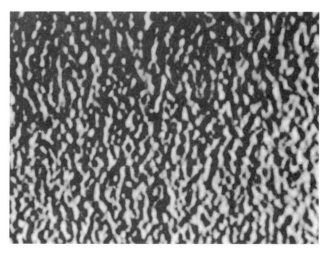

FIGURE 17–13. Excessive orange peel. (Courtesy of DuPont Company.)

Remedy. If the condition is slight, no remedy is necessary. If the condition is excessive, clean, sand (with fine sandpaper), compound, and polish the affected area.

OVERSPRAY (LACQUER AND ENAMEL)

Overspray is characterized by the appearance of a rough or dull paint finish, similar to dry spray.

Cause. Overspray is caused by the settling of semi-dry paint particles on adjacent finished and unprotected surfaces during spray painting operations.

Remedy. If the condition involves the same color, compound and polish the affected area when dry. If the condition involves two colors but is slight, compounding and polishing may eliminate the condition. If the condition is severe, sand and color-coat as required.

Prevention. Use proper masking and/or covering techniques.

PEELING (LACQUER AND ENAMEL)

Peeling is the separation of a paint film from the surface in sheet form (see Figure 17–14).

Cause

1. *Not cleaning and sanding* the original factory acrylic enamel thoroughly before painting
2. *Not using the recommended sealer* on the original factory acrylic enamel before color coating
3. *Not cleaning and sanding* the flexible exterior plastics thoroughly before color coating
4. *Not using the required primer* on polypropylene plastic parts before color coating
5. Incompatibility of the repair coat with the previous coat

Remedy. Remove the peeling paint completely. Prepare the metal and/or other surfaces as required to correct the cause, and refinish with a compatible paint system.

1. Thoroughly clean and treat old surfaces.
2. Use the required primer and/or sealer.
3. Follow the recommended refinish practices.
4. Use compatible repair systems.

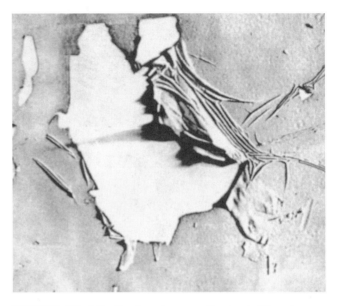

FIGURE 17–14. Peeling (poor adhesion). (Courtesy of BASF Corporation.)

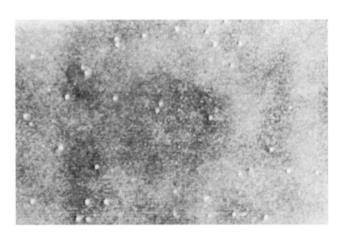

FIGURE 17–15. Pinholing. (Courtesy of General Motors Corporation.)

PINHOLING (LACQUER AND ENAMEL)

Pinholing is a series of tiny, fine holes or pits that give the surface a spotty, dull, or off-shade appearance (see Figure 17–15).

Cause. This condition is usually caused by trapping solvent or air in the paint film and then subjecting the paint to sudden high temperatures such as in force-drying.

Remedy. Sand down the surface with the appropriate sandpaper to remove the problem, and then color-coat as required.

PAINT CONDITIONS AND REMEDIES

Prevention. Allow plenty of flash time before subjecting freshly painted parts to force-drying. The amount of time depends on the solvents used and the shop temperatures.

RUB-THROUGH (LACQUER)

Rub-through or thin paint conditions are easy to see because the undercoat shows through the top coat (see Figure 17–16).

Cause. The usual cause of this condition is excessive compounding that removes the paint film. Too coarse a compound and/or negligent use of the polishing wheel can also cause this condition. The condition may be caused by insufficient color application during a repair. Most rub-through conditions occur at panel edges when panel edges protrude above the surface.

Remedy. For panels, clean the affected panel and color-coat as required. If edges or creaselines are thin, touch up with a brush as required.

Prevention. Before compounding, apply $\frac{1}{2}$- or $\frac{3}{4}$-inch masking tape to all high panel edges and sharp panel creaselines. After compounding, remove the tape and rub out these areas carefully by hand.

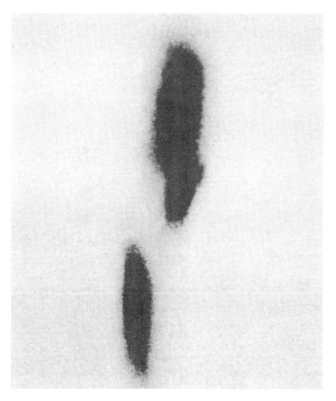

FIGURE 17–16. Rub-through (thin paint). (Courtesy of C.I.L. Paints, Inc., of Canada.)

RUNS OR SAGS (LACQUER AND ENAMEL)

Extra-heavy spray application of paint results in flooding the surface. Excess paint will first sag like a curtain and then turn into narrow runs that flow with gravity until excess paint runs out (see Figure 17–17).

Cause. This condition is caused by one or a combination of the following:

1. The spray gun is out of adjustment (see Chapter 4).
2. Holding the spray gun too close to the surface.
3. Not triggering the spray gun properly on the return stroke.
4. Not observing the applied color (due to poor lighting) and reapplying color to the same spot, flooding it.
5. Forgetting where clear coating was last applied and reapplying clear to the same spot, flooding it. Poor lighting also contributes to this condition.

Remedy

1. If the circumstances are right, brush out a minor run problem with a suitable brush and continue spraying.
2. If spraying with enamel, wash out the condition carefully. Blend the edges with a dry spray as required, and continue spraying (see Chapter 4).
3. If spraying with lacquer, allow to dry, sand out, and re-color-coat.

Prevention

1. Use the correct spray gun adjustment and application technique.

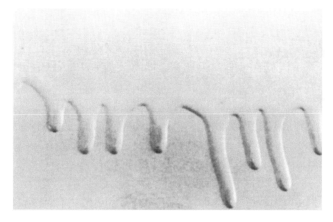

FIGURE 17–17. Runs or sags. (Courtesy of DuPont Company.)

2. Observe the application of every spray painting stroke and follow the progress of spray painting closely until the job is finished.
3. Allow sufficient flash time after each coat.

RUST SPOTS AND RUSTING (LACQUER AND ENAMEL)

Rust spots are usually accentuated by a rust-colored ring that forms at the affected area. Rusting beneath the film is usually made apparent by raised sections of film or blisters. Refer to Figure 17–18. After the film or blisters have broken, the rust begins to work back under the edges of the film. Since many primers are similar to rust in color, careful examination is necessary to identify minor rust conditions accurately. Major rust conditions are easy to identify.

Cause. Rust starts with moisture and chemicals attacking the metal through large or microscopic breaks in the paint film. An early stage of rusting is blistering, which, if unchecked, proceeds to peeling. Other causes are painting over rust that was not completely removed, painting over metal touched by bare hands, or painting over chemical deposits from sanding water.

Remedy. In minor cases where the paint is not blistered, wash the panel and clean the rust stain off with body polish or a mild rubbing compound, hand-applied; then protect the finish with an application of polish. In severe cases, remove the complete finish and rust from the affected area with sandblasting or a special cutting tool. **If the surface is rusted, sand the metal clean and bright to remove all traces of rust.** Treat bare metal surfaces with a two-part metal conditioner according to label directions. Prime the bare metal with the best primer system to prevent future rust. Complete the application of the primer–surfacer and color according to the system that is best for the car.

Prevention. See Chapter 18 for tips about preventing rust.

SAND OR FILE MARKS (LACQUER AND ENAMEL)

Sand or file marks appear after a metal and paint repair job dries and the surface is grained or scratched (see Figure 17–19).

Cause. Coarse file, disc, or sandpaper marks in the metal, solder, or plastic filler before painting. Also, the grit of the sandpaper used to sand the undercoat may have been too coarse. With present-day primer–surfacers, best results are achieved with No. 400 sandpaper.

Remedy. Minor sand marks or scratches on the color coat may be lightly sanded with No. 500 or 600 sandpaper and water and compounded to achieve a repair. In severe cases, sand and refinish as required. See Figures 8–11 and 8–12.

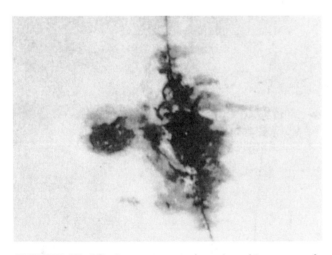

FIGURE 17–18. Rust spots and rusting. (Courtesy of C.I.L. Paints, Inc., of Canada.)

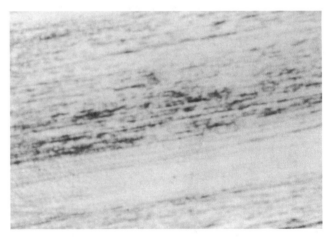

FIGURE 17–19. Sand or file marks. (Courtesy of General Motors Corporation.)

SANDSCRATCH SWELLING (LACQUER)

Sandscratch swelling appears as exaggerated sandpaper scratches (see Figure 17–20) and occurs mostly after spot repairs or panel refinishing are done over sanded original acrylic lacquer finishes. The condition is most apparent on dark colors.

Cause. The condition is caused by sanding acrylic lacquer surfaces with too coarse a sandpaper before color coating. The thinner in the fresh color coats swells the scratches.

Prevention. Do not sand acrylic lacquers unless required. When sanding with coarse sandpaper is required, finish sanding with extra-fine (No. 500, 600, Ultra Fine, 1200, or 1500) sandpaper. Other options include removing sandscratches by compounding, or applying an approved sealer according to label directions before color-coating.

Remedy. Remove minor sandscratches by rubbing and polishing. In certain instances, water-sanding with No. 500, 600, or finer sandpaper may be necessary before rubbing and polishing. Remove severe sandscratches by employing the steps above, "Prevention," and then color-coat as required.

SHRINKING AND SPLITTING OF PUTTY (LACQUER AND ENAMEL)

This condition appears as a pulling away and splitting of the top coat and the undercoat layers from a repaired area (see Figure 17–21).

Cause. Putty applied too heavily in thickness dries at the outer surface more quickly than at the under surface. As undercoats and topcoats are applied, solvents penetrate the complete paint film. On drying, the difference in shrinkage rates between the upper thickness and the lower thickness of the paint film causes the upper surfaces to split and crack open.

Remedy. Remove the putty from the affected area and reapply in a series of thin coats, with adequate drying time between coats (see the application section in Chapter 8). If the problem is considered deep, repair the condition with plastic filler (see Figure 8–8).

FIGURE 17–20. Sandscratch swelling. (Courtesy of C.I.L. Paints, Inc., of Canada.)

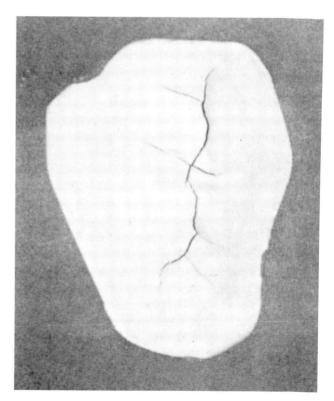

FIGURE 17–21. Shrinking and splitting of putty. (Courtesy of C.I.L. Paints, Inc., of Canada.)

STREAKS IN THE FINISH (LACQUER AND ENAMEL)

Dark or shaded streaks appear in the finish along the path of spray painting (see Figure 17–22).

Cause. This condition is caused by any one or a combination of the following:

1. Unbalanced spray pattern.
2. Poor spraying technique due to incorrect overlap.
3. The spray gun is dirty or needs repair.
4. Spraying metallic colors too wet (mottling).
5. Spraying any color too dry.
6. Substrate is too hot or too cold.

Prevention

1. Adjust the spray gun and check the spray pattern before starting each job (Chapter 4).
2. Use the correct spraying technique with sufficient overlap of strokes.
3. As necessary, clean the spray gun thoroughly.
4. As necessary, repair the spray gun.
5. Allow the metal to come to normal room temperature before spraying.

Remedy. Apply color according to the label directions while incorporating the previous prevention steps.

SWEAT-OUT OR BLOOM (LACQUER)

A dull paint repair sometimes appears after acrylic lacquer is applied and compounded and the repair has set for a day or two. Dullness in the paint is sometimes caused by

Cause. Compounding or polishing the color coat too soon after application, especially when retarder is used in the color coat. The condition is also caused by film shrinkage, which wrinkles the smooth surface with the final evaporation of solvents.

Remedy. Compounding after the film dries is usually sufficient to bring the gloss to an acceptable level. In severe cases, where film shrinkage results in a wrinkled or orange peel appearance, sand with No. 600 or finer sandpaper and water, and compound as required.

WATER SPOTTING (LACQUER)

This condition looks like small rings that surround each spot from which water has evaporated (see Figure 17–23). These rings appear to be etched into the paint finish and cannot be removed by normal washing or polishing.

Cause. The condition is caused by the evaporation of water droplets from a paint finish in bright sun when the surface temperatures are over 150°F. The condition becomes more severe as the chemical content of the water and the temperature are increased. A chemical reaction is believed to be caused by the evaporating water and the paint finish, resulting in the ring.

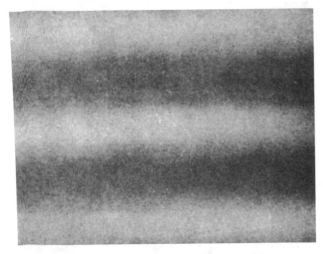

FIGURE 17–22. Streaks in finish. (Courtesy of BASF Corporation.)

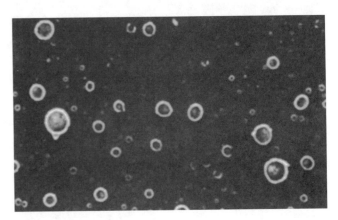

FIGURE 17–23. Water spotting. (Courtesy of BASF Corporation.)

Remedy. In minor cases, rub out and polish as required. Use a very mild or fine abrasive compound. In more severe cases, lightly sand with No. 600 or ultra-fine sandpaper and water, and compound and polish.

WET SPOTS (LACQUER AND ENAMEL)

The finish sometimes won't dry; it remains tacky in spots or on complete repair.

Cause. The condition may be due to any one or a combination of the following:

1. Application of finish over a surface contaminated with wax, silicones, oil, grease, fingermarks, or gasoline residue.
2. Poor ventilation in the drying room.
3. Air too humid and/or too cold.
4. Drier left out of the enamel when the color is prepared on the mixing machine.
5. The wrong type of thinner or reducer was used.
6. Extra heavy undercoats not properly dried.

Remedy. Sand or wash off the complete affected finish and refinish as required.

Prevention

1. Clean, sand, and reclean the surface as required with paint finish cleaning solvent.
2. Provide adequate ventilation and air movement in the drying area.
3. Allow the car to dry in average temperature conditions.
4. Include drier with the other ingredients when preparing color on the mixing machine.
5. Use the proper solvents.
6. Allow the undercoats to dry thoroughly.

WHEEL BURN (LACQUER)

A wheel burn is a dark, often rough smear on a panel surface.

Cause. Holding the polisher too long in one spot.

Remedy. Rub out with a cloth treated with paint finish cleaning solvent, and hand polish. In severe cases, water-sand with No. 600 sandpaper, and then rub out and polish.

WRINKLING OF ENAMELS

The wrinkling of an enamel paint finish looks like the surface of a dried prune skin (see Figure 17–24). The surface is wrinkled and puckered. At times, the wrinkling is so small that it cannot be seen by the naked eye. These surfaces appear dull and the wrinkling can be seen clearly with a magnifying glass.

Cause

1. Rapid drying of the top surface while the underneath remains soft
2. Application of excessively thick color coats
3. Spraying in hot sun, or exposing to sunshine before the enamel is thoroughly dry
4. Surface drying, trapping solvents
5. Fresh film subjected to force-drying too soon
6. Use of acrylic lacquer thinner in enamel

Prevention

1. Reduce enamels according to directions.
2. Apply as recommended.
3. Do not force-dry until solvents have flashed off.

Remedy. The best remedy is to remove the wrinkled film and repaint properly. The following operations constitute an alternate remedy if they can be done *while maintaining an 8- to 10-mil total film thickness*:

1. Force-dry or allow the wrinkled finish to harden thoroughly.

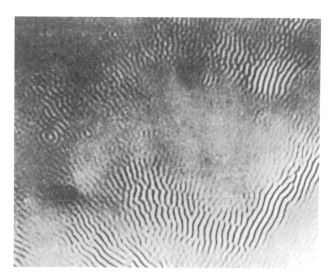

FIGURE 17–24. Wrinkling of enamels. (Courtesy of DuPont Company.)

2. Apply two or more coats of enamel-type primer–surfacer. Allow to dry.
3. Thoroughly water-sand with No. 400 sandpaper or finer.
4. Apply color coats as directed in the list for preventing wrinkling.

BINKS SUMMARY ON DIRT IN THE FINISH*

Dirt is one of the most prevalent problems encountered in the finishing industry today because it is just about the most elusive to solve. Dirt in the finish may be a result of one or a combination of the following conditions:

1. **The product is dirty:**
   a. All loose dirt must be removed from the product prior to entry into the spray area by proper blowing off, tack-wiping, or steam or pressure cleaning. The additional cost of this operation is made up many times in the reduction of rework.
   b. When masking operations are performed, tack off the product again after masking to remove any paper, dust, or lint.
2. **The paint is dirty:**
   a. It is quite easy to contaminate paint during mixing, thinning, and straining operations. Allowing dirt to build up on the mixing equipment and dirty containers adds to the problem.
   b. Some cheap materials may contain fillers that will not mix readily with the paint. Lumps, skins, and large particles must be strained out. Use paint supplier approved strainers (not cheesecloth or rags).
   c. Improper solvents may cause a precipitation in the paint that looks much like dirt. Use only the proper solvents recommended by the material supplier.
3. **The atomizing air is dirty:**
   a. Spray guns that are totally immersed in a solvent tank for cleaning quickly become "dirt generators." Dirty solvent leaves a residue in the air passageways, which will ultimately flake off into the air stream as the gun is used. Wash the outside of the gun with a solvent rag and flush the fluid passageways with solvent.

> **CAUTION**
> Do not submerge a spraygun in solvent.

   b. Piping used in air lines will usually rust or corrode as a result of moisture in the air. This rust will flake off and contaminate the air in the form of very fine particles and must be removed by filtration. These filters, called oil and water extractors, must be cleaned regularly.
   c. Check the condition of the air by holding a clean, white handkerchief over the end of the spray gun, with the air nozzle off and the fluid supply off. Trigger the gun so that the air will flow through the handkerchief for 1 minute. See the result in the handkerchief.
4. **The working area is dirty:** Good housekeeping is a requirement for a dirt-free paint job. Remove dust accumulation from walls, floor, equipment, and the like, regularly. Wet the floors to hold dust down. The spray booth should be of the proper type and size and equipped with an automatic damper to prevent back drafts. Do not grind, sand, or polish in the spray area.
5. **Improper operation of the booth:**
   a. The booth depends on clean air to function properly. Air intake filters must be kept clean to ensure this supply. These filters are often overlooked.
   b. Exhaust filters must be cleaned or changed regularly to ensure proper air movement through the booth.
   c. A booth equipped with doors should have interlocks on them so that the booth will not operate unless all doors are closed. The exhaust air will obviously draw dirt into the booth area.
6. **Poor operator techniques:**
   a. An operator who sprays at excessive air pressure will contribute to the dust and overspray accumulation in the spray area. The resulting high air velocities will not only contaminate the paint job, but often dislodge other dirt and dust particles that normally would not be a problem.
   b. Always start spraying the area of the product farthest from the exhaust fan. This precaution permits the overspray and offspray to move with the air flow over unpainted surfaces.

---

*Courtesy of Binks Mfg. Co.

PAINT CONDITIONS AND REMEDIES

7. **Lint on the clothing:** Operators should wear starched, lint-free clothing or clothing made of antistatic material.
8. **Static electricity:** Static electricity will attract dust particles like a magnet, especially when the humidity is low. Ground the product, the spray gun, and the operator when possible. For example, when painting a car, connect a wire from the ground terminal of the battery to a water pipe. The conducting air hose can ground the operator.

REVIEW QUESTIONS

1. What is the purpose of this chapter on paint conditions and remedies?
2. Why is it important to know the cause of a problem before repairs are made?
3. What is a prevention system?
4. What is meant by the cause of a paint condition?
5. What is meant by the remedy for a paint condition?
6. What is the remedy for excessive orange peel?
7. What is the remedy for a complete panel that has crazed severely all over and the panel must be repainted?
8. What is the remedy for rusting?
9. What is the primary cause of runs or sags?
10. What is the remedy for mottling?
11. Name several causes of dirt-in-paint problems.
12. What is the remedy for correcting blushing that has just occurred, and how can the condition be prevented as the color-coating operation continues?
13. What is the remedy for correcting severe acid and alkali spotting?
14. What is the conventional procedure for the painter to follow when fisheyes appear on application of the first coat and the entire panel still has to be painted?
15. To prevent crazing, how can the painter test an acrylic lacquer paint finish before making spot or panel repairs over it?
16. What paint problems can be removed from basecoat/clearcoat finishes without repainting?

CHAPTER 18

Rust Repairs and Prevention

INTRODUCTION

The purpose of this chapter is to familiarize automotive refinish and collision repair trade personnel with a sound approach and repair procedures for making rust repairs on cars that are guaranteeable. Making long-lasting, high-quality repairs over rusted areas involves:

1. **Removing all rust completely** from exterior car surfaces before paint operations are done.
2. **Using the best plastic-filling methods and materials** that are water resistant and/or waterproof.
3. **Using the best primer and primer–surfacer available** for durability purposes.
4. **Rust-proofing the reverse or hidden side of repairs** to keep moisture and condensation off interior rust-repaired surfaces.

This method, when done properly and thoroughly, guarantees that rust will not return. When workmanship is poor in any of the repair steps and water gets to the steel, however, the rust condition will return.

"STOP RUST" PRODUCTS

Several products are now available at paint jobbers and auto accessory stores that "stop rust upon application." When used properly, these products actually stop rust in its present condition and prevent it from spreading. Label directions advise how much original rust must be removed before application. These products change iron oxide (which is rust) into a compound that is compatible with topcoats and is not affected by penetrating moisture. For best results, follow label directions.

One example of a product available through the Permatex label is called Extend (a Loctite Corporation product). Its label directions read as follows:

1. a. Use when the temperature is between 50° and 90°F.
   b. Use gloves to prevent staining of hands.
2. Scrape or wipe off dirt and oil. Sand or wire brush rusted surface.
3. Shake well before using. Pour a small amount of Extend into a small disposable container. Do not return unused material to the bottle or dip the applicator into the original bottle because the contents may be activated prematurely.
4. Brush or sponge on a thin layer using long, thin strokes. Try to contain Extend to rusted area being treated because the product will stain the surrounding paint. Remove splatters with detergent and water immediately.
5. A black coating will begin to appear in 15–30 minutes. Apply a second thin coat within 20 minutes to produce a uniform color.
6. Allow 24 hours protected from moisture before topcoating. Acrylic latex finish paints require a solvent based primer–sealer before painting. Lacquer/enamel base paints may be used without a primer.
7. Clean brushes and other tools immediately after use with detergent and water.
8. For automotive applications, allow an extra 24 hours drying time (or infrared drying equivalent) to allow trapped moisture to evaporate before topcoating. For a smooth finish, apply a sandable primer–surfacer before topcoating.

> **NOTE**
>
> This product has not yet been approved by any specific car factories or by any paint companies. However, many professional painters, auto accessory stores, including the author, have tried this product out for a number of years and we find the product works as stated in the label directions. The parent company, Loctite Corporation, is a major supplier of many products to a number of car companies, including GM.

CAUSES OF RUST

Several factors contribute to the severity of rust on cars:

1. Car environment
2. Lack of car maintenance
3. Age of the car
4. Design of the car
5. Incorrectly repaired collision damage

A car may be affected by two or more of these factors, in which case the corrosion is that much more severe.

Car Environment

The environment is the single biggest cause of rust on cars. Highly industrialized areas aggravate and promote corrosion conditions. The smoke and pollutants from chimneys of coal-burning furnaces, like those in steel mills, contribute heavily to corrosion conditions. The northeast quadrant of the United States, most of Canada, and other countries that use salt to melt snow and ice on streets and highways during cold weather have the most corrosion problems on passenger cars and trucks. Another rust-problem area is along the Florida and Gulf coasts, where the air is highly humid and salty.

Another factor that promotes rust conditions in these areas of the world are car-wash businesses that use recycled "salt" water. These businesses use recycled water as a means of cutting water costs and for national water conservation. As cars from salty streets are washed, the salty water runs into a water-saving tank and is reused. Some of these car washes have a freshwater final rinse but many do not. Washing cars with salty water promotes corrosion, particularly in water-trap areas on cars.

Lack of Car Maintenance

Dirt, salt, chemicals, and water are the principal ingredients that cause rust on cars when these ingredients come in contact with the steel through microscopic or larger breaks in the paint finish. Any car that is not washed, cleaned, and waxed properly and periodically has a tendency to rust faster when exposed to rust-producing elements.

Age of the Car

Once started and not checked or repaired, rust conditions become more deeply embedded in the steel. The older the car, the more severe the corrosion.

Design of the Car

Many cars, by styling or incidental design, have water pockets or water traps built into them. A water trap is a location on a car so designed that it holds water against steel painted parts after runoff of water following a car wash, a rainfall, or heavy dew. Water traps are invariably the weakest spots on a car from a corrosion standpoint and are among the first to rust through when the car is exposed to severe salt and chemical environments.

Incorrectly Repaired Collision Damage

Insurance companies pay collision repair shops to restore each collision-repaired car to the original condition. This restoration includes application of rust-proofing and undercoating. Due to speedy repair and/or negligence, however, some repair shops fail to rust-proof the interior construction of affected panels or areas of cars as written on repair orders. This failure results in serious premature rust problems.

In some cases, collision repairpersons use an improper panel-to-panel overlap joint, which results in an unprotected rust trap. In such cases, cars start to rust at the time of collision repair. In other cases, metal and paint repairs are made carelessly, as follows:

1. All known rust is not removed completely from the repair area before painting.
2. The metal is not cleaned or conditioned before painting.
3. A weak primer and/or primer–surfacer system is used before color coating.

TYPES OF RUST

Rust can be divided into three general categories: surface rust, corrosive pitting and scale rust, and perforation rust.

Surface Rust

Surface rust is the first level of rust. This condition can be seen and repaired completely and satisfactorily on an exterior surface if done early with correct repair techniques. Figure 18–1 shows corrosion in the beginning stages. In a surface rust condition, rust is just starting and there is no serious blistering. Surface rust is the beginning of corrosive pitting. The condition is mild and is not very deep. This rust is the easiest type to repair because the condition can be corrected by sanding, metal conditioning, and painting.

To repair surface rust:

1. Sand to shiny bright metal.
2. Apply a two-part metal conditioner (Table 6–2) to dissolve the remaining rust in surface pits (Figure 18–1).
3. Wash the surface with water; blow-dry with compressed air to dry the surface.
4. Apply the best primer and primer–surfacer system before color-coating.

Corrosive Pitting and Scale Rust

Corrosive pitting is an advanced form of rusting. **If left unchecked, rusting proceeds laterally, under a paint film, to cause rust blistering, as well as in depth, into the thickness of the metal.** Rust deposits on a surface later turn into scale rust. Usually, a large number of pits are affected, which turns into scale on the surface as corrosion progresses.

To repair scale rust:

1. Remove rust from each pit thoroughly and completely.
   a. Sandblasting is the best and quickest method of rust removal.
   b. If sandblasting is not possible or available, use a pointed cutting tool and $\frac{1}{4}$-inch drill motor to grind each pit completely. Pointed abrasive tools are available through paint jobbers and are fitted with a $\frac{1}{8}$-inch shank for use with a small drill motor. If the number of corrosive pits is too great, the only answer is sandblasting or sectional panel replacement.
2. After removing the rust, treat each pit and surface area with metal conditioner. Rinse with water and blow-dry.
3. Apply the best primer and/or primer–surfacer, as required.

The painter can see by studying Figure 18–2 why **hand or power sanding alone cannot remove corrosive pits without removing excess metal.** If not removed properly and completely, corrosive pits continue to grow in depth and diameter and the rust problem continues.

The best test to prove when corrosive pitting is removed completely is to tin the affected spot. Tinning is a metal repair operation that is required just before solder filling. Solder will not stick to a rusted surface, so when a spot area can be tinned correctly, the painter can be assured that there is no rust on the surface.

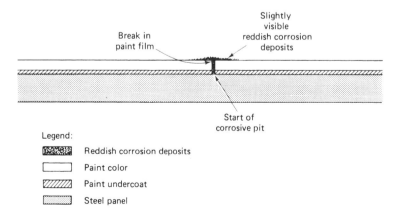

FIGURE 18–1. Cross-section of beginning stages of corrosion.

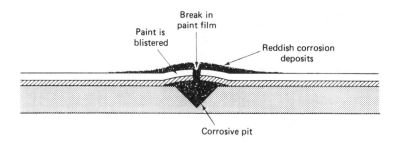

FIGURE 18–2. Cross-section of severe corrosive pitting.

> **CAUTION**
>
> Tinning on large flat surfaces should be done only by a technician experienced in this phase of metal repair. Improper tinning on a large flat panel will cause metal distortion and, in turn, will spoil the surface appearance of a panel. Before tinning a spot on a panel, the entire surrounding area must be heated uniformly by the gradual application of heat with a torch until a ballooning expansion is noted in the area metal. At this time, tinning can be done effectively and safely. Allow to air cool only. Do not quench.

Perforation Rust

Perforation rust is the most advanced form of rust and consists of complete perforation of a panel. The condition can originate on either side of a panel and is also known as a rust-through or rust out. The repair of perforation rust involves repair or replacement of the affected panel section or panel.

DETERMINING RUST REPAIR COSTS

Straight-time work operations are estimated to make sectional metal repairs. Paint materials and labor are estimated and added to make up the total. Panel replacement work operations are determined through flat-rate schedules. Also, paint material and labor are determined from flat-rate schedules. Both are added together to determine the total cost.

RUST REMOVAL

The following is a procedure for removing rust from a complete car. It includes a patch repair for perforation rust.

1. The most efficient way to remove rust, including corrosive pitting, is by sandblasting. As noted in Figure 18–2, corrosive pitting and other advanced forms of rust cannot be removed efficiently by disc sanding or by other means of sanding because the depth of the condition means that too much metal will be removed.
2. If sandblasting is done in cold or rainy weather, below 50°F, the operation can be done safely and efficiently in a paint spray booth with the air exhaust on. Also, sandblasting can be done in a paint spray booth on a year-round basis.
3. If sandblasting is done in warm and dry weather, above 50°F, it can be done safely on a driveway or other suitable location outside the repair shop.
4. When sandblasting to remove rust at specific locations, perform the necessary parts-removal operations. For specific removal and installation instructions, refer to the applicable car factory service manual.
    a. **Rust at moldings** (steel, aluminum, or plastic):
       (1) Remove moldings.
       (2) Store parts, clips, and screws for later reinstallation.
    b. **Rust at nameplates:**
       (1) Remove nameplates.
       (2) Store parts.
    c. **Rust at stationary glass windows:**
       (1) Remove perimeter moldings and clips.
       (2) Store parts. When necessary, use emergency repair parts.
       (3) **Mask adjacent window glass with two to three thicknesses of cloth-back body sealing tape (2 inches wide, minimum).**
       (4) Where necessary, cut and remove glass adhesive sealant to expose the rust condition. Replace the sealant when the painting is done.
    d. **Rust along the edge of vinyl top:**
       (1) Remove the moldings.
       (2) Detach and fold back the vinyl top.

FIGURE 18–3. Patch repair of rust-through condition.

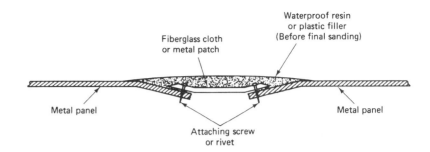

e. **Rust under the vinyl top** (heavy, hard blisters push up the top cover):

(1) Remove the vinyl top.

(2) **When installing a vinyl top, use only factory-approved vinyl-top adhesive.**

5. Protect the adjacent area parts with several layers of masking paper.
6. Sandblast all rusted surfaces until the steel is bright and clean. No discoloration spots should remain. Adjust the air pressure for clean and thorough sandblasting as required. Observe the safety precautions outlined in Chapter 6.
7. If the area being repaired is "blind" (no access to the back side), apply anticorrosion compound through the outer perforation(s) to protect the bared metal of the inner construction cleaned during the sandblasting operation. If the back side of the repair area is accessible, apply anticorrosion compound to the repair surfaces and the inner construction as the last step of the repair.

NOTE

For the availability of anticorrosion compound, check with the service department of the car dealer or zone office or with a local rust-proofing shop. Most sprayable, heavy-bodied anticorrosion compounds are basically similar in purpose, construction, and use. These compounds are designed to keep air and water away from the metal. Without water and air, steel does not rust.

---

8. Carefully "sink" metal (on an incline) around the edge of the perforation for a distance of 1 to $1\frac{1}{2}$ inches, as shown in Figure 18–3.
9. Clean the exterior bare metal surfaces thoroughly by wiping with Part I of a metal conditioner.
10. Make a repair patch of suitable metal, such as aluminum, copper, or steel. Cut and fit the repair patch to bridge the perforation and secure temporarily in place with two or three screws. Check the patch alignment (Figure 18–3). Be sure that the heads of the screws are positioned below the final surface line of the repair as shown. Do steps 11 and 12 before securing the patch permanently with screws or rivets.
11. Use one of the following products (or equivalent products) to make rust-proof plastic repairs:

   a. Rust Out (No. ROF-1, Unican); see the paint jobber.

   b. Duraglas (U.S. Chemical & Plastics, Inc.); see the paint jobber.

Rust Out and Duraglas make use of cream hardeners and contain glass fibers that contribute to strength. For correct use and application, follow the label directions. Both products are moisture-resistant or moisture-proof and corrosion-resistant and are available in all parts of the country.

CAUTION

The ordinary or commonly available polyester fillers are not recommended for rust repairs because they tend to be water absorbent and promote rather than retard corrosion.

---

12. Remove the repair patch. Prepare a suitable amount of plastic. Apply about a $\frac{1}{16}$-inch coating of plastic to the *back side* of the patch and install the patch with screws or suitable rivets. Wipe the repair surface smooth.
13. Prepare and apply the first coat of plastic filler to the complete repair area. Using a rubber or plastic spreader, apply an initial thin coating of plastic firmly to the repair surface. Apply the remainder of the prepared plastic to the initial fill and level out for general smoothness.

14. Prepare and apply all additional plastic filler to complete filling the repair area.
15. As soon as the plastic becomes solidified to the touch, use a cheese-grater-type rasp and file the repair surface to the approximate exterior contour. This filing is done in the early stages of curing before the plastic hardens. Then sand the surface smooth to the approximate final contour with No. 80 sandpaper after the plastic cures. Finish sanding with No. 220 sandpaper.
16. If necessary, apply final glaze coat(s) of plastic to the surface to produce a final smooth surface free of imperfections. More than one glaze coat may be required to achieve final smoothness.
17. Allow to cure. Sand smooth with No. 220 sandpaper.

CAUTION

Do not apply paint finish cleaning solvent on plastic filler. The solvent is absorbed by the plastic and dries very slowly, which leads to blistering of the paint finish.

---

18. Apply metal conditioner *to the bare metal* as described in Chapter 6 and according to product label directions.
19. Apply a primer system.
20. Apply a primer–surfacer.
21. Apply color as required.
22. If access to the back side of the repair is possible, apply anticorrosion compound to it. Plug access holes as required.

CHIP-RESISTANT COATINGS

The purpose of chip-resistant coating is to protect the lower underside metal panels against stone and gravel chipping when cars travel on gravel roads. The tires pick up and throw stones against the car with great force, and the abrasive action is similar to sandblasting. While the paint may chip away, the protective coating remains and protects the metal (Figure 18–4).

The early factory chip-resistant coating was made of a plastisol material similar to a vinyl top coating. The latest factory material is a tough urethane plastic that has less orange peel than earlier materials. The thickness of a good protective material has been in the range of 15 to 20 mils. Table 18–1 lists several good chip-resistant coatings available to the painter, and more are available on the market.

In the event of collision damage to panels covered with chip-resistant coating, the coating must be removed to bare metal before metal repairs and repainting can be done. **The protective coating is designed for full panel repairs only.** The material is not suitable for spot repairs. Use the following repair procedure:

1. Remove the coating from the damaged panel.
   a. Some early coatings can be removed by using a heat gun on the product and then scraping the material off with a sharp putty knife.
   b. If the material is not readily removable as described in step 1a, the fastest method of

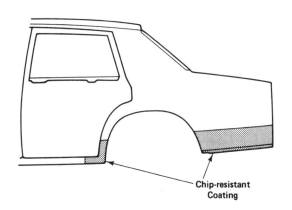

FIGURE 18–4. Location of chip-resistant coatings. Some styles include lower areas of doors. (Courtesy of General Motors Corporation.)

TABLE 18–1 Chip-Resistant Coatings

Purpose of Product	Rinshed-Mason	PPG	3-M Company
Chip-resistant material applied to low body panels prior to topcoating	891 (see label) and acrylic lacquer or M-2 enamel and proper solvent	DX-54	Aerosol 05911

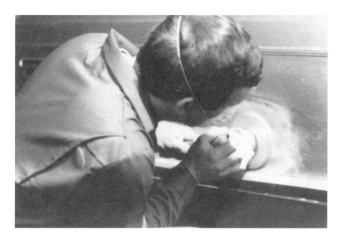

FIGURE 18–5. Sanding off chip-resistant coating. (Courtesy of BASF Corporation.)

FIGURE 18–6. Spraying straight primer followed by primer–surfacer. (Courtesy of BASF Corporation.)

coating removal is with a disc sander and a No. 36 or 40 grit disc. Finish with a No. 80 grit disc (Figure 18–5).
2. Make all necessary sheet metal repairs according to current recommended procedures.
3. Clean the surface with paint finish cleaning solvent.

CAUTION

Keep solvent off any plastic filler repairs.

---

4. Clean and mask the car for paint repair:
   a. If the car is equipped with a lower body side molding, use it as the upper breakline. Keep masking off the sheet metal.
   b. If the car is not equipped with a lower molding, mask as follows:
      (1) Position the masking paper and tape $\frac{1}{4}$ to $\frac{3}{8}$ inch above the natural body crease-line.
      (2) Position $\frac{1}{2}$-inch 3-M Fine Line masking tape over the conventional masking tape and position the edge of the tape on the center of the crease-line. This tape edge forms the upper border.

NOTE

The center of a body panel crease-line is a good breakline for this purpose. Use Fine Line tape because it leaves a sharp, clean edge with the thick vinyl coating.

---

   c. Mask the balance of the car, such as the wheels and bumper.
5. Apply metal conditioner to the bare metal as recommended by the paint supplier and then apply straight primer.

NOTE

If R-M's 834 or DuPont's 615S primer is used, do not use a metal conditioner. Allow the primer to dry per label directions.

---

6. Apply the best type of primer–surfacer, allow to dry, and scuff-sand (Figure 18–6).
7. Secure chip-resistant coating according to Table 18–1, or secure an equivalent product.
8. Prepare the material and equipment per label directions.
   a. Prepare R-M products as follows:
      1 Part 891 flex agent
      2 Parts Alpha-Cryl lacquer or Miracryl-2 enamel. Do *not* reduce before application.
   b. The following three companies' products are ready to be used as packaged for application with a suction-feed spray gun:
      (1) PPG's DX-54, Road Guard
      (2) Martin-Senour's 6850 or 6851, Vinyl-Tex
      (3) Sherwin-Williams's G1W295, vinyl Gravel Guard
   c. 3-M Company's 08874, Rocker Schutz, is ready to be used as packaged, but 3-M recommends that the product be applied with an undercoating gun, as shown in Figure 18–7.

FIGURE 18–7. 3M textured protective coating aerosol application. Three coats are required. Follow label directions. As with any lacquer-type product, do not apply over refinish enamel. (Courtesy of 3M Automotive Aftermarket Division.)

FIGURE 18–8. R-M's chip resistant coating is applied with a conventional spray gun. (Courtesy of BASF Corporation.)

9. Spray out a sample test panel per label directions for full coverage and compare the orange peel level to the car. Make adjustments to the air pressure (and material feed on guns so equipped) as follows:
   a. For coarse orange peel, apply with lower air pressure while leaving the material feed wide open.
   b. For finer orange peel, apply with higher air pressure while reducing the material feed slightly (if so equipped).
10. Apply protective coating to the car (see Figure 18–8) using the same technique that produced an acceptable finish on the test panel. Allow to dry as follows:
    a. If acrylic lacquer topcoating is used, allow at least 1 hour for drying at normal temperature (70°F).
    b. If acrylic enamel topcoating is used, allow at least 2 hours for drying (70°F).
11. Apply the finish color as follows:
    a. If no molding bordered the upper area of the repair, remove the original strip of $\frac{1}{2}$-inch Fine Line masking tape before the protective coating dries. To remove the tape, pull up and away from the panel.
    b. Place a new strip of Fine Line masking tape just slightly above, instead of right on, the upper edge of the coating. This slightly relocated strip of tape ensures full coverage of the protective coating.

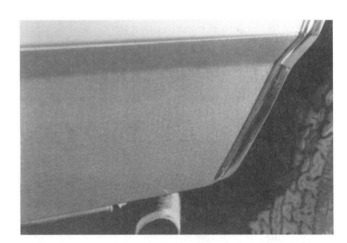

FIGURE 18–9. After the drying period, apply color as required to achieve the final color match. (Courtesy of BASF Corporation.)

12. Apply final topcoats according to label directions or as directed in Chapter 12. Increased abrasion resistance is obtained by using a catalyst in standard acrylic enamel or by using a final enamel clearcoat that is catalyzed (Figure 18–7).
13. Clean equipment immediately after painting with medium lacquer thinner or equivalent. The finished job is shown in Figure 18–9.
14. Allow the paint finish to dry per label directions before putting the car into service.

RUST REPAIRS AND PREVENTION

REVIEW QUESTIONS

1. What is the purpose of this chapter?
2. List several factors that contribute to the severity of rust on cars.
3. What is the single biggest cause of rust on cars?
4. Describe surface rust and give a remedy for its repair.
5. What is corrosive pitting?
6. Can corrosive pitting or other severe forms of rust be removed completely with conventional disc-sanding or hand-sanding methods?
7. What is the best method of removing corrosive pitting and all forms of rust completely?
8. What is the best test to use to prove that corrosive pits are removed completely?
9. What is perforation rust?
10. Where is the best place to do sandblasting inside a building in cold weather and still be in OSHA compliance?
11. Explain the procedure for repairing perforation rust by replacing a section of panel.
12. Name three different brands of plastic products, each of which can be used to make rust-proof plastic repairs.
13. Why are the commonly available polyester plastic fillers not recommended for rust repairs?
14. Explain the procedure for removing OEM chip-resistant coating from a car.
15. Explain the procedure for applying chip-resistant coating with a suction-feed spray gun.
16. How much time is recommended for drying of chip-resistant coating before applying standard acrylic lacquer or enamel?

GLOSSARY

abrasive (aluminum oxide): The toughest and most durable abrasive for use on metal surfaces; not quite as sharp as silicon carbide, but it does not break down as easily.

abrasive (silicon carbide): The hardest and sharpest abrasive known; shiny black in color; best abrasive for sanding paint; breaks down fast on metal surfaces.

ABS plastic (acrylonitrile-butadiene-styrene):
1. The hard ABS is used on door arm rest bases and other areas where hardness is a factor.
2. The soft ABS is used in combination with vinyl (PVC) to form instrument panel covers and similar parts.

acrylic enamel: An enamel coating derived from an acrylic polymer containing hydroxyl functionality.

adhesion: The ability to stick to the surface due to interfacial forces.

adhesion promoter: A coating applied over both an existing unsanded topcoat and the coated area immediately adjacent to the unsanded topcoat to promote the adhesion of a subsequent topcoat. No topcoat, primer, primer–sealer, or primer–surfacer shall be classified as an adhesion promoter.

air dry: The drying hard of a lacquer or enamel applied paint material at ordinary room temperatures and without the aid of artificial heat.

alkyd enamel: An enamel coating derived from any of several different synthetic resins made by heating together a polybasic acid, such as phthalic or maleic acid, and a polyhydric alcohol, such as glycerin or a glycol.

antiglare safety coating: A coating that is formulated to eliminate glare for safety purposes on the interior surfaces of a vehicle and that shows a reflectance of 25 or less on a 60° gloss meter.

arcing: Pivoting the spray gun at the wrist or by using an arm movement pivoting at the elbow. Causes an uneven (thin) application of paint at the beginning and end of each pass. Used in spot repairs and blending.

atmospheric pressure: 14.7 psi represents the pressure a column of air 12 inches by 12 inches exerts at sea level, although the column of air extends the entire thickness of the earth's atmosphere. From sea level, the earth's atmosphere extends upward more than 5 miles.

atomize: The breaking up of a paint material into small droplets using a paint spray gun and compressed air by counteracting forces of air at the air cap.

baking: The application of heat at temperatures of 200°F and higher. Technically, baking is not used in refinishing. At this temperature, glass would break and plastic parts would deform. *See* force dry.

barrel sanding blocks: Small, handheld sanding blocks of round construction that are ideal for small, isolated defect removal.

basecoat: A pigmented topcoat that is first applied as part of a multistage topcoat system.

basecoat/clearcoat topcoat system: A topcoat system composed of a basecoat portion and a clearcoat portion. When applying a basecoat/clearcoat system, the combined VOC total of both systems determines compliance.

binder: The resin portion of a paint that holds the pigments and other ingredients of a paint together.

blending: Combining different varieties of materials by thoroughly intermingling so that a line of demarcation cannot be distinguished. Producing a harmonious effect. Mixing.

capillary action: The electrical attraction that occurs between a liquid (the primer) and a metal surface when the surface is prime coated. The electrical forces are so strong that the primer is drawn into full contact with all of the metal surfaces, even along the walls of all sandscratches.

catalyst: A substance that causes or speeds up a chemical reaction when mixed with another substance but does not change by itself.

chip resistance: The ability to withstand chipping when exposed to normal stone and sand abrasion, and to withstand normal contact with a neighboring car when opening the door in a parking lot.

chroma: The strength or intensity of a color. The departure from white, gray, or black from the neutral axis of the color tree. Often referred to as saturation/desaturation.

clean air district: A portion of a state comprised of several counties that have similar pollution problems and that adopt a common set of clean air regulations.

clean air national rule: The clean air national rule consists of two parts. Part 1 of the rule applies to the entire country, and VOC limits are placed on the manufacturer of all paint materials. Part 2 of the plan applies to specific states and counties that have the most pollution problems. The VOC limits in these states are more strict than the national rule. The clean air rules are enforced by the federal Environmental Protection Agency (EPA).

clearcoat: A topcoat that contains no pigments or only transparent pigments; it is the final topcoat applied as part of a multistage topcoat system.

clear urethane enamel: A water-clear urethane finish with plasticizer and ultraviolet screener, but lacking pigments and metallic particles. Clear urethane must be mixed with a catalyst before use.

coating: A material that is applied to a surface and that forms a film to beautify and/or protect that surface.

cold cracking: Cracking of an excessively thick paint finish when subjected to cold and then to warm temperatures. As metal expands, the paint does not expand with it due to internal stresses in the paint and the paint cracks. Also, flexing an improperly painted flexible plastic part in cool or cold temperature will cause cracking. See "Plasticizer," Chapter 9.

color: A color is determined by the light waves an object reflects while it absorbs all other light waves. An object is blue if it absorbs all light waves except blue, which it reflects. An object is black when all light waves are absorbed and none are reflected. An object is white when all light waves are reflected.

color coat: To apply only the top color and no undercoat to a spot, large area, panel, or the complete vehicle.

color drift: Color variances on original equipment manufacturer (OEM) cars from the same factory.

combination polish and compound: A popular class of polishes that contain a certain amount of specific cutting abrasives in conjunction with chemical compounds of wax and/or chemical composition. They are popularly known as single-product polishes and clean and polish in one operation.

compatibility: The ability to become blended with other materials. For example, the ability of the color of one company to be repaired with a similar type of material from another company.

compounding: The use of abrasives to level a coarse surface before polishing and/or to make a smooth surface coarser before topcoating for adhesion.

condensation: When a cold object, like a car, is brought into a hot, humid room, water condenses from the air on the cool surface. Fast-evaporating solvents lower the surface temperature on a car and cause condensation on humid days.

crocking: The rubbing off or transfer of color from a painted object onto a person's clothing.

diluent: A liquid that makes the material to which it is added more liquid. In other words, it thins or dilutes. Diluents in lacquer thinner also serve as a solvent for the resin.

dissolve: Cause to pass into solution; separate into component parts.

double coat: One sprayed single coat of paint followed immediately by another sprayed single coat.

durability: The ability of a paint to maintain protection and appearance when subjected to the elements for long periods of time.

elastomeric materials: Coatings specifically formulated for and applied over coated or uncoated flexible plastic parts for the purpose of adhesion.

enamel: A coating that cures by chemical cross-linking of its base resin. Enamels can be distinguished from lacquers because enamels are not resoluble in their original solvent.

evaporation: The escape of solvents from the paint into the air.

evaporation rate: Basically, the amount of time it takes for solvents to evaporate. Similar products from different companies evaporate at slightly different times. Generally, solvents are classed in four categories: fast, medium, slow, and retarder.

exempt compounds: Any of a specific list of compounds that have been proven to be nonprecursors (noncausing agents) of ozone or pollution. *See also* nonexempt compounds.

face: The direct observation of a color at a right angle.

face of metallic color (tinting): The appearance of a color when viewed from the perpendicular (90° angle).

fading: A gradual decrease in the brilliance of a color.

featheredging: The tapering of the broken edges of paint during sanding so that the featheredge is hardly felt or is not felt at all.

feathering: Tapering to a very fine application of color at the beginning and end of a stroke by arcing and triggering. Color goes on like the cross-section of a feather. An excellent technique for spot repair.

film build: The total thickness of the undercoat and topcoat finish of a painted panel, which is measured in mils.

fisheyes: The appearance of small, round, craterlike openings in the finish immediately after color application. Caused by surface contamination with silicones.

flash: Allowing solvents from a newly painted surface to evaporate from a glossy to a more dull surface.

flattening agents: A special combination of talc powder and/or other ingredients and solvents that can be added to paints to dull the gloss of the finish or to increase the metallic sparkle in the base color of basecoat/clearcoat metallic paints.

fluorescent light: The emitting of visible light from a tubular electric lamp having a coating of fluorescent material on its inner surface and containing mercury vapor whose bombardment by electrons from the cathode provides ultraviolet light.

fog coat: Use in metallics. Following a wet coat in which mottling or streaking occurs, move the gun back two or three times the normal distance and apply a fog coat with continuing fluid flow and circular motion until the condition is corrected. Then move to another area. This technique unifies the metallic and pigment orientation. If necessary in enamels, apply just enough mist-coat solvent to enhance the gloss.

force dry: The application of infrared heat to a painted surface to speed drying. Normally, temperatures to 180°F maximum are safe on vehicles.

glaze: In this book, the term *glaze* has the same meaning as *polish*. Some products are designed for hand use and may be applied manually. Machine glaze products are designed to be applied by machine.

gloss (or luster): The brilliance of the reflection of a finish. Gloss is measured by percentage; 100 percent gloss is perfect. In general, there is high-gloss, gloss, and semigloss. Zero gloss is a flat finish.

guide coat: A reference coat. The color of the topcoat is different from the color of the undercoat, and the system is used to serve as a guide coat in sanding to control the sanding depth. The system aids in determining when a smooth surface has been reached.

hardener: *See* catalyst.

hiding: Has the same meaning as covering. Most standard colors achieve hiding at 2 to 2.5 mils. Proper hiding is important in color matching.

high volume, low pressure (HVLP) spray: A piece of equipment used to spray a coating by means of a gun that operates between 0.1 and 10 psi air pressure.

hue: An excitation of the sense of sight created by beams of light that allow humans to distinguish one color shade from another.

humidity: Water vapor in the air. It is measured by percentage. High humidity slows the evaporation of solvents, and low humidity speeds evaporation.

hydrocarbons: Any of a vast family of compounds containing carbon and hydrogen in various combinations; found especially in solvents and fuels. Most of the hydrocarbons are major air pollutants, and they may be active participants in the photochemical process.

impact resistant coating: Any coating applied to the rocker panel areas of cars for the purpose of chip resistance against road debris.

intermediate color: A mixture of an adjacent primary and a secondary color. Examples are red-orange, yellow-orange, blue-green, blue-violet or blue-purple, and red-purple or red-violet.

iridescent: Rinshed-Mason uses this term to denote their metallic color lines.

isocyanate resin: This resin is a principal ingredient in urethane hardeners. Because this ingredient has toxic effects on the painter, he or she should always wear a correct respirator approved by NIOSH (see Chapter 5).

lacquer: A coating that dries primarily by solvent evaporation and hence is resoluble in its original solvent.

mass tone (tinting): The color of a tinting base prior to mixing with other bases.

metallic color: Acrylic paints made by adding several sizes of metallic particles to the basic acrylic ingredients: acrylic resin, low opacity pigments, plasticizer, and solvent.

metallic particles: Most generally, aluminum flakes with iridescent and reflective properties used in various combinations of sizes in paint colors. When added to the paint material, it is commonly called a metallic color.

metamerism: A phenomenon whereby the spectral reflectance curves in a color match under one light source but do not match under a second light source.

mica: An element found in nature. It has many characteristics giving it pigment qualities, especially when coated with various chemicals. Micas aid in producing the brightest and greatest variety of colors with transparent depth. A pearlescent color is but one example.

micron: One-twentieth of a mil. Twenty microns equals one mil. The size of sandscratches is measured in microns.

midcoat: A semitransparent topcoat that is the middle topcoat applied as part of a three-stage topcoat system.

mil: One one-thousandth of an inch. Automotive paint thicknesses are measured in mils. See Chapter 9 for paint finish thicknesses.

mist coat: A light spray coat of high-volume solvent for blending and/or gloss enhancement.

mottling: Dark-shaded or off-color areas or streaks within a paint finish.

multistage topcoat system: Any basecoat/clearcoat topcoat system or any three-stage topcoat system manufactured as a system and used as specified by the manufacturer.

National Fire Protection Association (NFPA): This organization makes and monitors the fire safety regulations for the United States.

National Institute for Occupational Safety and Health (NIOSH): An agency that works with OSHA to do two important functions:
1. Makes recommendations for safety.
2. Does all testing and certifying of all tools and equipment designed and built for industry.

National Rule Clean Air Act: *See* clean air national rule.

nonexempt compounds: Any of a specific list of compounds that are known to be, or are suspected of being, toxic or potentially toxic or to cause other environmental problems.

Occupational Safety and Health Act (OSHA): The purpose of the act is "to assure as far as possible every working person in the nation safe and healthful working conditions and to preserve our human resources."

OEM: Original equipment manufacturer.

opaque: Impervious to light or not transparent: light cannot be seen through it.

orange peel: A natural occurrence in refinishing in which the resulting finish has uneven formations on the surface similar to those of the skin of an orange. A certain amount of orange peel is normal and acceptable. Coarse or rough orange peel is unacceptable and is caused by improper refinish techniques and/or practices that need correction.

overlap: That specific area of coverage in which one spray pattern application is extended over and is partly covered by the next application.

oxidation: The combining of oxygen from the air with the paint film. One principal cause of acrylic enamel drying.

paint thickness gauge: Instrument used to measure the thickness of a coat of paint. Two types of paint thickness gauges are used in the refinish trade: magnetic and electronic.

phosphate coating: A chemical bond on a steel surface that provides the best adhesion for undercoats. Produced by a metal conditioner or a primer.

photochemically reactive solvents: Common strong paint solvents such as aromatics, the branch chain ketones, diacetone alcohols, and trichlorethylene. Solvents may contain up to 20 percent toluene, 8 percent xylene, 20 percent methylisobutyl ketone, or 20 percent trichlorethylene. Emits pollutants into the atmosphere that do not dissipate.

pigment: A finely powdered, relatively insoluble substance that imparts black or white or color to coating materials. Some pigments are derived from natural sources and some are produced synthetically.

pitch of metallic color (tinting): The appearance of a color when viewed from any angle other than face, and usually 45° or less.

pitch or side tone: The observation of a color from an acute angle, possible from a 45° angle or less.

plasticizer: An oily type of substance that adds flexibility to an otherwise brittle substance. Used in acrylic lacquers but not in enamels. The resins of enamels have a built-in quality of flexibility.

polishing: The application of a product, a substance, made especially for the purpose of creating gloss on a vehicle. Usually of a wax or chemical compound construction. In this sense, polishes and sealants serve the same purpose. A true polish contains no abrasive, as indicated in the label directions.

polymerization: A change of state occurring chemically when certain ingredients are combined to form another compound with different physical properties, such as in making concrete. Ingredients cannot be returned to their original state. Enamels dry and form by this process. Lacquers do not because they can be redissolved.

polypropylene: A tough, lightweight rigid plastic made through a mechanical and chemical process. It is used extensively in the manufacture of interior trim parts. It is a difficult plastic to paint and requires a special primer.

polyurethane resin: The resin system used in urethane products. *See* urethane enamel; urethane plastic.

polyvinyl chloride: A thermoplastic material composed of components of vinyl chloride. It has outstanding resistance to the elements and can be made into a hard or soft plastic. PVC, as this plastic is known, has many uses, including use in vinyl tops, instrument panel covers (with ABS), and exterior plastic filler panels.

pot life: The amount of time a painter has to apply a plastic or paint finish to which a catalyst or hardener was added. Six hours is usually the pot life for catalyzed enamels.

precoat coating: A coating applied to bare metal primarily to deactivate the metal surface for corrosion resistance to a subsequent waterbase primer.

pressure drop: The difference in hose pressure at the transformer (on the wall) and at the spray gun (at the car). Caused by the size of the inside hose diameter. Friction caused by the walls of the hose and the length of the hose.

pressure feed gun: A paint spray gun with the fluid nozzle being flush with the air nozzle. Paint material is force-fed to the spray gun using a pressure cup, tank, or material pump.

pretreatment coating: A coating that contains no more than 12 percent solids, by weight, and at least $\frac{1}{2}$ percent acid, by weight. It is used to provide surface etching and is applied directly to bare metal surfaces to provide corrosion resistance and adhesion. The trade knows this item as metal conditioner.

primary colors: Red, yellow, and blue.

primer: A coating applied for the purposes of corrosion resistance or adhesion of subsequent coatings.

primer–sealer: A coating applied prior to the application of a topcoat to produce color uniformity or to promote the ability of an underlying coating to resist penetration by the topcoat.

primer–surfacer: A coating applied for the purpose of corrosion resistance or adhesion, and to promote a uniform surface by filling in surface imperfections.

psi or PSI: Regarding air pressure, pounds per square inch.

GLOSSARY

refinish: The replacement of undercoat and top color coat; the complete finish of a spot, large area, panel, or complete vehicle.

refraction: The breakup of a light beam into its component parts. In sunlight, light rays include all the colors in a rainbow: red, orange, yellow, green, blue indigo, violet, etc.

repairability: The ability to be repaired satisfactorily in service.

retarder: Prevents blushing. Slows down evaporation of solvents from the paint material. Can be used for fine blending in enamels. Primarily, retarder makes possible the wet-on-wet painting system, which prevents overspray from showing on complete paint jobs.

rocker panel area: The panel area of a car that is no more than ten inches from the bottom of a door, quarter panel, or fender.

rub through: The inadvertant removal of a colorcoat primarily at the panel edges on high misaligned panels during compounding and/or polishing operations because the abrasive of the polishing agent cuts away the colorcoat.

sandscratches: Frequent condition in paint finishes that is measured in microns. Specific sandscratches caused by specific sandpaper can be removed effectively from paint finishes with the same size abrasive compound/polish.

saturation/desaturation: *See* chroma.

sealant: The final product used in the process of polishing a car to a high gloss.

secondary colors: Mixture of two primary colors: orange, green, and purple or violet.

service parts identification label: A paper decal with a protective plastic coating and attached in designated locations of the vehicle. It provides the following for the painter: vehicle identification number, WA number, two-digit paint code number, and paint technology (type). Used by GM since mid-1984.

settling: The gradual sinking of the heavier pigments, binders, and metallic particles that make up a paint material when it is allowed to remain in the liquid state.

single coat: Spray painting once over a surface, with each spray pattern pass overlapping the previous pass by 50 percent.

single-stage topcoat: The colorcoat is of a single film construction, although it is applied in more than one coat. The film thickness of single-stage colorcoats is about two mils.

smog: A contraction of "smoke" and "fog," it describes the mixture of pollutants composed primarily of ozone and oxidants.

solids: Colors that contain no metallic flakes; must be tinted within the confines of a color hue and to near exactness. Consist of highly opaque pigmentation.

solution: A homogenous mixture; when the solids are distributed evenly in a mixture of reduced paint.

solvent: A liquid substance capable of dissolving or dispersing one or more other substances. Provides a solution.

specialty coatings: A special class of coatings consisting of a long list of products that serve special needs on a car. They constitute a low percentage of a paint shop's business or usage. Examples are chip resistant coatings, flexible plastic coatings, low-gloss finishes, etc. Because they are low usage items, a higher VOC limit is permitted.

spectrophotometer: A specially devised electronic tool that, when placed on an automotive finish, automatically determines and portrays a formula for that color. This formula will closely approximate the color being checked.

spot repairs: Repairs to cars in which the damaged area to be repaired is limited only to a portion of any given panel so that an entire panel need not be repaired.

suction-feed gun: A paint spray gun in which paint material is fed into the spray gun by atmospheric pressure due to a partial vacuum created by the design of the air and fluid nozzle.

swirl mark: A semiround scratch pattern in a paint finish that follows the path of a polishing wheel or a DA sander. Swirl marks are caused by the tools of compounding and polishing. Removal requires the selection and use of a proper polishing system.

thermoplastic plastics: Plastics that are flexible; that is, they can bend or flex and recover their shape quickly. They are affected by high heat.

thermoset plastics: Plastics that are hard and rigid in construction, and they are not affected by high heat.

three-stage topcoat system: A topcoat system composed of a basecoat portion, a midcoat portion, and a transparent clearcoat portion. The VOC content of a three-stage topcoat system is determined by a special formula explained in Chapter 7.

topcoat: A coating applied over another coating for the purpose of appearance, identification, or protection.

topcoat removal limit: When making repairs on topcoats, no more than $\frac{1}{2}$ mil of clearcoat should be removed to maintain film integrity and durability on basecoat/clearcoat finishes, and no more than $\frac{1}{3}$ mil of colorcoat should be removed on single-stage finishes. When these limits are exceeded, appropriate colorcoat repairs are required. *Note:* The above limits are required when topcoating is to be avoided.

top-feed spray gun: For this type of spray gun, material is fed to the air cap by gravity.

touch-up coating: A coating applied by brush or nonrefillable aerosol to cover minor surface damage. Dispensed in containers of no more than eight (8) ounces.

transfer efficiency: The ratio of the weight of the coating solids deposited on an object to the total weight of the coating solids used in a coating application step, expressed as a percentage.

translucence: The property of allowing light to pass through but the objects beyond cannot be clearly distinguished; partly transparent. A property of metallic colors.

transparency: The property of allowing light to pass through so that objects can be identified clearly. The opposite of opaque. A property of metallic colors.

ultraviolet screener: An ingredient added to paint finishes that is designed to cut off or reduce ultraviolet light penetration into a paint film. This ingredient is needed primarily in metallics and in clears. Provides durability to paint.

undercutting: Removing too much primer–surfacer when sanding.

under tone (tinting): The color produced by a tinting base when it is mixed with white or aluminum.

universal blender (PPG-830): A custom blend of solvents that allows for use in all types of shops and conditions and with various products.

urethane enamel: Requires an isocyanate hardener for curing. The ingredients have toxins that can affect a person coming in contact with them. Protective clothing and a NIOSH-approved respirator are required when applying materials containing isocyanate hardeners. The material is available in colors and in clear form, and provides a glass-smooth finish approximately two and a half times harder than ordinary acrylic enamels.

urethane plastic: As used on bumpers, filler panels, and quarter extension panels, urethane plastic is of the thermoplastic type. It is tough and flexible. This material can be plastic-welded. Parts made of it require flexible plastic paint finishes.

vacuum: The absence of air.

value: The lightness or darkness of a color.

vaporization: The conversion of solvents into gases during spray painting.

vehicle: The liquid portion of a paint.

ventilation: The correct movement of air during spray painting and drying of the finish. One of the variables of spray painting.

vinyl: The common plastic name for polyvinyl chloride (PVC), as in "vinyl tops" and "vinyl trim." Vinyl plastics require a special type of paint in refinishing. For best refinish results, it is best to use factory-recommended brands of vinyl paints.

viscosity: The flow characteristics of a paint material that determine how well it will atomize, how well it will "flow out" on the work, and the type of equipment necessary to move it.

VOC regulations: Rules and laws of the clean air act that seek to reduce the amount of VOC in the air and thereby limit pollution. The regulations vary from state to state, county to county, and district to district. For a general description, see Chapter 1. For specific rules of specific areas, check with the local clean air district.

volatile organic compound (VOC): An organic compound that readily evaporates at normal temperatures; some are smog-forming. Government regulations seek to limit the amount of VOCs in the air.

waterbase enamel finishes: Each of the major paint manufacturers have a VOC waterbase paint system available in all areas. Included are complete undercoat and colorcoat systems.

waterproof: Sheds water completely; none penetrates.

water resistant: Sheds most water; some penetrates.

INDEX

A

ABS (flexible) parts, 230-31
Abrasives, 94, 95. *See also* Compounding, Sanding, Sandpaper
Acid rain, 203–205
Acrylic enamel,
   application, 161–62
   single stage
      color application, 190
      color preparation, 190
      spot repairs, 189–90
Acrylic finishes
   and cleaning solvents, 91
   and hardeners, 155
Acrylic–lacquer, 161
   single–stage
      application, 162–63, 189
      repairs, 188–89
      testing for, 148
      and sealers, 134
Acrylic urethane enamel
   single stage
      application, 180–81
Adhesion promoter, 134, 226, 234, 237
Adjustments
   spray gun, 37–41
Air cap, 23
   cleaning, 51
Air compressors
   capacity, 73
   diaphragm–type, 74, 75
   installation, 74, 75
   maintenance, 74–75
   piston–type, 73–74
   pressure switch, 74
Air filter and regulator, 75–76
   installation, 76
   pressure drop, 77
   refinish hoses, 76–78
Air inlet, 26
Air pollution, 5–8, 121
Air pollution control districts (APCDs), 8
Air pressure, 157
Air quality management districts (AQMDs), 8
Air system
   for HVLP use, 45–46

Air valve, 25
   packing, 25
   spring, 26
Aluminum, 91
   flakes in paint, 142, 143
American car factory, paint use, 146
Apprentice painter, 3
   review of refinishing trade, 11–14
Apron tapers, 112–13
Arcing, 48
Atomize, 21
Automotive refinishing, theory and skill development, 37
Automotive refinishing trade, 1, 2
   associated businesses and organizations, 2

B

Baffle, 26
Basecoat/clearcoat finish
   application, 175, 176–77, 178, 193
   blending, 191–92
   repairs to, 171–74, 198
   testing for, 148
Binder, 142
Binks summary of dirt in the finish, 258–59
Blending, 169–70, 174
   and spot repair, 185
BRYG color wheel, 212–13
Buffing, 203
Bull's eye, 187

C

Capillary action, 132
Car components, and paint application. *See also* Exterior plastic parts, Interior plastic parts, Panel repair, Vinyl tops
   door, 192–94
   fenders, 159–60
   hood, 159–60
   luggage compartment, 231
   rear deck, 160
   roof panel, 159
   sides, 160–61
Chroma, 212
Chrysler
   paint and trim codes, 17, 18

Clean Air Act, 3
Clearcoat
    and acrylic urethane enamel (single stage), 181
    application, 175, 177, 180, 192, 193–94, 227, 237
    blending, 170
Cold strippers, 105, 107
Color, 141
    application, 10–12
    blending, 169–70, 174
    charts, 17, 19–20
    drift, 169
    exterior, 17, 19
    inspection, and lighting, 209–11
    interior, 17, 20
    mixing, 154–55
    reduction, 155
    shade, 207
Color blenders, 122, 170
Color matching, 85, 167–69, 234
    and problem solving, 214–17, 218
    variables affecting, 207–209
Color plotting, 218, 220
Color systems
    field repair, 147–48
    of vehicle manufacturers, 11
Color theory, 211–14
Compatibility of paint, 148
Compliant areas, 6
Compounding, 195
    hand rubbing, 197
    machine, 197–98
    problems and remedies, 198
Costs, determining, 2
Crack resistance, 234
Crocking, 225
Cup, paint viscosity, 124–28

D

Dirt, 201–202
Door aperture refinishing tape (DART), 109, 110
Drying equipment, 82–83
Drying temperature, 9
Dusting gun, 72

E

Emergency services, 8
Enamel paint systems, 134, 161
    testing for, 148
Environmental Protection Agency (EPA), 5–6, 68
Estimator, 2
Exempt solvents, 121, 142
Extenders, 142
Exterior colors, 17, 19
Exterior plastic parts (flexible)
    bumper cover, 237–40
    elastomeric paint finishes for, 233
    identifying, 235
    location of, 233
    paint repair systems for, 234–35

    painting, 233–36
        procedure for, 236–37
    vinyl tops, 240–42

F

Factory painting and color use, 9
Featheredging, 97–99, 136, 186, 187
Feathering, 48
Field paint repair materials, 146
Finesse repair systems, 149
Finish, tests to determine type, 148
Fire prevention, 8
Fisheyes, preventing, 164, 202–203
Flattening agents, 142
Flooding test, 38–40
Fluid feed valve, 24
Fluid inlet, 26
Fluid needle, 23
    packing, 24
    packing nut, 23–24
    spring, 24
Fluid tip, 23
Ford
    paint and trim codes, 16–17
Full jet spray pattern, 10
Full–open spray gun adjustment, 40

G

Gauges
    banana, 151
    paint thickness
        electronic, 149, 150–51
        magnetic, 149, 151–52
    pencil, 151
General Motors
    paint and trim codes, 14–16
Gloss, 234
    flattening compound, 231
Government regulations, 5–8, 12–13, 22
    and compliant areas, 6
    and dusting guns, 72
    and HVLP equipment, 32
    and least regulated areas, 6–7
    and pressure hoses, 78
    and spray booths, 78
    and VOC concentration, 122
Graduated Glasurit paint mixing paddle, 127

H

Handy maskers, 112–13
Health services, 8
Hiding, 162
Hose control, 158
Hot wire test, 223
Hue, 212, 213

I

Industrial fallout damage, 203–205
Insurance companies, 5

Interior colors, 17, 19, 20
Interior body components, plastic, color keying, 223–24
   paint
      application, 226
      preparing to, 226
      procedures, 226–27
   paint standards, 224–25
   testing paint durability, 225–26
Interior body components, vinyl,
   products for painting, 228–31
Intermediate colors, 212

L

Labor costs, 2
Lacquer–based products,
   and government regulations, 6
Let down panel, 173
Light, for color inspection, 209–11

M

Manual spraying, 10–11
Masking, 110
   aerials, 118
   crease–line or reverse, 118–19
   door handles, 117
   door jambs, 119
   exterior plastic parts, 118
   filler, 116
   front ends, 118
   general purpose units, 112–13
   lock cylinders, 18
   and imprinting, 112
   materials for, 111
   of molding, 114, 115
   newspaper for, 114
   paper for, 111
   of perimeter areas, 114–16
   prefolded drape plastic film, 111–12
   tape, 111
   techniques, 113–14
   tools for, 113
   wheels, 118
   windows, 116–17
Material costs, 2
Material safety data sheets (MSDSs), 13–14
Media blasting, 108–109
Metal conditioning, 9–10
   booth, 10
   for refinishing, 91–92
Metallic color
   and aluminum particles, 142
   shade of, 207
   single stage, description, 147–48
   tinting colors, 218, 219
   topcoats, 143
      multicolored single–stage, 144
      single–stage, 144
   variables affecting, 125, 207–209
Mica flakes, 142, 143–44, 171–72

Micron, 195
Midcoat application, 177, 180, 193
Mil, 195
Mottling, 163
Multicoat finishes, 190–91
Multistage topcoats, 144
   multicolored, 145
Munsell color tree, 213–14

N

National Fire Protection Association (NFPA), 78
National Institute for Occupational Safety and Health
   (NIOSH), 67–68
Nonattainment areas, 6
Nonexempt solvents, 121, 142

O

Occupational Safety and Health Administration
   (OSHA), 8, 12, 67
Original equipment manufacturer appearance (OEM)
   standards, 225–26, 234
Office of Emergency Services (OES), 8
Orange peel, 202
Overspray, 48

P

Paint
   adhesion, 225
   application procedure, 158
   analyzing, 119–20
   cup, 125–28
   disposal, 87
   fade resistance, 225
   and fisheyes, 164
   flexibility, 225
   formulas, 217
   ingredients, 141–42
   insects, 164
   low VOC, 148
   paddles, 125–28
   preparing to, 178
   reduction ratio guide, 126–28
   smudges in, 164
   thickness, 149–52
Paint companies, refinish, 3
   regional training centers, 3
Paint identification, 14–17
Paint jobbers:
   and color orders, 19–20
   miscellaneous materials from, 4
   and regulations, 21
   responsibilities, 4
Paint mixing room, 9
   and equipment, 83–85, 153–54
Paint problems and remedies
   acid and alkali spotting, 245
   blistering, 246
   bloom, 256
   blustering, 246

bull's eyes, 246–47
chalking, 247
checking, 247
cracking, 247
crazing, 248
dirt, 248, 258–59
dry spray, 249
etching, 249
fisheyes, 249–50
lifting, 250
micro-checking, 247–48
mottling, 250–51
off-color, 251
orange peel, 251
overspray, 252
peeling, 252
pinholing, 252–53
putting, shrinking and swelling, 255
rub-through, 253
runs, 253
sags, 253
rust, 254
sand and file marks, 254
sandscratch swelling, 255
streaks, 256
sweat-out, 256
water spotting, 256–57
wet spots, 257
wheel burn, 257
wrinkling, 257
Paint removers, 105–106
   categories of, 106
   methods for using, 106, 107–108
Paint shops
   conventional, 1, 5
   custom, 1, 5
   equipment for, 87
   high-volume, 1, 5
   operation of, 2–3
   regulation of, 7–8
   safety rules, 12–13
Painter, 1–2, 3
Panel repair
   and color blending, 169–70
   and color matching, 167–69
   procedure, 174–75
   sectional, 178–81
   spot and partial
      multicoat finishes, 190–94
      single-stage acrylic enamel, 189–90
      single-stage acrylic lacquer, 188–89
      single stage finish, 188
      surface preparation, 186–87
   surface preparation, 174–75
   and tricoat blend, 176
Panel replacement, 176–78
Pearl luster effects, 171–72
Performance settings charts, 40, 41, 43, 45
Pigments, 142

Plastic
   ABS (flexible) parts, 230–31
   fillers, 91
   sanding, 99–100
   rigid parts, 227
      painting, 228, 229
   thermoplastic, 223
   thermoset, 223
Polishing, 196–98
   buffing, 203
Pollution, and finish damage, 203–205
Primary colors, 212
Primer
   application, 10, 226
   conventional, application, 135
   and converter, 134
   flexible, 237
   guide coat filling and sanding, 136–37
   -sealer, 134
   straight, 132
   -surfacer, 133, 187, 238
      improper use of, 137–38
      conventional, application, 135–36
   waterborne epoxy (Glasurit 76–22), 131–33
   waterborne filler (Glasurit 76–92), 133
Production line, description, 9
Putty, 134–35
   glazing, 135
   polyester, 138

Q
Quadcoat finishes, 145

R
Refinish hoses, 76–77
   maintenance, 77–78
   regulations, 78
Refinishing
   color reduction, 155
   color preparation, 154–55
   paint application, 158–61
   paint systems, 161–63
   precoating checklist, 154
   silicone additives, 163–64
   spray gun check, 156–57
Refraction, 144
Repair, 185
Repair order, 2
Resin, 142
Respirators,
   air supplied mask or hood, 68, 70–71, 173
   compressors for, 71
   dust particulate, 68, 71
   general spray painting (TC–23C), 68–70
   negative pressure test, 69
   positive pressure test, 69–70
   qualitative fit test, 69
Retarders, 123–24
Rinshed-Mason vinyl top paint system, 241

INDEX

Robotic spraying, 10, 11
Runs, 201
Rust
   causes of, 262
   and chip resistant coating, 266–68
   corrosive pitting, 263
   cost of removal, 264
   perforation, 264
   products to stop, 261
   removal of, 264–66
   scale, 263
   surface, 263

S

Safety rules, 12–13
   for paint removers, 107
   for power sanding, 105
   for sandblasting, 109
Sags, 201
Sandblasting, 108–109
Sanding
   and backup block, 97
   common techniques for, 97–99
   disc technique, 102–105
   featheredging, 97–99
   and fisheyes, 202
   Green Corps Dust–Free (DF) products for, 101–102
   by hand, 96, 99–100
   health precautions, 96
   materials, 93–94
   wet, 198
   with power tools, 100–102, 105
   STIKIT system, 99–100, 101
Sanding blocks, 93
Sandpaper, 93–94
   abrasives in, 94, 95
   dry–type, 95
   how to use, 96–97
   waterproof, 95
Sandscratches, 138
Saturation, 212, 213, 216
Screener, ultraviolet, 142
Secondary colors, 212
Shop towels, 92
Silicone
   additives, 163–64
   removal of, 90–91
Solid color
   single–stage, 144
   tinting colors, 218, 219
   topcoats, 143
Solvent recycler, 86–87
Solvents, 142
   color blenders, 122, 170
   exempt, 121, 142
   fast–drying, 123
   medium drying, 123
   nonexempt, 121, 122–24, 142
   and record keeping, 128
   retarders, 123–24
   slow drying, 122–23
   and drying temperature, 124
   types of, 121
   water, 142
Spectrophotometer, 85
Sponge pads, 93
Spot repair. *See also* Panel repair
   single stage, 185–86
   spray gun adjustment, 40–41, 44
   spray gun technique, 65
Spray booths, 78, 154
   air replacement, 82
   cross–draft, 79
   cure cycle, 80
   down–draft, 79, 80, 81
   lower cost, 81–82
   maintenance of, 82
   selecting, 79–80
   spray cycle, 80
   waterwash, 79
Spray fans, 38
Spray gun,
   adjustments, 38, 63
   cleaning, 50–52, 52–53, 54
      with Mattson Quick Clean procedure, 56–58
      with Mattson Quick Switch cartridge, 53, 55–56
   companies, 4
   conventional high–pressure, 21–22, 33–36
   daily performance check, 59
   flooding test, 38
   fluid containers for, 28,
   high–volume, low–pressure (HVLP), 12, 22–23, 30, 32–36
   lubrication, 52
   maintenance, 50–52
   parts of, 22–26
   performance comparison, 33–36
   preapplication check, 156–57
   pressure feed, 22, 28, 51
      cups, 30, 31–32, 51
   spray patterns, 38
   suction–feed, 12, 22, 25
      cups, 28, 30, 31
   top–feed, 22, 26–27, 29
   troubleshooting guide, 60–61
   washers, 52–53, 86–87
Spray–out panels, 211
Spray–out test record, 221
Spray painting, methods,
   corners, 50
   curved surfaces, 50
   edges, 50
   horizontal surfaces, 50
   long panel, 50
   panel, double coat, 50
   rectangular panel, exercise, 64
   single coat, 48–49
   spot repair technique, exercise, 65

Spray painting stroke
  arcing, 48
  banding, 49–50
  distance, 47
  exercises to develop, 58, 62, 63–65
  feathering, 48
  fundamentals, 62–63
  grip for, 46–47
  gun position, 47
  movement of, 47
  overlap practice, 63–64
  for panel, 48–49, 50
  speed, 47–48
  triggering, 48, 63
Spray pattern
  and flooding test method, 38–39
  normal, 38
Spreader control valve, 24
Squeegees, 92–93
Steel, 91
STIKIT sanding system, 99–100, 101
Striping, repainting, 181–83, 184
Surface condition, 119–20
Surface preparation, 11
  car washing, indoors, 89–90
  car washing, outdoors, 90
  equipment for, 92–93
  general cleaning, 90
  masking, 110–16
    areas of vehicle, 116–19
  metal conditioner, 91–92
  paint removers, 105–108
  power sanders and files, 100–105
  refinish tape, 109–10
  sandblasting, 108–109
  sandpaper and sanding, 93–100
  silicone removal, 90–91
  paint finish cleaning solvent, 90–91
Surface preparation work station, 85–86

T
Tack rags, 92
Temperature, and finish, 124
Thermosetting, enamel topcoats, 146
Thinners, 123–24
3M Perfect-It Paint Finishing System, 196, 198–201

Tinting
  guide, 217–18
  measurements, 216
  procedure, 215
Topcoats, 143–44, 226–27, 237
  classifications of, 144–46
  thermosetting, 146
Training centers, 3
Transfer efficiency, 33, 35
Translucence, 144
Transparency, 144
Tricoat finishes, 145
  blending, 176
  repairs to, 171–74, 192–94
Trigger, 25

U
Undercoat
  application, 135–38
  categories of, 131–38
  history, 131
  purpose of, 131–32
Undercutting, 136
Urethane enamel paint system,
  applying, 161–62

V
Value, 212, 213
Vinyl, products for painting, 228–31
Vinyl tops
  paint systems for, 240–41
  procedure for painting, 241–42
Viscosity, 9
  checking, 124–25
  cup, 125–28
  measuring, 9
Volatile organic compounds (VOCs), 5, 6, 121

W
Wall regulator, calibrating, 45–46
  pressure drop, 77
Warning label, 70
Warranties, 170–71, 190
Water, 121
Wet sand operations, 10

ABOUT THE AUTHOR

Harry T. Chudy earned his B.S. degree in education from Wayne State University in 1938. He taught in the Detroit Public School System. He was a member of the Ferris State University Autobody Advisory Board from 1965 to 1983. He also served as an instructor of Automotive Refinishing at Ferris State University in 1975 and 1976.

Although the author has held several positions with General Motors over a 33-year period, his primary duties consisted of teaching automotive refinishing to newly hired body repair persons who later became instructors to teach in General Motors Training Centers. The author also served as instructor at several General Motors Training Centers.

The author is well aware of the needs of both the experienced painter and the apprentice.